The Military Architecture of Jordan During the Middle Bronze Age

New evidence from Pella and Rukeis

P. Bruce McLaren

BAR International Series 1202
2003

Published in 2016 by
BAR Publishing, Oxford

BAR International Series 1202

The Military Architecture of Jordan During the Middle Bronze Age

ISBN 978 1 84171 567 4

BAR Publishing is the trading name of British Archaeological Reports (Oxford) Ltd. British Archaeological Reports was first incorporated in 1974 to publish the BAR Series, International and British. In 1992 Hadrian Books Ltd became part of the BAR group. This volume was originally published by Archaeopress in conjunction with British Archaeological Reports (Oxford) Ltd / Hadrian Books Ltd, the Series principal publisher, in 2003. This present volume is published by BAR Publishing, 2016.

Printed in England

BAR
PUBLISHING

BAR titles are available from:

BAR Publishing
122 Banbury Rd, Oxford, OX2 7BP, UK
EMAIL info@barpublishing.com
PHONE +44 (0)1865 310431
FAX +44 (0)1865 316916
www.barpublishing.com

CONTENTS

LIST OF TABLES

LIST OF FIGURES

Acknowledgments

This book is based upon my doctoral dissertation, entitled *The Military Architecture of Middle Bronze Age Jordan*, completed in the Department of Near Eastern Archaeology at the University of Sydney, Australia. I would like to thank my supervisors, Dr A.V.G. Betts and Prof. D.T. Potts, for their assistance and support in completing this work. I would particularly like to thank Dr A.V.G. Betts for the opportunity to work on material from Rukeis and Dr S.J. Bourke for the opportunity to work on material from Pella, which provided the core data used in this study. Work in Jordan was made possible by the Department of Antiquities of the Hashemite Kingdom of Jordan. I was given financial support through an APRA (Australian Postgraduate Research Award) from the *Australian Research Council,* and special grants courtesy of the *Near Eastern Archaeology Foundation* and the *Catherine Southwell-Keely Fund.* I would like to thank David Davison for ready assistance and advice in the publication of this book. Thanks to Stuart Cullinan for providing computer support.

Colleagues who have offered important advice and insights into aspects of this study, or who have aided me in the completion of this work, include: Dr. A.V.G. Betts (Univ. of Sydney), Dr. S.J. Bourke (Univ. of Sydney), Dr. D. Clark (Walla Walla College, WA), Dr. S. Eames (Univ. of Sydney), Prof. I. Finkelstein (Tel Aviv Univ.), Dr. S.W. Helms (Univ. of Sydney), Dr. D. Ilan (Hebrew Union College, Jerusalem), Prof C. Lamberg-Karlovsky (Peabody Museum, Harvard), Prof. M. Maqdissi (Damascus), Prof. A. Mazar (Hebrew Univ., Jerusalem), Prof. G. McMahon (Univ. of New Hampshire), Dr. P.R.S. Moorey (Ashmolean Museum, Oxford), Dr. K. Nashef (Birzeit Univ., Nablus), Prof L. Stager (Semitic Museum, Harvard), Prof. J. Strange (National Museum, Copenhagen), Prof. I. Theusen (Carsten Niebuhr Inst., Copenhagen), Dr. N. Qadi (Yarmouk Univ., Irbid); and my colleagues in the Department of Near Eastern Archaeology at the University of Sydney.

Special thanks and gratitude go to the McLaren Family and the Lay Family for their patience and support. I would like to extend special thanks to Dr. S.J. Bourke for his limitless advice, encouragement and friendship. Finally, this work would not have been possible without the tireless proof reading and love of my wife, Cynthia.

Abbreviations

EBA	Early Bronze Age
EBI	Early Bronze Age I
EBII	Early Bronze Age II
EBIII	Early Bronze Age III
EBIV	Early Bronze Age IV
MBA	Middle Bronze Age
MBI	Middle Bronze Age I
MBII	Middle Bronze Age II
MBIII	Middle Bronze Age III
MB-LB	Middle Bronze Age - Late Bronze Age Transition
LBA	Late Bronze Age
AAAL	Annals of Archaeology and Anthropology, Liverpool
AAAS	Annals Archeologiques Arabes Syriennes
AASOR	Annual of the American Schools of Oriental Research
ADAJ	Annual of the Department of Antiquities, Jordan
AJA	American Journal of Archaeology
AJSLL	American Journal of Semitic Languages and Literature
APEF	Annual of the Palestine Exploration Fund
Archaeology	Journal of the Archaeological Institute of America
ARM	Archives Royales de Mari
AS	Anatolian Studies
Atiqot	Journal of the Israel Antiquities Authority
BA	Biblical Archaeologist
BASOR	Bulletin of the American Schools of Oriental Research
BAR	British Archaeological Reports
Berytus	Journal of the American University of Beirut
BMB	Bulletin. Musee de Beyrouth
CAH	Cambridge Ancient History
EI	Eretz Israel
ICANE	First International Congress on the Archaeology of the Ancient Near East, Bietak, M. *et al* (eds). Rome. 2000
IEJ	Israel Exploration Journal
ICAC	Acts of an International Colloquium on Absolute Chronology, Astrom, P. (ed), Goteburg. 1987
Iraq	Journal of the British School of Archaeology in Iraq
JEA	Journal of Egyptian Archaeology
JPOS	Journal of the Palestine Oriental Society
JSSEA	Journal for the Scientific Study of Egyptian Antiquities
Kush	Journal of the Sudan Antiquities Service
Levant	Journal of the British School of Archaeology in Jerusalem and the British Institute at Amman for Archaeology and History
Mari	Mari, annales de recherches interdisciplinaires
NGR	National Geographic Research
Paleorient	Journal of the CNRS
PEQ	Palestine Exploration Quarterly
Qedem	Monographs of the Hebrew University, Jerusalem
QDAP	Quarterly of the Department of Antiquities, Palestine
RB	Revue Biblique
SICAC	Acts of the Second International Colloquium on Absolute Chronology', Aegypten und Levante III. 1992.
SHAJ	Studies in the History and Archaeology of Jordan
Sumer	Journal of Archaeology and History in the Arab world
Syria	Syria. Revue d'Art Orientale et d'Archeologie
Tel Aviv	Journal of the Tel Aviv University Institute of Archaeology
VT	Vetus Testamentum
ZDPV	Zeitschrift des Deutschen Palaestina-Vereins

Chapter 1: Background

The Military Architecture of MBA Jordan

This study analyses the military architecture of Middle Bronze Age (MBA) Jordan. Although military architecture is one of the defining features of urban development in the MBA Levant, the military architecture of Jordan has never been adequately assessed. The study aims to redress this imbalance and to consider the military architecture of MBA Jordan within the trends of urban and regional developments. The result will be a greater understanding of the relationship between Jordan and the Greater Levant during the MBA. The core data for the study was excavated at Pella and Rukeis under the supervision of the author and is complemented by data from other sites in Jordan (figure 1).

The MBA in the Levant represents one of the most expansive and prosperous phases of early urban development (figure 2). According to Wright, this period was 'the classic age of the sovereign city' (Wright 1985a:172). Rapidly growing cities were the result of a major shift towards an increasingly centralized society. Monumental fortification systems were common features of these urban centres. As Oren explains, 'perhaps the clearest manifestation of the highly organized social and political structure of Middle Bronze Age Canaan is the immense fortification systems, comprising ramparts, moats, *glacis*, city walls, and city gates' (Oren 1997b:256).

The purpose for these elaborate building works has been heavily debated. The primary function was almost certainly defensive (Mazar 1997:250), although these architectural features may also have provided level platforms for temples and palaces (Finkelstein *et al.* 1993:378) and served as physical representations of power (Bunimovitz 1992). Due to their sheer physical size, fortifications are often well preserved and provide one of our most useful data sources for studying MBA urbanism. According to Oren, fortifications 'perhaps more than any other aspect of the material culture, could help to elucidate the mechanism of state formation in southern Canaan' (Oren 1997b:256).

Fortifications are an important indicator of urban development and settlement patterns. This premise relies on two important factors that provide the platform for the study. The first factor involves the development of standardised features of military architecture during the MBA. The second factor involves the interpretation of standardised features of military architecture as evidence for the movement of people, ideas, and trends in urbanism.

Standardisation

A central theme of Bronze Age Levantine architecture is the development of standardised features of planning and construction. The development of standardised construction features is most evident in religious and military architecture, but may also be observed in features of town planning (Callaway 1980:63; Harif 1979:167; Yassine 1974:133). It is the development of standardised features of architecture that provides the most useful insights into patterns of urbanism, regionalism, and the movement of people and ideas.

The standardisation of military architecture first becomes evident during the EBA. At this time, there was 'uniform architectural development' (Helms 1976a:9), most notably in straddle towers, semi-circular towers and bent-axis gates. Straddle towers (figure 3a) were a common feature at Tell Dothan (Helms 1977a:103), Tell Ta'anach (Lapp 1969:2), Jericho (Sellin and Watzinger 1913), Ai (Callaway 1980:26), Tell el-Far'ah North (de Vaux 1962:212), Rosh Hanniqra (Tadmor and Prausnitz 1959:72), Terqa (Buccellati and Kelly-Buccellati 1983:47) and Tell Khoshi (Lloyd 1940:13). Semi-circular towers (figure 3b) were also a standard feature of the EBA, at sites such as Ai (Callaway 1980:157), Jericho (Kenyon and Holland 1981:14), Me'ona (Braun 1989:96-97), Arad (Amiran 1978:13; Amiran and Ilan 1996:19-23), Lerna (Caskey 1964:15-16), Demirchihuyuk (Korfmann 1977:37), Arslan tepe (Palmieri 1984:208), Tell Agrab (Delougaz and Lloyd 1942:220) and Lerna (Amiran 1978:13). Similarly, standard building practices are reflected in bent-axis gates (figure 3c) at Rosh Hanniqra (Tadmor and Prausnitz 1959:72-88) and Tell Dothan (Helms 1977a:112).

The trend towards increasingly standardised architectural features continued during the MBA, most notably in gates and earthen ramparts (see pp.34-35). Triple-entry gates (figure 4a) are commonly found in MBI-III contexts at numerous sites throughout the Levant and are widely regarded as a chronologically important hallmark of the MBA (Gregori 1986)[1]. Earthen ramparts were also constructed using increasingly standardised and sophisticated techniques (Pennels 1983:61). Ramparts were constructed using layers of earth deposited in a deliberate fashion to facilitate drainage and ensure stability (Finkelstein *et al.* 1993:300). Layers of chalk bonded the rampart with a *glacis* surface and provided a base for the regular deposition of earthen layers (figure 4b). This construction technique has been referred to as the sandwich method of rampart construction and is evident at a range of sites (see p.38).

The examples of towers, gates and ramparts discussed above demonstrate that a particular architectural feature may be considered representative of a specific period. It is important to note, however, that these features are not *exclusive* to a specific period. For example, semi-circular towers may also be noted at Old Babylonian period Tell Kannas (Finet 1977:83), Roman period Lejjun (Parker 1982:6), Bar Kokhba period Beitar (Carroll 1925:103), Parthian period Kish (Langdon 1935:121) and Islamic period Qasr al-Hayr al-Sharqi (Graber 1965:107-122) and Samarra (Northedge *et al.* 1990:136).

[1] Similar developments of standard gate types may also be noted in contemporary Middle Kingdom Egypt, where gates formed a distinctive class of their own at Buhen (Emery 1958; 1959; 1960), Uronarti (Reisner 1960:13) and Mirgissa (Wheeler 1961:97).

Triple-entry gates are also known at LBA Megiddo (Finkelstein *et al.* 2000:122); and, Iron Age Tell el-Kheleifah (Glueck 1939c:8-21), Tell Dor (Stern 1983:261), Tell Batash (Mazar 1997:258), Tell el-Nasbeh (Zorn 1997:53), Hazor (Yadin 989:31), Megiddo (Loud 1948:46), Tell Dan (Biran 1994:236; Biran *et al.* 1996:12, 25), Gezer (Yadin 1958a:80-86) and Beersheba (Aharoni 1973 Pl 83; Detlef Jericke 1992:122-148).

Although standard features of military architecture are not exclusive to a single period, evidence strongly indicates that specific features of military architecture were consistently employed in defensive constructions at numerous sites throughout the Middle East during the MBA. Certain features were so commonly utilised that they may be considered characteristic and representative of the MBA. Based on this, any parallels between standard features of military architecture at sites in Jordan and well-dated sites elsewhere may be indicative of patterns of regionalism and urbanism.

Architecture as an Indicator of Movement

Because military architecture became increasingly standardised during the MBA, it is possible to establish, through accurate dating, where a certain feature was first constructed and when that feature was employed at sites elsewhere. This information helps to elucidate the areas of technological innovation and the time period and routes associated with the dispersal of the new technology.

The value in using fortifications to understand the movement of peoples and ideas was realized during the earliest stages of archaeological scholarship in the Levant. Following his work at Tell el-Yehudiyeh during the early 20th century, Petrie suggested that the earthen rampart fortifications at the site should be associated with the Asiatic *Hyksos* dynasty, because the ramparts had more in common with earthen ramparts in the Southern Levant than with those in Egypt (Petrie 1906). Petrie expanded this theory to suggest an origin for the *Hyksos* in northern Syria and Central Asia (Petrie 1930:22; 1931:3; *et al.* 1952:2). Petrie's work represents the first attempt to view military architecture as evidence for the presence of a specific cultural group (Wright 1968:3,7).

Petrie's association of earthen rampart fortifications with the *Hyksos*, and his belief that the *Hyksos* had a northern origin, strongly influenced scholarship in the early 20th century. His views were adopted by Woolley at Carchemish (Woolley 1921:39; 1952:222-224) and by Albright at Tell Beit Mirsim (Albright 1930:4; 1932b:19). Albright considered that the earthen ramparts at Tell Beit Mirsim provided 'the first concrete proof of the barbarian occupation ... in the period of the great *Hyksos* interruption about the middle of the 18th century B.C' (Albright 1932a:8). Albright attributed the origin of the earthen rampart to 'the plains of Eastern Europe and Turkestan, whence it was brought into western Asia in the 18th-17th century B.C' (Albright 1932a:8)[2].

Following the work of Albright, scholars such as Kenyon (1952; 1969:56), Wright (1956:17), Tufnell (1958:45-47), Lapp (1964:6-8), and Bull and Campbell (1968:3) adopted the theory that earthen ramparts are directly related to the arrival of the *Hyksos* from the north. This theory would later be disputed by Parr, who demonstrated that MBA earthen ramparts were conceptually a development of EBA fortifications (Parr 1968:41-42). Other scholars, who have dismissed a *Hyksos* origin for earthen ramparts, include Dever (*et al.* 1970:43 fn7), Tubb (1983:59) and Gregori (1986:83).

Fortifications have figured prominently in other important debates regarding the origins and movements of people. For example, the origin of EBI period fortifications in the Southern Levant has been attributed to the arrival of people from the Northern Levant (Kenyon 1979; de Vaux 1970; 1971; Lapp 1970). This theory has been disputed by Miroschedji 1971; Schaub 1982 and Mabry 1996. The origin of casemate walls and triple-entry gates in the MBA Southern Levant has been attributed to the arrival of the *Amorites* from Mesopotamia (Kaplan 1975). This theory has been disputed by Gregori (1986:96-97). More recently, the origin of freestanding earthen ramparts in the MBA Southern Levant has been attributed to the arrival of the 'coastal *Amorites*' (Stager 1999:237).

Although features of military architecture have been interpreted as evidence for the movement of people and ideas though the Levant, there has been no comprehensive analysis of the MBA fortifications of Jordan. This study analyses the standardisation and employment of MBA military architectural features, their relationship to the movement of people and ideas, and their usefulness in understanding regionalism and urban development in MBA Jordan. The outcome is a more detailed and accurate assessment of the relationship between MBA Jordan and the Greater Levant.

Geography

The core data for this study comes from sites within the region of the modern Hashemite Kingdom of Jordan, bordered by Syria to the north, Iraq to the East, Saudi Arabia to the south, and Israel and the Palestinian territories to the west. The borders of Jordan present an artificial geographical framework, as there are few significant geographic divisions between Jordan and her neighbours. Even the Jordan River would not have significantly restricted the movement of people, ideas, and cultural influences in antiquity. For the purposes of this study, Jordan is divided into the regions of the East Jordan Valley, the Jordanian Plateau, the Hauran and the Desert (figure 5a).

[2] Albright's views were influenced by the discovery of similar earthen ramparts in Central Asia (see Pumpelly 1908). Albright assumed that these ramparts predated the MBA ramparts of the Levant when in fact they were of a much later period (Wright 1969:32).

The geography of the Greater Levant is a relatively open landscape; one region flows into the next without impassable topographic barriers (Magness-Gardiner 1997:304). The Southern Levant is more particularized than the north, being 'compacted between sea and desert' (Wright 1985a:5), but served an historically important function as a land bridge between Africa and Asia (King 1999:94). A complex system of primary and secondary trade routes pass through the area, facilitating relatively easy movement from one region to another. During the MBA, the absence of major geographical barriers allowed for a ready movement of people, trade items, and ideas and resulted in the emergence of homogeneous cultural traits in terms of architecture, metalworking and pottery production (Dever 1987). The relatively open topography of the Levant thus played a central role in the development of urbanism and military architecture in MBA Jordan.

The East Jordan Valley

The determining geographic feature of the region is the Rift Valley, which comprises the Beq'a Valley, the Jordan Valley, the Dead Sea, the *Arabah* and the Gulf of Aqaba. This long valley defines the western border of Jordan. The monumental rift is geographically related to the formation of the Jordanian Plateau to the east. The plateau runs north-south and combines with the Jordan Valley to produce a series of longitudinal climatic (figure 5b) and vegetation regions (el-Eisawi 1985; Long 1957; Magness-Gardiner 1997:304; Shehadeh 1985). During the MBA, the climate of the region was similar to today (Magness-Gardiner 1997:304; van Zeist 1985:201), and much of the natural environment of the region was yet to be significantly impacted (Gophna *et al.* 1986:81; van Zeist 1985:201; Mabry and Palumbo 1992:69). The north-south orientation of the Jordan Valley and the Jordanian Plateau affected agricultural practices, settlement patterns, and the movement of trade.

The Jordan Valley runs from Lake Tiberias to the Dead Sea. This region receives 50-350mm annual precipitation and is a warm steppic weather zone (Shehadeh 1985:33-35), charaterised by Mediterranean and Irano-Turanian soils and vegetation (el-Eisawi 1985:50-51). The valley is conducive to agriculture, receiving water from perennial *wadis* and the Jordan River. The Jordan Valley is also connected to major trade routes, notably the Plain of Esdraelon, which opens to the Mediterranean in the west, and the Rift Valley, which continues into the Beq'a Valley to the north. To the south of the Dead Sea, the Rift Valley becomes the desert of *Arabah* as far as the Gulf of Aqaba. This region is characterised by sandy, saline soils and minimal rainfall. Vegetation is Sudanian. The region is not conducive to agriculture or large urban settlement.

Although only the east bank of the Jordan Valley is located in modern Jordan, the entire valley must be viewed as a single geographical entity. The climate, agricultural potential and proximity to trade routes made the Jordan Valley an important area for human settlement during the Bronze Age. Some of the most significant MBA cities in Jordan are located in this region, including Pella which provides some of the central data for this study.

The Jordanian Plateau

The Jordanian Plateau runs parallel to the Jordan Valley and the Central Highlands to the west. The plateau varies in altitude from 800-1250m above sea level and is characterised by a series of *wadis* running from east to west down to the Jordan Valley. The Jordanian Plateau receives cool and temperate weather and an average annual precipitation of 300-500mm. The region has Mediterranean and forest vegetation and is suitable for pastoralism and dry farming.

The plateau may be divided into a series of sub-regions. To the north is the Irbid region, which extends from Wadi Yarmouk in the north to Wadi Zerqa in the south. This area is a geographical extension of the Hauran and the Damascene and is a productive region characterised by good soil, reliable rainfall, and an open landscape (Magness-Gardiner 1997:315). To the south is the Amman region, which encompasses the plateau south of Jerash and the Baqah Valley. This is a fertile, hilly region that receives good rainfall (Magness-Gardiner 1997:316). South of the Amman region are the Madaba plains, which stretch as far as the Wadi Wala. Climatically, this is a transitional zone on the desert frontier (Magness-Gardiner 1997:317). South of the Madaba plains, rainfall decreases and the environment becomes less conducive to agriculture and sedentary human settlement. The region of Kerak stretches from the Wadi Hasa to the Wadi Mujib and receives marginal annual rainfall (less than 200mm) that allows for dry-farming (Miller 1991:2-3).

The Hauran

To the north and east of the Jordanian Plateau, there is a slight descent into the plains of the Hauran. There is no geographical barrier between the plateau and the Hauran; although, the plains form a distinct region characterised by red soils and a black basaltic landscape. The region is fertile, receives 200-300mm annual precipitation, and is conducive to cereal dry farming and steppe pastoralism (van Liere 1963:116). This region stretches north and east from Irbid into southern Syria and the Damascene. Therefore, it is necessary to view the Hauran as a geographic region that stretches beyond the modern Syro-Jordanian border.

The main geological formation associated with the Hauran is the Jebel Druze. These basaltic hills rise as high as 1860m above sea level at Jebel el-Arab. By comparison, the plains of the Hauran are approximately 900m above sea level. As a result, a series of *wadis* form watercourses that cut onto the plains. Rukeis, a core data source for this study, is located on such a *wadi* system.

The Desert

The desert stretches to the east of the Jordanian Plateau and to the south of the Hauran. The desert forms the greater extent of modern Jordan and is charaterised by sporadic annual rainfall (less than 100mm) and Saharo-Arabian vegetation. The desert is not conducive to sedentary urban activity and none of the sites discussed in this study are located in this region.

The Greater Levant

The other important regions of the Southern Levant are located to the west and north of Jordan (figure 5a). The Central Highlands, to the west of the Jordan Valley, are comprised of a narrow limestone ridge. The western slope of the ridge is long and gradual while, the eastern slope drops sharply into the Jordan Valley. The range rises as high as 1000m near Bethel before declining into the Negev in the south. Routes through the range are limited because of a lack of water sources. To the north-west of the range, the Carmel Ridge extends to the coast, effectively forcing overland trade routes to head inland. To the west of the Central Highlands is the *Shephelah*, a region characterised by rolling hills.

The coastal plain and the coast are west of the *Shephelah.* The plain varies in width from 200m to 40km and is appropriate for grain production. Woods, marshlands and sand dunes characterise some areas. The coastline lacks natural ports, and significant coastal centres were not established until the MBA. The Esdraelon Plain provides the major passage inland from the coastal routes, east to the Jordan Valley and north to Lebanon and Syria. The Esdraelon includes the Plain of Akko, the Plain of Megiddo, and the Jezreel Valley.

To the north of Jordan, the Hauran Plains merge with the Damascene of southern Syria, effectively linking the Northern and Southern Levant. The Anti-Lebanon mountains (an extension of the Jordanian Plateau) and the Beq'a Valley (an extension of the Jordan Valley) are located west of the Hauran and Damascene. On the western side of the valley, the Lebanon Mountains rise sharply before dropping into the coast. This area has tremendous topographical contrasts and environmental diversity (Marfoe 1979a:3-5). The Northern Levant, located to the north of the Damascene and Lebanon, is not directly linked to Jordan geographically.

Chronological Framework

The periods that are directly relevant for this study are the EBIV, MBI, MBII and MBIII. It is not the purpose of this study to enter the detailed debates regarding the most accurate chronology and appropriate terminology. I adhere to the following framework for the reasons described below. All dates given throughout the study are in years B.C.

EBI-III	**3500-2300**
EBIV	**2300-2000**
MBI	**2000-1775**
MBII	**1775-1650**
MBIII	**1650-1550**

The chronological framework of the MBA is heavily debated. Opposing schools of thought argue for either a high chronology (Goetze 1952, 1957; Kitchen 1967, 1987a, 1987b; Tuman 1987; Kempinski 1992; Dever 1985, 1991, 1992 and Weinstein 1992:27) or a low chronology (Albright 1952a, 1952b, 1956, 1957, 1973; Gates 1987; Astrom 1992; O'Connor 1983, 1985; Malamat 1992; Bietak 1987a, 1991, 1997). The unresolved issue in this debate is which dating scheme should be applied to Egyptian and Mesopotamian king-lists (Gates 1987:76; Hoffmeier 1989:182; Huber 1987:16; Kempinski 1992:69; Tuman 1987:197; Ward 1992:61,2).

All of the important archaeological finds that are relevant to providing an absolute date for the MBA, such as the Montet Jar (Bietak 1987a:97; 1991:54; Porada 1966; O'Connor 1983; 1985; Tufnell and Ward 1966; Ward 1978) or the cartouches of Pepy I and Hotepibre found at Ebla (Matthiae 1984:23; 1987:113;166; Ryholt 1998:1), are problematic. Although carbon dates are limited for this period (Gerstenblith 1980:74)[3], they provide the most reliable dating material. Available carbon dates indicate that the beginning of the MBA should date to c.2000 B.C (Oren 1973:34,35). It is for this reason that I adhere to the high chronology for the purposes of this study. The SCIEM (*Synchronization of Civilizations in the Eastern Mediterranean*) project of the Austrian Academy (Bietak and Kleinsgutl 2000) will hopefully provide a more concrete MBA chronology in the near future.

The period of the MBA is characterised by cultural continuity (Kenyon 1973:88,90; Bienkowski 1989:170; Bietak 1991:43) with archaeological evidence for three sub-phases of development (Amiran 1969b; Dever 1992:10; Mazar 1968:91-93). The tripartite division is most clearly expressed by the terms MBI, MBII and MBIII (Dever 1992; 1997a; Finkelstein 1993). This terminology is adopted for the purposes of this study.

The term EBIV is adopted to emphasize this period's greater degree of continuity with the EBA (Wright 1937:3; 1961:103; Lapp 1970:115-117; Oren 1973; Dever 1980:35,46; 1992; Richard 1980) than with the MBA (Philip 1989:207). The term EBIV is more appropriate than Copper Age (Petrie 1931:3), Caliciform Period (Tufnell 1958:31), MBI period (Albright 1932b:8-18; 1962:38; Amiran 1960:213; 1974; Dever 1973a:56,57) or Intermediate EB-MB (Kenyon 1951:106; Prag 1974:69; Strange 2000:1571).

[3]More carbon samples have been tested from EBA contexts (Callaway and Weinstein 1977:1-16; Rast and Schaub 1980:21-62).

Historical Background

The EBI-III periods

Many of the urban developments that define the MBA find their origins in the EBI-III periods. The EBI-III is characterised by the centralisation of society within city-states; the expansion of international trade; an increase in the use of bronze; and, the construction of urban fortifications (de Jesus 1978; Kempinski 1983:237; Muhly 1977:73; Renfrew 1972:440-475; Wright 1985a:39; Yakar 1984a; 1985:35).

Trade
During the EBA, there is evidence for expanded trade contacts between Syria and Egypt (Matthiae 1982:89); Mesopotamia, Syria and Anatolia (Behm-Blancke 1980:215; Ozguc 1960:25); and Egypt and the Southern Levant (Amiran 1965:30-34; 1969a:50-54;1978:114; Ben-Tor 1991; Esse 1991:103-109; Finkelstein and Gophna 1993:14; Harrison 1993:83-86; Hennessy 1967b:49; Levy *et al.* 1995:26; 1997:45; Ward 1991; Weinstein 1984:62; Yeivin 1960:193-200). The effects of expanded trade were also felt in Jordan. One example is the presence of copper ore mined in the Wadi Feinan (Hauptmann and Weisgerber 1992:61-3; Najjar *et al.* 1995:519) at Jericho and Numeira (Hauptmann and Weisgerber 1992:63) and Arad (Hauptman *et al* 1999:1-18).

Urbanism
Expanded trade networks resulted in a growth in prosperity and a centralisation of society into fortified urban centres (Foster 1977:39; Klengal 1990:94). The development of city fortifications is one of the most important developments of the EBA. Although there is some evidence for fortifications preceding the EBA (Abu al-Soof 1968:3-16; Caneva 2000:165; Giannessi 2000:555), walled cities first appeared with relative uniformity throughout the Levant during the first phases of the EBA. A range of theories have been proposed to explain the emergence of fortifications in the Southern Levant. These include an invasion by an already urbanised population from the north (Kenyon 1979; de Vaux 1970; 1971; Lapp 1970); an indigenous development (Schaub 1982; Mabry 1996); and, a mixture of both (Callaway 1972, Hennessy 1967a).

The developments in urbanism that characterised the EBA are also evident in Jordan. During the EBI period, urban centres were established at a number of sites in Jordan, such as Jawa (Helms 1987a:43,44; Helms 1987b:54-9), Bab edh-Dhra (Schaub 1982:67-75), Tell Umm Hamad (Sauer 1986:3) and Pella (Bourke 1997:99; Bourke *et al.* 1998:182). Research undertaken in the Hesban region (Ibach 1987:151-3), the Wadi Hasa (MacDonald 1988:166), the Madaba Plains (Harrison 1997:1), the southern *Ghor* and north-east *Arabah* (MacDonald 1992:41) and on the Kerak Plateau (Miller 1991:307; Steele 1990:26) demonstrate that the EBI was a significant period of sedentary activity.

The trend towards larger, more elaborate fortification systems continued in the EBII-III periods. Although the number of sites decreased, the size of the sites increased, reflecting the trend towards increased nucleation of society (Finkelstein and Gophna 1993:6; Gophna and Portugali 1988:14-16; Esse 1991:150-152). In Jordan, EBII-III large fortified centres have been revealed at Bab edh-Dhra and Numeira (Schaub 1982:67-75), Khirbet Zeraqun (Ibrahim and Mittman 1997:388), Tell el-Handaquq North (Mabry 1996:150), Khirbet Iskander (Richard 2001:441), Dhiban (Sauer 1986:3), al-Lajjun, Mudawarah and Rujm Birjes (Schaub 1992:167) and possibly at Tell Mustah (Prag 1992:155).

The EBIV

The EBIV period marks a break in tradition. The period is characterised by a decline in trade and sedentism and a renewed emphasis on pastoral and semi-pastoral activities. There have been many theories regarding the reasons for this change, including: Egyptian pressure on the Southern Levant during the 5th-6th dynasties (Mazar 1968:66; Callaway 1978:46-58); the collapse of the Egyptian Old Kingdom (Kempinski 1978:36); the cessation of international trade (Richard 1980:10-11); climatic changes (Harlan 1985); and, destruction by invading nomads (Glueck 1955; Albright 1961,1962; Falconer and Magness-Gardiner 1984; Kenyon 1966b, 1971; Lapp 1966, Hennessy 1967b; Mazar 1968; Prag 1985:81).

The EBIV urban hiatus had a dramatic effect throughout much of the Middle East; although, the extent of the impact varied from region to region. Egypt and Mesopotamia entered a period of decline, and an occupational break occurred at sites on the Khabur and Assyrian plains (Weiss 1993:999). The cities of the richer agricultural areas of Syria continued to exist on a reduced scale (Klengal 1990:95; Kramer 1977:59; Matthiae 1982:89-90). In the Southern Levant and Jordan, the trend away from urbanism and centralised society was more marked. Most fortified sites were abandoned, and there was a sharp decline in the number of urban settlements (Finkelstein 1991:23-26; Gophna and Portugali 1988:16).

One notable development of this period is increased human activity in the marginal zones along the desert frontiers of the Negev and Jordan (Finkelstein 1989a:130-134; Haiman 1996:1; Finkelstein and Perevolotsky 1990:76; MacDonald 1992:71). Although settlement in Jordan was sporadic during this period (Ibach 1987:33-9; MacDonald 1988:11; Yassine *et al.* 1988a:171; Yassine *et al.* 1988b:196), there is evidence for EBIII-IV continuity at Khirbet Iskander, Bab edh-Dhra, Iktanu, Aro'er, Tell Umm Hamad, Ader, Tell el-Hayyat and Tell el-Umeiri (Richard 1987:242-5).

The MBA

The MBA represents a re-emergence of the same urban trends that characterised the EBI-III periods (Wright 1985a:43,53). The MBA was characterised by a number of specific traits, including widespread use of tin bronze; increased use of the potter's wheel and standardisation in ceramic production;

expanded trade; and, increasing urbanism and centralization of society in the form of large, fortified centres.

Trade

MBA textual material describes an expansion of trade throughout the Levant and the Eastern Mediterranean (Georgiou 1979:86; Malamat 1971:31-38), with intensive trade between Babylonia, Assyria, Syria and Anatolia (Dalley 1977:155; Gerstenblith 1983:8; Klengal 1990:95; Larsen 1976; Walker 1980:15; Veenhof 1977:109). The archaeological record provides evidence for MBA trade between Egypt, the Levant, Anatolia, the Aegean and Cyprus (Gittlen 1981:49; Mazar 1968:77; McGovern and Harbottle 1997:141-158; Mee 1978:121-156). The expansion of trade resulted in an increasingly homogenous material culture, as is evident in pottery (Kochavi *et al.* 1975:164; McGovern *et al.* 1994:31; Sevin 1984:222); temple architecture (Matthiae 1990:113); ivory work (Scandone Matthiae 1990:146-160); scarabs (Ben-Tor 1997:188; Tufnell 1956:73); and metal work (Philip 1989:207-10; 1995:523). In the Southern Levant and the Egyptian Delta, the material culture evolved into a blend of elements from both regions (Bietak 1991:27-72; Dever 1985:79).

Urbanism

The establishment of a city-state society accompanied the expansion in trade during the MBA. In the Southern Levant, a highly nucleated political structure and urban society, characterised by fortified cities, developed (Finkelstein 1989b:140; Finkelstein 1991:26-27; Gophna 1978:144; Gophna and Portugali 1988:17; Oren 1997a:xxiii). This development is attested in Egyptian textual material (Albright 1941:19; Dothan 1976:1; Mazar 1968:75,81; Rainey 1994:83).

In Jordan, developments in MBA urbanism are evident in the East Jordan Valley (Funk and Richardson1958:84 Najjar 1992:152,3; Prag 1992:155; Smith 1987:55; Yassine *et al.* 1988a:183; 1988b:196); the Hauran (Betts *et al.* 1995;1996; Eames 2000); the Irbid region (Braemer 1987; Kafafi and Knauf 1989; Lenzen and McQuitty 1984; 1989; Lenzen *et al.* 1985; Mare 1991; Mittman 1970); the Amman region (Abu Dayyeh *et al.* 1991; Greene 1992; Ibrahim 1987;1989; Lehmann *et al.* 1991; Northedge 1983; Ma'ayeh 1960; McGovern 1989; Simmons and Kafafi 1988; Wimmer 1989; Zayadine 1973; *et al.* 1987, 1989); and the Madaba region (Boling 1989:188; Geraty *et al.* 1987;1989; Herr *et al.* 1991a; 1991b; 1999). There is only limited evidence for MBA settlement south of the Madaba Plains and the Dead Sea (Ibach 1987; MacDonald 1988; 1992; Miller 1991).

Political Development

The expansion of trade combined with the increasing power of the city-state society to produce the first true period of international interaction at a political and cultural level (Oren 1997a:xxiv; Ozguc 2000:1247). Textual material from Mari, Alalakh and Ebla is indicative of a heightened level of political interaction between Syria, Egypt and Mesopotamia (Albright 1940:23; Buccellati 1988:43-62; Weinstein 1975:12). In Syria and Mesopotamia, individual city-states competed for political and economic control, occasionally taking military action to expand their dominions (Bittel 1958:24, Dornemann 1987:276, Klengal 1990:96; Matthiae 1982:90; Mazar 1968:78-9; Munn-Rankin 1956:74).

During the MBA, there was also a significant increase in political, commercial and cultural relations between Egypt and the Levant. During the MBI period there is limited evidence for Egyptian military campaigns into the Southern Levant, leading some scholars to argue that the MBI was a period of Middle Kingdom rule in the region (Albright 1922:110-139). Weinstein has countered this theory by demonstrating that less than fifty Egyptian objects have been discovered in the Southern Levant (Weinstein 1975:9). Much of the Egyptian material consists of funerary amulets brought from Egypt rather than genuine trade items (Ben-Tor 1994:11; Ben-Tor 1997:187).

During the MBII-III period, new dynasties were established at Avaris in the Eastern Nile Delta. This development was either the culmination of a rapid conquest or a gradual, peaceful influx of Asiatics (Oren 1997a:xxii). These dynasties are well-attested textually and archaeologically and are unique because the rulers were Asiatics of Levantine origin rather than Egyptians. Textual sources refer to these rulers as *Hyksos* or 'rulers of foreign lands' (Oren 1997a:xx). Egyptian power was threatened by the emergence of the *Hyksos*, and the ruling dynasties retreated to Thebes.

Although Egyptian chronology is unclear for this period, two parallel ruling dynastic lineages existed at this time, the Theban Dynasties 13 and 17 and the *Hyksos* Dynasties 14 through 16. The period of two competing dynasties came to an end with the destruction of Avaris by Ahmose, who pursued the *Hyksos* to the Southern Levant stronghold of Sharuhen, which is commonly equated with Tell el-Ajjul, Tell Haror, or Tell Far'a South (Oren 1997b:253). These events coincide with the end of the MBA in the Levant and mark the beginning of the New Kingdom in Egypt.

The episode of *Hyksos* history represents a dramatic increase in the level of cultural and political interaction between the Levant and Egypt. For the first time, Egypt was dominated and threatened by a foreign culture, forcing her into economic difficulty and military conflict (O'Connor 1997:62). The question of *Hyksos* origin has dominated debate. Some scholars support a Syrian origin (Bietak 1987b:52; Scandone Matthiae 1997:420), and others emphasize the close links with southern Palestine, based on pottery (McGovern and Harbottle 1978:151), scarabs (Kempinski 1997:328), temple architecture (Oren 1997b:264) and burial practices (Wapnish 1997).

The strong cultural link between the *Hyksos* and the Southern Levant is emphasized by the *Hyksos* retreat to a stronghold in that region. The closest cultural ties seem to be in south-west Palestine (Oren 1997b:255). The MBII-III period in south-west Palestine is characterised by a rapid urbanization process and the construction of similar earthen ramparts at Tell Haror, Tell el-Far'a South, Tell el-Ajjul, Tell Jemmah and Tell Masos. The fortifications at these sites are similar to *Hyksos* ramparts in the Egyptian Delta at Tell el-Yehudiyeh and Heliopolis (Wright 1968). The development of these urban centres is closely associated with the 'political and economic reorganisation' of the *Hyksos* period (Oren and Yekutieli 1996:87). MBII-III period Jordan falls outside of the sphere of *Hyksos* activity. Evidence for contact with the *Hyksos* culture is limited to a few luxury trade items found at Pella and Amman (Magness Gardiner 1997:321).

The MBA ends with a widespread destruction of most urban centres in the Southern Levant. There is little continuity in urban activity from the MBA to the LBA (Gonen 1984:69-70). Theories for these events include Egyptian invasion (Wright 1956:17; Dever 1990:75-77; Weinstein 1991:105-15) and internecine warfare or earthquake (Hoffmeier 1989:183-90; 1990:86; 1991:117-9). Convincing arguments have been made against an Egyptian invasion (Shea 1979:1-5; Redford 1982:117). It is likely that the widespread destruction of so many cities was the result of an earthquake, possibly associated with the Thera eruption (Goedicke 1992).

Textual and Iconographic Evidence

Textual material from late 3rd Millenium Southern Mesopotamia (Civil 1994:128) and Neo-Assyria (Mallowan 1950:158) provides insights into the nature and appearance of military architecture. Unfortunately, textual material relevant to the MBA in the Levant is scarce. Some textual sources record the construction of defensive works, such as those undertaken by Hantili I at Boghazkoy (Gurney 1966:4), but specific details of military architecture are lacking. Most military references are limited to descriptions of weaponry, such as the battering ram or the bow (Sukenik 1947:11).

Iconography is an important source of information for reconstructing the vertical appearance of fortifications. As with textual evidence, iconographic material relating to the MBA is scarce. A rare example is a siege scene from MBI period Beni Hasan, which shows a vertical elevation of a fortification system (figure 6a). Much of the iconographical material that depicts military architecture is associated with earlier or later periods and can only be related to the MBA only through inference. Predynastic examples from Egypt include the 'Narmer Palette' (figure 6b), the 'Bull Palette' (figure 6c) and the 'Cities Palette' (figure 6d), each of which depict the ground plan of a fortified enclosure. A slightly later example is the 5th-6th Dynasty wall painting from the Tomb of Anta at Deshashe, which depicts the plan of a Levantine city defended by a wall and tower system (Petrie 1898 Plate IV).

These early iconographic representations are simplistic, stylised, and are essentially limited to a plan of a defensive circuit. The iconography of New Kingdom Egypt and Neo-Assyria provides valuable evidence for the vertical appearance of city ramparts; although, these representations post-date the MBA. Ultimately, iconographic representations are only of limited use, because they are schematic representations. This is most evident in the fact that iconographic representations of a single city may vary considerably (Jacoby 1991:117-120). Iconography does not provide a genuine representation of the city, rather the 'artist records architecture with an attempt at imitating reality' (Tucker 1994:115).

Archaeological Evidence

Our knowledge of MBA military architecture is almost exclusively the result of archaeological excavation. Excavation has been extensive and has resulted in a detailed corpus of defensive features. The major categories of defensive features are wall systems, tower systems, gate systems and earthen ramparts. Each of these categories has a range of related architectural features.

The central features of military architecture first appear during the EBA and are well attested in the archaeological horizons. These features are again utilized in the MBA, but in a more complex, varied and elaborated fashion. Wall systems are slighter and more curvilinear than during the EBA (Wright 1985a:173). During the EBA, tower systems were one of the most standardised and defining features of military architecture, but during the MBA they are only rarely employed (Wright 1985a:191). More complex and standardised gate systems develop during the MBA, notably the triple-entry gate which is 'the most evolved architectural entity' of the period (Wright 1985a:191). The earthen rampart, which was primarily used to protect the exterior base of a wall during the EBA, emerges in the MBA in a range of forms (Wright 1985a:173). In general, the MBA is characterised by 'an increasing diversification of building forms and programmes' (Wright 1985a:514).

It is not the purpose of this study to discuss the entire corpus of military architecture from the Bronze Age Levant. A comprehensive summary of the known corpus of MBA military architecture may be found in Wright (1985a). Table 1 presents a list of important MBA fortified sites that are directly relevant to this study and reveals the extent of archaeological data on this subject see (pp.11-12). These sites provide comparative material for the MBA military architecture of Jordan (figure 1).

Glueck undertook the earliest significant inquiry into Bronze Age settlement in Jordan during the 1930's (Glueck 1934; 1935; 1939a; 1951). Glueck concluded that the EBA was well attested in the archaeological horizon, but found that 'between the eighteenth and thirteenth centuries B.C. there is a complete gap in the history of settled communities' (Glueck 1935:139). Glueck associated the MBA with a 'sharp decline of permanent settlement' (Glueck 1951:423). This view was generally accepted until the 1950's when increased excavation began to reveal concrete evidence for MBA urbanism. Since then, the argument for a MBA urban hiatus has been disputed by Zayadine (1973:20), Thompson (1974:65-66), Ibrahim (1987:77), Mabry and Palumbo (1992:69), and Najjar (1992:152).

As Glueck had correctly observed, the EBA in Jordan was an important period of urban development represented by a nucleation of society in the form of large, fortified cities. Fortified urban centres of the EBA have been revealed in the major geographical regions of Jordan, including: Pella (Bourke 1997:99; Bourke *et al.* 1998:182), Tell Abu al-Kharaz (Fischer 2000:448) and Tell el-Handaquq North (Mabry 1989:65-67; 1996:123-127) in the Jordan Valley; Khirbet Zeraqun on the northern Jordanian Plateau (Ibrahim and Mittman 1997:388); Tell Jamid on the Yarmouk river (Leonard 1992:24); Tell Jawa on the Madaba Plains (Geraty *et al.* 1989:144); Bab edh-Dhra (Rast and Schaub 1978; 1980; 1981) and Numeira (Coogan 1984; Rast and Schaub 1981:37) to the south-east of the Dead Sea; and, Jawa in the Hauran (Helms 1973; 1975; 1976b; 1977b; Betts 1991). These fortified sites date both to the EBI (Jawa and Tell Handaquq North) and EBII-III periods (Khirbet Zeraqun, Tell Jamid, Tell Jawa, Bab edh-Dhra, Numeira) and exhibit the full range of defensive features, including city walls, towers, posterns and gates. EBA urban development in Jordan thus closely reflects patterns of urbanism elsewhere in the Levant.

Evidence for extensive MBA settlement has been revealed through survey in the East Jordan Valley (Yassine *et al.* 1988a, 1988b), on the north-west Jordanian Plateau between the Wadi Yarmouk and Wadi Zerqa (Mittman 1970), in the Amman region (McGovern 1989; Abu Dayyah *et al.* 1991; Simmons and Kafafi 1988) and on the Madaba Plains to the south of Amman (Ibach 1987; Boling 1989). Survey work undertaken further south in the region of Kerak has revealed very little evidence for MBA urban activity (MacDonald 1988; Miller 1991). MBA settlement was most intense in the Jordan Valley and on the Jordanian Plateau. With the exception of Pella and Rukeis, the excavation history of the most significant fortified urban centres of MBA Jordan is outlined below.

The East Jordan Valley

The East Jordan Valley survey revealed a total of 27 sites dating to the MBA, a decline from the number of sites in the EBA (Yassine *et al.* 1988a, 1988b). The majority of these sites were located to the north of Deir Alla. Of these sites, Pella, Tell Deir Alla and Tell Nimrin were significant fortified urban centres during the MBA.

Tell Deir Alla

Tell Deir Alla is located 40km south of Pella, in the vicinity of the Wadi Zerqa, and is strategically located on the major trade route between the Jordan Valley and the Jordanian Plateau. The site is small (4ha) and is best described as a walled town (Magness-Gardiner 1997:315) of the MBII-III period (Franken and Ibrahim 1989:203). Excavations in Trench D revealed brick on stone fortification walls that resemble 'a casemate wall, but may in fact have been part of a gate' (Franken 1961:364). Against the exterior face of the wall a thick layer of earth had been dumped during the period of the wall construction, possibly the remains of an earthen rampart. Associated ceramic evidence indicates a date in the MBII period for the possible rampart (Franken and Ibrahim 1978:74).

Tell Nimrin

Tell Nimrin is located 40km south of Deir Alla, on the Wadi Shueib, and is strategically placed on an important trade route between the Jordan Valley and the Jordanian Plateau. Tell Nimrin was a small walled town (1.5ha) with substantial deposits for each phase of the MBA (Dornemann 1990).

Excavation in Area N40/W20 revealed possible fortification walls with two major phases of construction (Flanagan *et al.* 1990:144; 1992:90; 1994:218). The first construction phase dates to the MBII period and consists of two large stone walls meeting at a juncture. The wall running north-south (L222) is comprised of at least ten courses of small boulders and is preserved to a height of 4.0m, though the true base of the foundations was never revealed. The wall is at least 3.0m wide (Flanagan *et al.* 1994:217). The wall running east-west (L230) has at least nine stone courses and a brick superstructure that is approximately 2.0m high. The excavators suggest that these walls 'constituted elements of a tower or other large structure associated with the wall' (Flanagan and McCreery 1990:144).

The second construction phase dates to the MBIII period and consists of a new superstructure constructed above Wall L222 and Wall L230 (Flanagan *et al.* 1994:17). A soil deposit that was placed against Walls L222 and L230 possibly represents part of an earthen rampart. The deposition of the soil and the new wall superstructures belong to a single, deliberate phase of construction (Flanagan *et al.* 1994:17).

Jordanian Plateau: Irbid Region

Survey work on the north-western Jordanian Plateau has revealed considerable evidence for MBA settlement. The decline in the number of sites from the EBA is less sharp here than in the East Jordan Valley and regions of the Jordanian Plateau to the south (Mittman 1970).

Tell Irbid

The dominant site is Tell Irbid (22ha), located at a point where the Plains of the Hauran and the Damascene meet with Wadi Yarmouk, a major trade route into the Jordan Valley. Excavations at the site revealed a monumental basalt wall circuit surrounding the *tell* (Lenzen and McQuitty 1984:297; Lenzen *et al.* 1985) associated with a *glacis* (Ibrahim 1987:76). The fortifications have been dated to the beginning of the MBI period (Lenzen and McQuitty 1989:300; Lenzen 1997:181).

Abila

Excavations at Abila to the north-west of Tell Irbid have revealed MBA material (Mare 1991:206). Magness-Gardiner suggests that the fortifications at the site may also date to the MBA (1997:316).

Jerash
To the south, evidence for MBIII settlement has been revealed at Jerash (Braemer 1987). Jerash was an important site in antiquity and exhibit some evidence of EBA military architecture (Glueck 1939b:26). It is possible that the site was also a fortified settlement during the MBIII, but further excavation is required to address this question. A possible MBA settlement has been revealed at nearby Dahr Medineh (Kafafi and Knauf 1989).

Jordanian Plateau: Amman Region

The Amman region comprises the Jordanian Plateau between the Wadi Zerqa in the north and the Madaba Plains in the south. As in the East Jordan Valley, there is a notable decline in the number of sites from the EBA to the MBA. Survey in this region suggests that MBA settlement was sparse, focused mainly in the urban centres of Amman, Tell Safut and Sahab (McGovern 1989; Abu Dayyeh *et al.* 1991; Simmons and Kafafi 1988). MBA residential architecture has also been revealed at Abu Snesleh, located 15kms east of Amman (Lehmann *et al.* 1991).

Amman
The first concrete evidence for MBA fortifications was revealed at the Amman citadel during the 1960's. In Area III, excavation revealed two MBA city walls associated with the remains of an earthen rampart (Dornemann 1983:18-19; Northedge 1983:440 and Zayadine 1973:20). At a later date, a deep sounding was excavated at the west end of the lower terrace (Sqs M26-27) revealing MBII-III period occupation. Although MBI pottery was discovered, no occupational evidence of this period could be identified (Zayadine *et al.* 1987:305-308). Excavation of the sounding was continued by Greene and 'Amr, who revealed large stone walls of the MBII-III period 'perhaps forming part of the tower or gate' (Greene 1992:116-117).

Further excavation undertaken on the south slope of the lower terrace (Area A) revealed a city wall (2005), abutted by an earthen rampart, which in turn is supported by a retaining-wall (2015). Wall 2005 is over 2.0m wide and preserved to a height of 1.85m. Wall 2015 is 1.6m wide and preserved to a height of 1.4m. The earthen rampart between the two walls had been prepared with a layer of earth 1.5m thick and covered by another layer of chipped flint, approximately 0.3m thick. The rampart is covered with a *glacis* consisting of 'head sized rubble with a greyish clay'. The walls and rampart date to the MBII-III period (Zayadine *et al* 1989:359).

Sahab
Sahab is located 12kms south-east of Amman at the interface of the Jordanian Plateau and the Eastern Desert. The 5ha site is defended by a monumental brick wall (Area HIII) and a 'typical *glacis* rampart' (Area GII) dating to the MBII-III periods (Ibrahim 1975:76; 1987:76; 1989:516-518). The relationship between the two defensive elements was never established (Ibrahim 1997:450). LBA fortifications were also revealed at the site (Ibrahim 1987:76-77; 1989:519).

Tell Safut
Evidence for MBA architecture has also been revealed at Tell Safut, located approximately 15kms north-west of Amman (Magness-Gardiner 1997:317). During the 1950's, Ma'ayeh noted 'a sloping plastered *glacis* revetment, resting on natural rock and crowned by a wall. The *glacis* was constructed of different layers of sand, *huwwar*, and earth, beaten into a kind of terre pise' (Ma'ayeh 1960:115). Sauer dated the *glacis* to the MBII-III period (Sauer 1985:200). More recent excavations by Wimmer suggest that the earthen layer is a natural geological formation rather than a rampart (Wimmer 1989:513). As at Sahab, the perimeter defensive wall is of a LBA date (Wimmer 1997:449).

Jordanian Plateau: Madaba Plains Region

Evidence for MBA settlement declines to the south of the Amman region. Survey work undertaken in the vicinity of Tell Hesban revealed eleven sites with MBA pottery, although no evidence has been found for MBA settlement (Ibach 1987). A similar pattern has emerged in survey work undertaken in the vicinity of Kerak (MacDonald 1988; Miller 1991). In the hills above the Dead Sea, an MBA *glacis* has been reported at Khirbet el-Mekahayyat, although evidence for this is limited (Saller and Bagatti 1949). The most substantial evidence for MBA settlement has been revealed at Tell el-Umeiri in the Madaba Plains region to the south of Amman and north of the Wadi Wala (Boling 1989:188). Future excavation at nearby Tell Jalul may also reveal an important MBA settlement.

Tell el-Umeiri
MBIII period fortifications have only been revealed at the 6ha site of Tell el-Umeiri, which is located 10kms south of Amman. Excavation in Area B revealed a freestanding earthen rampart (Geraty *et al.* 1987:192; 1989:12; Herr *et al.* 1991a:11; 1991b:159; 1999:104). There is no evidence for a crowning wall along the central portion of the rampart, however, a possible corner tower has been revealed in the north-west corner of the *tell*. The relationship between this structure and the rampart is yet to be established (Clark 2003 *pers. comm.*). LBA and Iron Age defences were founded directly upon the rampart (Clark 1994; Geraty *et al.* 1989:245-247; Herr *et al.* 1991a:53-57).

Khirbet Iskander
Another relevant site is Khirbet Iskander, located to the south of the Madaba Plains on the Wadi Wala. Although originally fortified during the EBII-III period, the site is unique in the Southern Levant, because the fortifications were rebuilt and reused during the EBIV period (Richard *et al.* 2001:441). The fortifications comprise casemate walls, towers and a gate (Richard and Boraas 1984; 1988; Richard 1983; 1990:36-48).

The Hauran

The fertile plains of the Hauran extend beyond the northern border of Jordan into southern Syria. Survey work (Braemer 1984;1988;1993) and excavation of large *tells* such as Tell Zheir

(Braemer 1991) and Khirbet Umbashi (1994:124) have revealed that the MBA was a high period of urban development in the Syrian Hauran. This pattern is also evident in survey work and excavation undertaken in the Wadi al-Ajib on the Jordanian side of the border (Betts *et al.* 1995; 1996; Eames 2001). The Hauran reflects the strong patterns of MBA urbanism that are evident in the Irbid region of the Jordanian Plateau. These two regions are geographically connected.

Rukeis is the most extensively excavated fortified urban centre of the MBA in the Jordanian Hauran. The only other excavated MBA site in the Jordanian Hauran is Jawa, located 50kms to the east of Rukeis. Jawa was primarily occupied during the EBI period; although, there is some evidence for MBI period resettlement involving the rebuilding of the EBA fortifications (Betts 1991:26; Helms 1977b:24; 1989:141). The central structure of the MBI resettlement was a pillared building measuring 28x26m, which probably functioned as a caravanserai (Helms 1975:26).

Table 1: Important MBA Fortified Sites in the Coast, Coastal Plain, Shephelah, Northern Negev, Central Highlands, Esdraelon Plain, Jordan Valley and Syria

Coast and Coastal Plain	Period	Description	References
Ashkelon	MBI-III	Earthen ramparts, crowning wall, tower, arched gateway	Stager 1991; 1999:238 Schloen 1997:221
Kabri	MBI-III	Earthen rampart, multiple walls	Prausnitz and Kempinski 1977 Kempinski and Miron 1987 Dever 1997c:261
Aphek	MBI	Walls, tower, 'mini-rampart'	Ory 1938 Kochavi 1975; 1997 Kochavi *et al*. 1979; 2000
Tell Poleg	MBI	Walls, towers, earthen ramparts, double-entry gate	Kochavi *et al*. 1979: 13 Wright 1985a:203
Tell Akko	MBI	Walls, earthen ramparts, single chambered arched gate	Dothan 1973a; 1974; 1975; 1981 Dothan and Conrad 1978; 1979; 1983; 1984 Dothan and Raban 1990
Tell Poran	MBI	Earthen ramparts	Gophna 1992
Tell Burga	MBI	Wall, towers, earthen rampart, *fosse*	Kochavi *et al*. 1979:142-3
Tell Zeror	MBI	Wall, tower, earthen rampart, *fosse*	Kochavi *et al*. 1979 Ohata 1970
Tell Hefer	MBI	Walls	Paley *et al*. 1984
Ain Zurekiyeh	MBI	Walls	Gophna and Ayalon 1982
Achzib	MBI	Earthen rampart, *glacis* and *fosse*	Prausnitz 1963; 1975
Tell Kisan	MBI-II	Earthen ramparts, wall	Albright 1938a: 24 Seton-Williams 1980: 390
Tell el-Ajjul	MBII	Triple-entry gate, wall, earthen rampart, *glacis*, *fosse*	Petrie 1931; 1932; 1933; 1952 *et al*; Albright 1938b Wright 1985a:202 Stewart 1974:11
Tell Nagila	MBII	Wall, *glacis*	Amiran and Eitan 1963a; 1963b
Jaffa	MBII	*Glacis*	Kaplan 1964
Tell Mevorakh	MBII	Earthen rampart	Stern 1973; 1984
Tell Jemmah	MBII-III	Earthen rampart	Petrie 1928 Van Beek 1992; 1997
Tell Jerishe	MBII-III	Wall, *glacis*	Geva 1983
Yarneh-Yam	MBII-III	Earthen ramparts, double-entry gate, triple-entry gate	Kaplan 1967; 1969
Tell el-Far'a South	MBII-III	Walls, triple-entry gate, earthen rampart, *glacis*, *fosse*	Petrie 1930: 15-17 Weinstein 1997:305
Ashdod	MBIII	Earthen rampart *glacis*, double-entry gate	Dothan 1969; 1971; 1973b
Shephelah and Negev	**Period**	**Description**	**References**
Tell Beit Mirsim	MBI-III	Walls, towers, *glacis*, possible triple-entry gate	Albright 1926; 1928; 1930; 1932a; 1932b; 1938a
Tell Masos	MBII	Earthen rampart, retaining wall, *fosse*	Aharoni *et al*. 1974; 1975 Fritz and Kempinski 1983
Beth Shemesh	MBII-III	Wall, triple-entry gate	Mackenzie 1913 Dever 1997b:311
Lachish	MBII-III	*Glacis*, rampart	Tufnell 1958 Ussishkin 1978
Gezer	MBII-III	Walls, towers, earthen ramparts, triple-entry gate	MacAlister 1911 and 1912 Dever 1970; 1973b; 1974 Wright 1985a:202
Tell Batash	MBII-III	Rampart	Kelm and Mazar 1979; 1982; 1985; 1986; 1987 Kelm 1991 Mazar 1997
Tell Haror	MBII-III	Earthen rampart, *glacis*, *fosse*	Oren and Yekutieli 1996
Tel Malhata	MBIII	Walls, tower, *glacis*	Kochavi 1967

Table 1 (cont.): Important MBA Fortified Sites in the Coast, Coastal Plain, Shephelah, Northern Negev, Central Highlands, Esdraelon Plain, Jordan Valley and Syria

Central Highlands	**Period**	**Description**	**References**
Beth-Zur	MBII-III	Walls, tower	Albright and Sellers 1931 Funk 1958 Sellers *et al.* 1968
Jerusalem	MBII-III	Walls	Weill 1920; 1947 McAlister and Duncan 1926 Kenyon 1962; 1963; 1966a; 1967; 1968 Shiloh 1984a; 1984b
Shechem	MBII-III	Solid and casemate walls, earthen ramparts and retaing walls, triple-entry and double-entry gates	Wright 1956; 1957; 1985:202 Sellin 1926; 1927; 1928 Welter 1932
Tell el-Far'ah North	MBII-III	Walls, towers, earthen ramparts and double-entry gate	Mallet 1973: 29-46; 1987 Wright 1985a:201
Tell Ta'anach	MBIII	Earthen rampart	Lapp 1964; 1967; 1969 Sellin 1904, 1905
Tell Shiloh	MBIII	Walls, earthen ramparts	Albright 1923 Buhl and Holm-Nielson 1969 Finkelstein 1985 Finkelstein *et al.* 1993
Bethel	MBIII	Wall, earthen rampart *glacis*,	Albright 1934
Tell Dothan	MBIII	Wall	Free 1958
Tell Marjamah	MBIII	Wall, Possible tower	Zohar 1980
Esdraelon Plain	**Period**	**Description**	**References**
Megiddo	MBI-III	Upper town walls, earthen rampart and *glacis*, possible triple-entry gate. Lower town freestanding earthen rampart with retaining wall	Loud 1948, fig 8 Kempinski 1989:58; Plan 4 Herzog 1997:150-151 Ussishkin 1997:462 Finkelstein *et al.* 2000
Jordan Valley	**Period**	**Description**	**References**
Tell Dan	MBI	Earthen ramparts, two triple-entry gates	Biran 1980; 1984; 1994:75-90 Biran *et al.* 1996: 58-61
Jericho	MBI/II-III	Walls, earthen ramparts, tower, possible gate	Sellin and Watzinger 1913 Kenyon and Holland 1981 Marchetti *et al.* 2000
Hazor	MBII-III	Upper town walls. Lower town	Yadin 1958b; 1960; 1989
Tell el-Hammah	MBII-III	Walls, earthen rampart, *glacis*	Cahill *et al.* 1988
Syria	**Period**	**Description**	**References**
Mari	EBII-MBI	Freestanding earthen rampart and *glacis*, crowning wall with salients	Margueron 1997:414; 2000:910-912
Tell es-Sweyhat	EBIII-IV	Walls	Holland 1976; 1977
Hammam et-Turkman	EBIII-IV	Casemate wall	Van Loon and Meijer 1983
Tell Jidle	EBIV	Wall	Mallowan 1946:134
Carchemish	MBI-III?	Walls, gates, towers, earthen ramparts	Woolley 1921; 1952 Parr 1968: 32
Qatna	MBI-III?	Earthen rampart with crowning wall, triple-entry gate	du Mesnil du Buisson 1935: 40-41 Wright 1985a:202
Tell Hamidi	MBI	Walls, towers, gate	Wafler 1990
Ebla	MBI	Earthen rampart with retaining walls, fortress towers on summit, four combined triple and double-entry gates	Matthiae 1967; 1968; 1970; 1997a; 1997b; 2000 Peyronel 2000
Mari	MBI	Earthen rampart	Margueron 1983 figure 1: 2001:45
Tell Bi'a	MBI	Wall, *glacis*	Strommenger 1983; 1990
Qadesh	MBI	Casemate wall	Pezard 1931: 4-12 Parr 1983; 1997:114 Bourke 1993: 162
Tell Rimah	MBI-II	Earthen rampart	Howard Carter 1965 Oates 1967
Tell Kannas	MBI-II	Walls, towers, *glacis*	Finet 1977
Tell Hadidi	MBII	Walls, possible tower and *glacis*	Dornemann 1979
Bosra	MBII	Earthen rampart abutting circuit wall	Braemer *pers. comm.*
Tell Tuqan	MBII	Triple-entry gates, towers, walls	Guardata 1990; 1996
Tell Umm el-Marra	MBII	Walls, earthen ramparts, retaining walls, gate, possible tower	Schwartz *et al.* 2003

Chapter 2: Pella

Introduction

Pella (*Khirbet Fahl*) is a large 12ha site situated in the lower foothills of the East Jordan Valley (figure 7). The environs of Pella host a combination of riverine forest vegetation (van Zeist 1985:199), steppic grasses, and pine forest with oak and pistachio (Mabry and Palumbo 1992:67). During the MBA, the region was supported by a multi-crop, intensive agricultural industry (Flanagan *et al.* 1992:106). Pella is at sea level, receives an annual precipitation of 280-350mm and is adjacent to a perennial water source in Wadi Jirm. The site is located at the confluence of important trade routes, including: the route north through the Jordan Valley to the Beq'a Valley and Syria; the route west through the Esdraelon Plain to the Mediterranean Sea; and, the route east to the Jordanian Plateau (McNicoll and Smith *et al.* 1982:12; 1992:xiii; Smith 1987:53).

Pella exhibits one of the longest continuous sequences of urban occupation in the Levant, from the Pre-Pottery Neolithic to the Islamic periods. During the EBA, the site was a large fortified urban centre. During the MBI-II transition, the inhabitants of Pella 'considered it necessary to provide the place with defences against enemy attack' (McNicoll *et al.* 1992:35). Pella remained an important urban centre throughout the MBA, functioning as a city-state surrounded by a series of small, undefended settlements such as Tell el-Hayyat (Falconer and Magness-Gardiner 1983; Knapp 1989).

The picture of Pella as a flourishing city during the earliest phase of the MBA is consistent with the reference to Pella in Egyptian Execration text E8 (Albright 1941a:19; Smith 1973:23). From the Egyptian perspective Pella was a troublesome city. As Magness-Gardiner points out, 'there must have been something substantial at the site early in the Middle Bronze Age for the Egyptians to worry about the ruler of a distant town in Jordan' (1997:312).

During the MBII-III periods, Pella profited from increased trade. Some of the best available evidence for foreign connections with MBA Jordan has been found at Pella. The range of cultural influences on Pella at this time is best reflected in the ivory box of the MBIII-LBI period (Potts 1985; Smith 1987). According to Magness-Gardiner, this item 'is one of the few pieces of evidence for possible interaction between the *Hyksos* and the Middle Bronze Age population of Jordan (Magness-Gardiner 1997:312). Magness-Gardiner concludes that, although the local economy was the most important factor of community life at Pella, the 'reference to Pella in the Execration texts, its fortifications, and its size and importance in later periods suggest that Pella was a major regional centre during the Middle Bronze Age' (Magness-Gardiner 1997:313).

The continued importance of Pella during the LBA is reflected in the Amarna letters and the annals of Thutmosis III and Seti I, and possibly those of Amenophis III, Haremhab and Rameses II (Funk and Richardson 1958:84). The Anastasi Papyrus records Pella as a supplier of wooden chariot parts for the Egyptian military (Funk and Richardson 1958:84). Excavation of LBA tombs at the site has revealed a rich corpus of Cypriot and Egyptian imports, indicative of the extent of trade activities at Pella during that period (Bourke and Sparks 1995:165).

Excavation History

The University of Sydney has excavated Pella since the 1970's. The core data presented in this study was excavated under the supervision of the author in Areas IIIF and XXVIIIC during the 1994-1997 seasons[4].

Area III (figure 8a) was the first part of the site to reveal evidence for MBA fortifications[5]. During the 1983-1992 seasons, excavation in Area III focused on the MBA fortification system[6]. Excavation in Areas IIIC and IIID revealed a major city wall (Wall 41) running north-south for over 30.0m through the excavation area. Wall 41 was part of the eastern fortification line, extending to the north along the eastern *tell* face and to the south towards the spring in the *wadi*.

Wall 41 was constructed using a combination of brick bands, with dark grey bricks at the top and bottom and green bricks in the middle. The brick superstructure was constructed above a stone substructure and protected by a layer of plaster. Four bored holes were revealed at the top of the brick superstructure, each hole measuring 0.28m - 0.35m, possibly designed to hold reinforcing beams (Potts *et al.* 1985:196-197; McNicoll *et al.* 1992:42). In the north of Area III, Wall 41 was 3.5m wide and preserved to a height of 7.0m. To the south of Area III, Wall 41 was heavily eroded (McNicoll *et al.* 1992:40).

Area IIIC is located to the interior of Wall 41. In 1992, a 0.5m wide foundation trench was identified at the base of Wall 41 in Area IIIC (Figure 8b). Ceramic material retrieved from the foundation trench was indicative of a construction date during the MBI-II transition. This phase (X) also included the construction of an exterior buttress (Buttress 100) and a second

[4] See Table 2 for site phasing (p14).

[5] Area III comprised three 10m x 10m trenches towards the *tell* centre (IIIN, IIIP, IIIQ), three 10m x 10m trenches in the south-east (IIIC, IIID, IIIE), and one 10m x 5m trench to the east (IIIF). The MBA fortifications are associated with Areas IIIC, IIID, IIIF.

[6] Earlier excavation during the 1950's (Funk and Richardson 1958) and 1960's (Smith 1973) had failed to expose any MBA material. An overview of the excavation of MBA material at Pella between 1980 and 1985 can be found in McNicoll *et al.* 1992:35-44. Bourke (1997) provides a good summary of MBA material excavated at the site between 1985 and 1995, and includes a table illustrating the general site phasing and the relative phasing of MBA material according to Area. Excavation undertaken in Area III from the third to the fifth seasons of excavation (1980-1982) revealed MBIII period domestic structures (McNicoll and Hennessy *et al.* 1982; McNicoll *et al.* 1984; Hennessy *et al.* 1983). No MBA material was excavated in the seventh season in 1985 (McNicoll *et al.* 1986), the tenth season in 1988 (Edwards *et al.* 1990), the thirteenth season in 1991 (which was a study season), and the fifteenth season in 1993 (Watson and Tidmarsh 1996). A preliminary account of the sixteenth and seventeenth seasons of 1994/1995, which includes some of the raw data used in this thesis, can be found in Bourke *et al.* 1998.

wall (Wall 52), possibly of a defensive character, running parallel to the interior of Wall 41 (figure 8a). A roadway originally ran between the two walls (Bourke *et al.* 1994:93; Bourke 1997:104). Four phases of MBA domestic architecture (IX-VI) abut the interior face of Wall 41, above the level of the roadway (Potts *et al.* 1985:198;1988:130; Walmsley *et al.* 1993:178). At the end of Phase VII, a break resulted in structural damage to Wall 41. A major reworking of the defences followed in Phase VI, involving the construction of cross-walls and buttresses (Bourke 1997:104; Potts *et al.* 1988:130). These buttresses include interior Buttress 83 and exterior Buttress 37 (McNicoll *et al.* 1992:42; Potts *et al.* 1985:197).

Phasing

There has been extensive excavation at Pella in a wide range of areas, each with a multiple number of trenches. The stratigraphy in each trench is given an individual phasing framework. The Trench phasing is then associated with the Area phasing, which in turn is associated with the Site Phasing.

This study is primarily concerned with the fortifications that were revealed during the excavation of Areas IIIF and XXVIIIC. Following the framework outlined above, each of these areas is phased individually. Because the occupational sequences differ between these two areas, the phasing terminology is inconsistent. For example, Phase B-C represents the first period of fortification construction in Area IIIF. Phase D represents the same period in Area XXVIIIC. The relationship between the phasing of the two areas is illustrated in Table 2.

Area IIIF

Area IIIF is located to the exterior of Wall 41. Excavation during 1983-1984 revealed a monumental brick wall (Wall 7), over 3.5m wide, running perpendicular to Wall 41 (Potts *et al.* 1985:197). The large size of the wall and its close proximity to Wall 41 indicate that Wall 7 also served a defensive function, either as a change in direction of the city wall or possibly as part of a gate (McNicoll *et al.* 1992:42).

In 1994, excavation resumed in Area IIIF to further examine the MBA fortifications, in particular the relationship between Wall 41 and Wall 7. The most important area of excavation is Locus 5, which comprised a 2.5m x 2.0m sondage placed at the point where Wall 41 and Wall 7 would most likely converge[7]. The sondage was carried through 6.0m of brick superstructure before reaching the substructure of Wall 41 and Wall 7. The stratigraphic sections and phasing for the sondage are illustrated in figures 9-12.

The sequence of developmental phases in Area IIIF occurred in relatively rapid succession during the MBI-II transition (Phases A-F), indicating that the MBI-II transition was a period of rapid growth and activity in this part of the site. It must be emphasized that the ceramic material from Phases E-F is limited. It is possible that these later phases could be associated with the more mature MBA.

Phase A

Phase A represents the earliest surviving settlement above sterile in this part of the site. It is comprised of a series of earthen deposits containing stones, pottery and ash layers. This phase of activity immediately precedes the construction of the MBA fortifications. Although there is evidence of EBI pottery associated with this phase (Bourke *et al.* 1998:189), the majority of forms are associated with the MBA. Sterile was reached only in the western half of the sondage where the foundations of Wall 41 (Phase B) were exposed. The comparative ceramic study for the Phase A sample (figure 13) places this phase in the MBI-II period transition. 69% of parallels are found with MBII assemblages and 22% of parallels are found with MBI assemblages. Over 78% of close parallels are with the MBII period.

Table 2: Pella Phasing

Area IIIC-D	Area IIIF	Description	Area XXVIIIC	Description
			A	Chalcolithic
			B	EBA
	A	MBA Pre Fortification	C	MBA Pre Fortification
X	B	MBA Wall 41	D	MBA Fortifications
IX	C	MBA Surface 5.17	E	MBA Post Fortification
VIII-VII	D	MBA Collapse	E	MBA Post Fortification
VI	E	MBA Wall 7	E	MBA Post Fortification
V	F	MBA Post Fortification	E	MBA Post Fortification

[7]Locus 5 was later changed to locus 7 during the course of excavations

Phase B-C

Phase B represents the construction of Wall 41. Wall 41 extends 0.3m into the north-west corner of the trench before disappearing behind the west section as it extends south. Wall 41 is preserved to a height of 6.0m (35 brick courses). A foundation trench had been excavated for the exterior substructure of Wall 41, as was evident on the wall interior (Area IIIC). The foundation cut in Area IIIF had been excavated through the earlier levels of Phase A into sterile. The cut had been lined with clay to provide a bed for the eight course stone foundation. Figure 9 demonstrates that the top level of the Wall 41 substructure stepped upslope from south to north. A similar system was employed to counter the slope in Area XXVIIIC.

A brick superstructure, comprised of three bands of brick types that vary in colour and consistency, is found above the Wall 41 foundation. The lowest brick band consists of twenty courses of ashy grey bricks fitted together with bands of orange mortar. The middle stage of the superstructure consists of a 1.2m thick band of green bricks with brown mortar. The highest surviving stage of the superstructure is constructed with the same grey bricks as the lowest stage of the superstructure. The use of different brick bands is consistent with earlier excavation undertaken on the interior of Wall 41 in Area IIIC. A similar use of bricks was also observed in Wall 7 (Phase E).

Reasons for the use of different brick types may be debated. The upper brick band could represent a reconstruction of Wall 41, contemporary with the construction of Wall 7 (Phase E). This explanation is unlikely, however, because Wall 7 abuts Wall 41. Damage to the superstructure resulting in a consistent cut back to a single level over such a large area is also unlikely. The changes in brick bands run consistently through the entire exposed length of wall.

The change in brick types is most likely a deliberate construction technique. A fortification system encompassing a city the size of Pella would require an enormous volume of bricks and a considerable period of time to build (see p.31, fn.20). Sections of the defences were probably built in stages using bricks from different clay sources. As one vertical section was completed using one type of brick, the next section could be constructed using another type of brick. Different clay types may also have been employed in a specific sequence in order to facilitate drainage and enhance the stability of the wall.

Phase C is comprised of a level of pise and plaster mixed with fist-sized stones (surface 7.12) that abuts the lowest brick courses of Wall 41. Due to the limited horizontal exposure within the sondage, it is difficult to determine the true function of the surface with absolute certainty. The surface may be a layer of randomly accumulated debris or it may represent a plastered *glacis*. There are multiple layers of plaster and stone, indicating that the slope has been deliberately resurfaced. If the feature is a *glacis,* it is a simple one. It is comprised of a plaster layer constructed above the existing slope rather than sealing an earthen rampart (see p.39)[8]. Whether surface 7.12 is a random feature or a true *glacis*, it immediately follows the construction of Wall 41 (Phase B). For this reason, Phases B and C are considered as two parts of a single phase.

Phases B and C are represented by a limited sample of pottery; therefore, construction of the Area IIIF defences is primarily dated by the pottery associated with Phase A, which immediately precedes the fortification construction, and Phase D, which immediately follows. Both Phases A and D are represented by substantial ceramic samples and are associated with the MBI-II transition.

Phase D

Phase D consists of an accumulation of debris following the construction of surface 7.12. The debris consists of a series of layers of slurry and brick collapse. Layers of different coloured bricks were identified, indicating that the debris is from Wall 41. Phase D represents a period of collapse when Wall 41 fell into disrepair and the possible *glacis* (surface 7.12) went out of use. The Phase D collapse corresponds to the occupational break between Phases VIII and VII in Area IIIC (Bourke 1997:104).

Phase D provides the largest pottery sample in Area IIIF. The results of the comparative ceramic study (figure 14) closely resemble those of Phase A. The Phase D sample finds 55% of parallels with MBII assemblages and 34% with MBI assemblages. Over 65% of close parallels are with the MBII period and 31% of close parallels are with the MBI period. This phase should thus be placed in the MBI-II transition or in the early part of the MBII period.

Phase E

Phase E represents the construction of Wall 7. This wall runs at a right-angle to Wall 41 and represents the final stage of fortification construction in Area IIIF. The lower 42 courses of Wall 7 consist of orange bricks bonded by dark brown mortar. The four surviving upper courses consist of dark red bricks bonded with white mortar. The foundation courses of Wall 7 stepped up from the east to the west, following the slope of level 5.6. No foundation trench or stone substructure was provided for Wall 7.

Wall 7 was constructed at a later date than Wall 41. This argument is based on the following factors. First, the foundations of Wall 7 are 3.0m higher than the foundations of

[8] There is no evidence for a similar *glacis* outside the city wall in Area XXVIIIC; although, it is worthwhile noting that earthen ramparts are not always constructed consistently around an entire site. Occasionally, ramparts are only constructed at the most vulnerable parts of the defensive circuit, as at EBII-III Tell Ta'anach (Lapp 1969:14), MBI-II period Carchemish (Woolley 1921:43) and possibly at MBII-III period Yavneh Yam (Finkelstein 1992:205) and at MBIII period Tell Shiloh (Finkelstein *et al.* 1993:52).

Wall 41. Second, a significant period of collapse (Phase D) had elapsed between the construction of Wall 41 (Phase B-C) and Wall 7 (Phase E). Third, Wall 7 abuts Wall 41 (see figure 9). Fourth, the bricks of Wall 7 are made from different coloured clay than the bricks of Wall 41.

The function of Wall 7 is debatable. It is unlikely that Wall 7 represents a return in the line of circuit Wall 41 (figure 15a), as Wall 41 continues to the north. Possibly, Wall 7 represents an external tower that was added to Wall 41 at a later date. Alternately, Wall 7 may represent part of a gate system, possibly a gate tower (McNicoll *et al.* 1992:42). This argument assumes that a break in the Wall 41 circuit is located not far to the north. The central problem lies in the difference in dates between Walls 7 and 41. If Wall 7 was part of a gate, the gate is a later addition to the Wall 41 circuit. Wall 7 can not have been part of an original gate associated with Wall 41. Also, if Wall 7 is part of a gate system, then the gate protrudes beyond the line of Wall 41 (as illustrated in figure 15b) rather than being contained within the line of the fortification trace. This would be highly unusual, as MBA gates were usually constructed largely within the defensive circuit (see figure 48).

Wall 7 may also represent an attempt to consolidate Wall 41 (figure 15c). The stratigraphy demonstrates that by the time Wall 7 was built, Wall 41 had partially collapsed (Phase D). It is possible that Wall 7 was constructed as an external buttress to shore up an unstable section of Wall 41. There is evidence for similar buttressing on the exterior (Buttress 37, Area IIID) and interior (Buttress 83, Area IIIC) faces of Wall 41. This work was undertaken during Phase VI in Areas IIIC and IIID. It is possible that Phase E in Area IIIF should be associated with this major period of reconstruction.

The ceramic sample that represents Phase E (level 5.6) is extremely limited. The ceramic material from Phase D (MBI-II) provides the most useful material for dating the Wall 7 foundations.

Phase F

Phase F represents 3.0m of collapse that accumulated to the east of Wall 41 and to the south of Wall 7. The collapse post-dates the construction of Wall 7. Although the pottery for this phase is limited (figure 16), the comparative ceramic analysis indicates that 67% of parallels are associated with MBII assemblages and 24% with MBI assemblages. Over 77% of close parallels are with the MBII period. Based on this data, Phases A-F should be associated with the MBI-II transition and MBII period.

Area XXVIIIC

Based on the presence of brick debris identified during a 1989 surface survey, a new trench area (XXVIIIC) was opened in the southern part of the site in 1995 (figure 7). Excavation continued in this area until 1997. The central feature of military architecture in Area XXVIIIC is a tower system flanked by city walls. The tower measures 8.0m east-west by 12.0m north-south. The entire south face of the tower has been exposed by erosion. The north face of the tower is revealed only 0.2m from the north baulk. To the west of the tower, a curtain wall is revealed for a distance of 11.0m. To the east of the tower, another curtain wall is revealed for a distance of 16.0m. In total, 35.0m of the east-west fortification circuit have been exposed.

Excavated locus areas are illustrated in figure 17. Loci 1-6 represent the original 3.0m x 6.0m trench area. Loci 7-9 represent a strategic cut that was made through the tower structure. Locus 10 represents an area inside the fortification circuit, to the west of the tower flank and abutting the interior of fortification Wall 9. Locus 10 revealed the deepest deposits of MBA material and is central to dating the fortifications[9]. The stratigraphic sections and phasing for Locus 10 are illustrated in figures 18-19. Locus 20 represents an area inside the fortification circuit, to the east of the tower flank and abutting the interior face of fortification Wall 10. This area provides important information on the nature of the MBA defences. Locus 20 is heavily eroded, which has resulted in shallow deposits and limited ceramic data[10]. Locus 30 comprises a chamber, running from north to south, within the core of Tower 1. The chamber is approximately 0.75m wide and of unknown length[11].

Phase A-B

Chalcolithic pits and burials represent the earliest evidence for human activity in Area XXVIIIC. This phase is only evident in Locus 10 (Phase A). The earliest evidence for human activity in Locus 20 is an EBA earthen deposit (Phase B).

Phase C

Phase C represents a MBA settlement preceding the construction of the fortification system. This phase is only evident in Locus 10. Phase C consists of a 1.5m deep deposit of alternating surfaces and earthen deposits. The only architectural feature of note is Wall 18, a 0.6m wide and three

[9] Locus 10 originally comprised an area measuring 8.0m (east-west) by 3.0-4.0m (north-south). As excavation progressed, the trench area was expanded to the west to gain a more substantial horizontal view of the Wall 9 interior, which includes Buttress 16. The western extension of Locus 10 measures 5.0m (east-west) by 3.0m (north-south) and is comprised of extremely limited remains compared to the eastern part of the locus. This is the result of heavy erosion on this part of the *tell* slope; a deep gully is situated to the west of locus 10. To the far western extent, curtain Wall 9 is completely eroded away.

[10] Locus 20 originally measured 8.0m east-west by 2.0-4.0m north-south. This locus area expanded with excavation.

[11] A 2.0m x 0.75m area was excavated to gain insight into the chamber composition. The southern wall of the chamber defines the southern extent of excavation. The northern extent of excavation consists of a section cut through the accumulated stratigraphy within the chamber.

course high wall, constructed using medium sized stones. Wall 18 is abutted and covered by the accumulated surfaces and earthen deposits. It is difficult to associate any definite function with Phase C in Locus 10.

Phase C provides the most substantial ceramic sample in Locus 10 and is clearly associated with the MBI-II period transition (figure 20-21). Over 58% of parallels are found with MBII assemblages elsewhere and 28% with MBI assemblages. Over 65% of close parallels are with MBII period assemblages. These results closely resemble those of Phase A in Area IIIF.

Phase D

Phase D represents the construction of the fortification system. Evidence suggests that the fortifications are the result of a single building effort. All aspects of the fortifications in Area XXVIIIC are discussed in this section.

The three major features of military architecture in this area include the tower (Tower 1), the western curtain wall (Wall 9) and the eastern curtain wall (Wall 10). These three features dominate the area. Secondary features include: Buttress 16 (10.39) built against the interior face of Wall 9; Buttress 12 (8.5) built against the exterior face of Wall 9 where it abuts the west flank of the tower; Wall 15 (20.10) built at a right angle to the interior face of Wall 10 and integral to the construction of the curtain wall; Revetment 10.26 abutting the western tower flank in Locus 10; and, the chamber built within the superstructure of the tower (Locus 30).

Excavation in Locus 10 revealed a 1.5m deep trench that had been cut for the foundations of the tower and city wall. The foundation trench cut through earlier material into sterile. Once the fortifications had been constructed, the remaining space between the fortification system and the vertical cut of the foundation trench was filled with alternating earthen, brick and ash deposits. The foundation cut for the fortification system (Phase D) was made from levels 10.29, 10.30, 10.37 and 10.41. Once the fortifications had been constructed and the fortification back-filled (10.47, 10.53 and 10.55) the area was sealed with a green brick surface (10.27, 10.28 and 10. 36). This surface defines the break between Phases C, D and E. The pottery sample from the foundation trench levels is confined to a few sherds and is too limited to be of use in a comparative ceramic analysis. The pottery from Phases C and D, that immediately precedes and follows construction of the fortifications, provides the foundation date.

Tower 1

Evidence for a stone substructure for Tower 1 (figure 22a) is revealed in Loci 9, 20 and 30. The stone substructure could not be seen in Locus 10, because it is covered by brick revetment 10.26 (figure 23a). In Locus 9 (figure 22b), the stone substructure is 1.0m deep at the exterior tower face and 0.5m deep towards the interior of the structure. In Loci 9 (figure 22b) and 20 (figure 25b), the stone substructure is deliberately stepped up the slope from south to north to counter the slope and provide a level base for the brick superstructure. The same technique is seen in Wall 41 in Area IIIF (figure 9).

The preserved height of the tower varies greatly due to the eroded slope. At its southern extent, only the stone substructure survived. To the north, the brick superstructure is preserved to a height of 5.0m (43 courses). The tower is eroded from west to east; the west flank is better preserved (24 courses) than the east flank (13 courses). Additional strength for the west tower flank is provided by the construction of revetment 10.26. This revetment consists of a brick skin, one course thick and eighteen courses high, built flush against the tower flank (figure 23a). The revetment is constructed within the tower foundation trench. There is no evidence for a similar revetment on the east tower flank.

The bricks used in the tower superstructure measure 0.4m x 0.4m x 0.1m and are laid with a running bond. Rectangular bricks, measuring approximately 0.7m x 0.4m x 0.1m, are used inside the core of the superstructure rather than on the tower faces or flanks. Along the line of the interior tower face, bricks measure 0.5m x 0.4m x 0.1m.

An unusual feature of the tower is a chamber located within the superstructure (figure 23b) The chamber measures 0.75m east-west. The north-south length of the chamber is not established; however, a 2.0m long section of the chamber was excavated from its southern wall. The chamber probably extends north through most of the width of the tower core. The vertical faces of the chamber are covered in lime plaster and had bowed inwards under the pressure of the brick superstructure. The stone substructure of the tower serves as the floor for the chamber, which is covered with a layer of plaster. A stone bench was revealed on the chamber floor. It is possible that other chambers exist within the tower superstructure.

The function of the chamber is debatable. Finds within the chamber, including complete ceramic vessels and the possible stone bench at the chamber base, indicate that the chamber originally functioned as an open area. The chamber may have been used for storage; although, the confined width of the chamber indicates that this would not have been practical. No obvious signs of stored material, such as storage jars or weaponry, were evident within the chamber. It is possible that the chamber held upright beams to support a further stage of the superstructure, but there is no evidence for burnt wood within the chamber. Instead, the material within the chamber consists of random layers of earthen fill that washed into the chamber once it was no longer used.

Curtain Walls 9 and 10

Curtain Wall 9 is a brick on stone wall, 2.5-3.0m wide, attached to the western face of the tower at an oblique angle (figure 24a). The wall is set back 1.5m from the south-west tower corner and is linked to the tower flank by a 1.0m deep stone buttress (Buttress 12). Wall 9 is well preserved to the east and heavily eroded towards the west, where it is preserved for a distance of 11.0m from the tower flank. The brick superstructure of Wall 9 is preserved to a height of 2.0m (18 courses) where the wall abuts the west tower flank (figure 24b). The brick superstructure of Wall 9 is built using a running bond and is integral to the tower superstructure. The exterior face of Wall 9 is covered with lime plaster.

Curtain Wall 10 is a brick on stone wall, 2.5-3.0m wide and built at a right angle to the east tower flank (figure 25a). The exterior face of Wall 10 is built on the same line as the tower exterior. Wall 10 is best preserved to the west and heavily eroded towards the east, where it is preserved for a distance of 16.0m east from the tower flank. The brick superstructure of Wall 10 is only preserved to a height of 0.4m (four courses) where the wall abuts the tower flank (figure 25b). The brick superstructure of Wall 10 is built using a running bond and is integral to the tower superstructure. A notable feature of Wall 10 is a sawtooth pattern of construction, approximately 7.0m east of the tower flank, where the line of the trace steps back 0.5m to the north. The wall line then continues to the east, though sloping slightly north to south.

The interior substructure of Walls 9 and 10 was constructed in a foundation trench. Due to heavy erosion, it is not possible to confirm whether the substructure exterior was also constructed in a foundation trench. Both Walls 9 and 10 have a stone substructure, as is evident in Locus 20 and to the west of Buttress 16 in Locus 10. The stone substructure is deeper at the exterior than the interior, a feature also noted in the tower substructure.

An unusual foundation feature may be noted in the section of Wall 9 between the west tower flank and Buttress 16. A platform of bricks was constructed over the entire area of the foundation cut, for this length of the wall. The line of the wall superstructure was then stepped in. The brick platform is three to five brick courses deep (0.3m - 0.5m) and represents a deliberate attempt to reinforce the foundation courses of the curtain wall and tower. Presumably, the brick platform is built above a stone substructure.

Buttress 16

Excavation of the Wall 9 foundation trench exposed Buttress 16 (10.39). Buttress 16 measures 2.5m east-west by 1.0m north-south and is preserved to a height of 1.0m (figure 24a). The foundations of Buttress 16 consist of three courses of green brick, topped with one course of medium sized stones. The stone course is on the same level as the stone substructure for Wall 9, to the west of the buttress, implying that the slope rose from east to west at this point. Above the stone course of Buttress 16 is a brick superstructure that is bonded with the brick superstructure of Wall 9. Buttress 16 and Wall 9 are thus part of a single phase of construction.

Buttress 12

Buttress 12 was built to a triangular plan within the oblique angle where Wall 9 abuts the west tower flank (figure 24a). Buttress 12 is 1.0m deep and links the west flank of the tower with the exterior face of Wall 9. The buttress consists of layers of stones consistent with the stone substructure of Tower 1 and Wall 9. Buttress 12 is sealed with a layer of lime plaster, as is also noted on the exterior face of Wall 9.

Wall 15

Wall 15 is a brick on stone wall, 1.5m wide, that runs at a right angle to the interior face of Wall 10. Wall 15 is structurally integral to the construction of Wall 10. Wall 15 has been heavily eroded and only 1.5m of the wall length (north-south) is preserved. A single brick course of the Wall 15 superstructure is preserved (figure 25a).

Wall 15 possibly functioned as a casemate dividing wall. Wall 15 is thinner than the curtain wall, as is typical of casemate defences. Stone packing between Wall 10, Wall 15 and the east tower flank is also indicative of a casemate fill (figure 25a). Although this packing is only preserved against the tower flank, it most likely extended as far as Wall 15. Wall 15 is also placed exactly where a casemate dividing wall should be, running parallel to the tower flanks, and forming a compartment between the exterior, and an assumed interior, curtain wall. There is some evidence for a double curtain (Walls 41 and 52) in Area III (see figure 8a), however, the existence of a roadway between these walls implies that there are no casemates in that area.

Phase E

Phase E represents the first phase of development following construction of the fortifications. The pottery associated with this phase is important for confirming the construction date of the defences in Area XXVIIIC.

Phase E is best represented in Locus 10. Preservation of Phase E in Loci 20 and 30 is poor. Following the construction of the fortifications, the entire area of Locus 10 was sealed with a brick surface (10.27, 10.28, 10.36, and 10.41). A series of accumulated earthen, stone and brick deposits and small north-south running walls (Walls 13, 14, 19) are found above this surface. The accumulation of occupational debris indicates that Locus 10 was intensively used at this time; although, it is difficult to say in what capacity. Each of the walls probably abutted the interior face of fortification Wall 9; although, this was never established due to heavy erosion. Even if the walls did abut fortification Wall 9, it is unlikely that they were structurally organic to the fortifications. The random placement of the walls indicates that they did not serve a casemate function.

The only substantial ceramic material for Phase E comes from Locus 10 (figure 26-27). The results indicate a date in the MBII period. Over 31% of parallels are found with MBI period assemblages and over 49% with MBII period assemblages.

Summary

• The fortification system revealed in Areas III and XXVIIIC was constructed during the MBI-II transition. Although these two areas are not related stratigraphically, evidence suggests that the fortifications are part of a single phase of construction.

• Excavation in Area IIIF exposed a monumental brick wall (41) that forms the main circuit trace (Phase B). This wall was constructed using different brick types to enhance stability. A simple *glacis* (Phase C) may have been constructed to protect the exterior wall face. Further evidence suggests that Wall 41 has a parallel internal trace (Wall 52); however, the second wall may have been part of a large building. The area between the walls was an open road. Phase B-C in Area IIIF equates to Phase X-IX in Area IIIC-D.

• In Area IIIF, a period of collapse (Phase D) was followed by a second constructional phase (Phase E). Phase E probably equates to Phase VI in Areas IIIC and IIID. A series of buttresses were constructed to provide added support to Wall 41. A large wall (Wall 7) was built at a right angle to the exterior of Wall 41. Wall 7 is either part of an external tower, a gate, or an external buttress for Wall 41. Further excavation is required to elucidate this issue.

• Fortifications were constructed in Area XXVIIIC during Phase D. The major defensive feature is Tower 1, which is the only excavated MBA tower in Jordan. City Walls 9 and 10 run off each tower flank. Wall 10 was constructed in the sawtooth manner and is possibly associated with a defensive casemate (Wall 15). Such a system would require a second trace wall to the interior of the site; however, the existence of an interior trace is not established. There is no evidence in Area XXVIIIC for an earthen rampart or *glacis* beyond the city wall.

Chapter 3: Rukeis

Introduction

Rukeis is situated on the Wadi al-Ajib in the Hauran (figure 1). The site is 1km to the south of the Syro-Jordanian border and 3km north of the village of Al-Dafyana. This area of the Hauran comprises the southern extent of the Jebel Druze, the prominent volcanic feature of the region. The Wadi al-Ajib runs south-west from the Jebel and is one of a series of *wadis*, which cut through the gently sloping region. The site is approximately 900m above sea level.

Rukeis is a small, 1.25ha site best described as a fortress or fortified town (figure 28). The fortification circuit that defended the site during the MBA dictates the round shape of the citadel. The entire wall circuit has been preserved, and the interior deposits have accumulated to the top of the fortification wall, as it exists today. Excavations have demonstrated that there was dense occupation in some places at least 25.0m beyond the main line of the visible fortification trace. The contours of the citadel suggest that there may have been a reservoir in the north-western part of the site.

Rukeis is situated on the wet-dry steppic interface and receives less than 200mm annual precipitation (Shehadeh 1985:31). Rukeis is located at the southern extent of the fertile plains of the Hauran, with the desert only a few kilometres to the south. To counter the lack of rainfall in this marginal zone, the inhabitants of MBA Rukeis constructed an elaborate hydraulic system. Similar systems may be noted at other sites in the Hauran, such as EBI period Jawa (Betts 1991), Byzantine period Umm el-Jimal (de Vries 1985:254) and Islamic period Berqa (Abujaber 1995:743). The basaltic soil of the region is fertile when irrigated and facilitates cereal production and pastoralism (de Vries 1985:252).

The major phases of occupation at Rukeis are associated with the MBA. Klengal argues that the Hauran was only thinly populated during this period (Klengal 1985:50), but evidence for major urban activity during the MBA has been established through more recent survey work (Braemer 1984; 1988; 1993; Eames 2001) and excavation (Braemer 1991; 1994). The importance of the Hauran during the MBA is reflected in the Egyptian Execration texts (Albright 1941b:30-35).

MBA Rukeis was not strategically located in relation to major trade routes. It was connected to the more verdant regions of the Levant through a system of secondary trade routes. The Hauran is geographically linked with the Damascene, and it is likely that trade generally moved in that direction. In the Roman, Byzantine and Nabataean periods, trade moved from Arabia to Azraq and onto Bosra and Syria (de Vries 1985:254). Rukeis may have figured in a similar trade pattern during the MBA. Although there is evidence for human activity in the Eastern Badiya from the Neolithic period onwards (Lancaster and Lancaster 1991:128; 1995:770; Betts and Helms 1989), sporadic rainfall makes it unlikely that an east-west trade passed through the desert, between the Hauran and the Euphrates.

Excavation History

From 1993-1999, the University of Sydney *Hauran Research Project* conducted four seasons of excavation at Rukeis[12]. The most significant excavation relevant to the fortifications was undertaken during 1994, 1995-96 and 1999 in Area 5 in the south, Area 8 in the east, and Areas 14 and 16 in the north (figures 28-29). Areas 5, 8 and 14 were excavated to sterile and provide the most substantial data for the dating and phasing of the fortification wall. Area 16 revealed the city gate. Details regarding Areas 5, 8, 14 and 16 are provided in the following sections of this chapter.

During 1999, Area 17 was established between Areas 14 and 16, in an attempt to stratigraphically link the two areas[13]. A wash gully cutting through Area 17 meant that this was not possible. As a result, work in Area 17 was subdivided into two areas, one integrated with Area 14 and the other integrated with Area 16[14]. The recording system at Rukeis follows a running number system, as opposed to the locus-level system used at Pella. The level numbers used in the text follow the trench numbers in Areas 5 and 8. Because Area 17 was a later extension of Areas 14 and 16, both the trench numbers and level numbers are used to describe the levels in those areas.

Phasing

The phasing schema at Rukeis differs from that employed at Pella. Rukeis is a smaller site, and occupation is limited to a few significant phases. It was therefore possible to establish a consistent phasing scheme that is applied to each area of excavation. The phasing associated with each area is illustrated in Table 3 (see p.21).

As this study is primarily concerned with the fortifications, only Phases II.1-II.2 are discussed. Phase II.2.i is the most relevant phase, because it represents the construction of the fortifications. Phases II.1 and II.2.ii-iii are also important for dating purposes, because those phases immediately precede and follow the construction of the fortifications.

---

[12] The site of Rukeis was first visited and described by Butler in the early 20th century. Butler noted the irregular dry-stone wall of large, uncut, basalt boulders which surrounded the *tell*. On the basis of Safaitic inscriptions, which he found at the site, he attributed the construction of the fortifications to that period (Butler 1919:116). During the 1980's, Braemer visited the site as part of his survey of the Hauran. Braemer also described the major features of the site and noted the predominance of MBA pottery (Braemer 1993:137). For more detailed background reference, see Betts *et al.* 1995 and 1996; Schroder 1997 and Eames 2001.

[13] Area 17 measured 17m x 3m.

[14] A 3m x 3m area was incorporated into Area 14. The stratigraphy and associated finds from Area 14 and the new addition from Area 17 have been phased together. Therefore, some level numbers begin with 14 while others begin with 17.

Table 3: Rukeis Phasing

Phase	Period and Description	Area 05	Area 08	Area 14	Area 16
I	EBIB. Domestic Occupation	*			
II.1	MBII. Pre-Fortification	*	*	*	
II.2.i	MBII. Fortification	*	*	*	*
II.2.ii	MBII. Post-Fortification	*	*	*	*
II.2.iii	MBII. Post-Fortification	*	*	*	*
II.3-4	MBII-III. Partial Site Abandonment	*	*	?	?
II.5	MBII-III. Reoccupation	*	*	?	?

Evidence for Phase I was only revealed in Area 5. Evidence for Phases II.3-5 was revealed in Areas 5, 8 and 14, and should be dated to the MBII-III period transition (Schroder 1997:145). Phase II.5 abuts the highest surviving point of the internal face of the fortification wall. Assuming that the city wall was still in use as a defensive feature, the wall was originally of a much greater height than it is today.

Area 5

Area 5 is located to the south of the site, abutting the exterior face of the fortification wall (figures 28-29). When excavation began in 1994, Area 5 was a 3m x 3m test trench. In 1995-96 Area 5 was expanded to 6.5m x 3m. The stratigraphy and phasing for Area 5 are illustrated in figures 30-32.

Phase II.1

The first MBA settlement at Rukeis preceded the construction of the fortifications. Evidence for this unwalled settlement was revealed in Areas 5, 8 and 14, indicating that the first MBA settlement was a similar size to the later fortified site. Phase II.1 in Area 5 is represented by a floor fragment and an earthen deposit. No architecture is associated with this phase. Phase II.1 is similarly represented in Areas 8 and 14, where it is never more than 0.5m in depth.

Ceramic data analysis indicates a MBII period date for Phase II.1 (figure 33). The fortifications of the previous and following phases are also associated with a MBII period date. The dating results indicate that the period of initial reoccupation and the subsequent development of a fortified settlement occurred in relatively rapid succession during the MBII.

Phase II.2.i

Phase II.2.i represents the construction of the monumental fortification circuit surrounding the site. The fortification wall (200) consists of monumental basalt blocks placed in approximately horizontal courses. The wall is preserved to a height of 4.0m (eleven courses); although, it was originally higher, based on the quantity of collapsed basalt boulders in the area. The width of the wall varies from 2.5-3.0m. The range in boulder sizes varies considerably; there is little evidence of worked stone. No evidence could be found for a plastered or rendered finish on the wall exterior.

A 1.0m deep trench was cut for the foundation courses of the wall. Once the wall substructure was complete, the foundation trench was filled with debris (levels 212, 213 and 215). It is unknown whether the foundation cut extends to the internal wall face or is confined to the exterior. In Area 8, no foundation cut was made for the internal wall courses. It is possible that the foundation cut was only made in parts of the site where the existing surface was considered unstable.

Area 5 has the most substantial pottery sample for Phase II.2.i (figure 34). 80% of close parallels are with the MBII period. Close parallels are only found with MBI and MBII sequences. No close parallels are found with MBIII period sequences. Phase II.2.i. should thus be associated with the MBII period.

Phase II.2.ii-iii

This phase, immediately following the wall construction, is well represented at the site. In Areas 5, 8 and 14, there are two architectural sub-phases. In Area 5, Phase II.2.ii is represented by levels 206-211 and Phase II.2.iii by levels 203-205. Although Area 5 is located outside of the fortification circuit, there are considerable occupational deposits associated with this phase, including a tabun, a series of plaster floors and various well-stratified deposits.

Only the diagnostic assemblage for Phase II.2.ii is used for dating in Area 5 (figure 35). The pottery sample for Phase II.2.iii is too limited to produce concrete results. Over 32% of parallels are found with MBI assemblages, 40% with MBII assemblages, and 20% with MBIII assemblages. Over 50% of close parallels are with the MBII period, while only 6% are with the MBIII period. This data indicates a date later in the MBII period.

Area 8

Area 8 is located in the east of the site, abutting the interior face of the fortification wall (figures 28-29). When excavation began in 1994, Area 8 was a test trench within Area 4. In 1995-96, Area 8 was expanded to 4m x 4m. The plan, stratigraphy and phasing for Area 8 are illustrated in figure 36-37.

Phase II.1

Phase II.1 represents the earliest evidence for human settlement in Area 8. In contrast to Area 5, there is no foundation trench for the fortification wall. The stratigraphic relationship between Phase II.1 and the city wall (Phase II.2.i) is clear, because Phase II.1 runs beneath the lowest foundation course of the fortification wall. The stratigraphy of this phase consists of a series of pits cut into sterile, sealed by fill layers and a plaster floor. Material within the pits is predominantly MBA. As in Areas 5 and 14, there is no architecture associated with Phase II.1.

The comparative ceramic analysis for Phase II.1 in Area 8 produced similar results to those in Area 5. Over 43% of parallels and 58% of close parallels are with MBII sequences (figure 38). There is only one close parallel with MBIII period sequences. Phase II.1 should thus be associated with the MBII period.

Phase II.2.i

In Area 8, the city wall (000) is preserved to a height of 4.5m (20 courses) and a width of 3.0-3.5m. In contrast to Area 5, the foundation courses are placed directly on the existing surface (Phase II.1). In Area 8, there is no evidence for a foundation trench for the city wall. Whether the wall exterior was provided with a foundation trench, as in Area 5, is unknown. No pottery is directly associated with this phase. As a result, a foundation date for the city wall relies on the ceramic samples from Phase II.1 and Phase II.2.ii-iii that immediately precede and follow the wall construction.

Phase II.2.ii-iii

As in Areas 5 and 14, there are two phases of MBA occupation in Area 8 that immediately follow the construction of the city wall. Phase II.2.ii is represented by over 3.0m of randomly deposited debris, consisting of layers of densely packed orange clay slurry, brown clay and grey ash. Phase II.2.ii is a deeper deposit in Area 8 than in Areas 5 and 14. These fill levels may be interpreted as accumulated domestic or industrial refuse following the construction of the MBA wall or as a deliberate attempt to build up and level this part of the site. No architecture is associated with this phase. Phase II.2.iii represents a direct continuity from Phase II.2.ii. In Area 8, this phase consists of occupational walls, floors and fill layers. All of these layers abut the interior face of the city wall.

A large amount of pottery is associated with Phase II.2.ii-iii in Area 8. Due to the size of the pottery sample, only a proportion of the sample was selected for the comparative ceramic analysis (figure 39-40). As the main aim was to confirm a date for the fortification system, pottery was selected from the Phase II.2.ii levels that immediately followed the construction of the city wall (levels 058-066).

The comparative analysis for this phase demonstrates continuity with Phase II.1. The largest number of close parallels is found with MBI and MBII period assemblages. More evidence for MBIII period parallels is found for this phase than elsewhere on the site. The majority of the MBIII parallels, however, are variable rather than close matches. The results suggest that the period between the initial MBA resettlement of the site and the development to a fortified site occurred relatively rapidly during the MBII period.

Area 14

Area 14 is located to the north of the site, abutting the interior face of the city wall. When excavation began in 1995-96, Area 14 was a 10m x 2m trench. During 1999, Area 14 was expanded to the west with a 3m x 3m trench[15]. The Area 14 plan, stratigraphy and phasing are illustrated in figure 41.

Phase II.1

Phase II.1 is immediately above sterile and consists of alternating earthen and stone deposits and surfaces. The phase ends with the construction of a surface over the trench area (levels 14038-39). There is no evidence of architecture during this phase, which implies that this part of the site was open to the elements. The same pattern of development is evident in Areas 5 and 8. Parallels for the Area 14 pottery associated with this phase indicates a date in the MBII period; however, the diagnostic material is too limited to offer substantial results in terms of dating this phase.

Phase II.2.i

As in Areas 5 and 8, Phase II.2.i in Area 14 is represented by the construction of the basalt city wall (14000). The phasing of Area 14 is more complex than in Areas 5 and 8, because a monumental brick wall (14033) directly abuts the interior face of the basalt city wall (14000). The brick wall, in turn, is abutted on the interior by another stone wall (14030). Walls 14033 and 14030 separate the city wall (14000) from the occupational stratigraphy to the interior of the site, causing difficulty in identifying phase changes in relation to the city wall. Despite this, it is likely that the three walls formed part of a single architectural unit integral to the defences.

---

[15]The Area 14 extension was originally part of Area 17 (see p.20).

Wall 14000
Wall 14000 forms the northern extent of Area 14. This wall has been traced around the entire perimeter of the site, is of a consistent construction in all excavated areas, and is part of a single phase of construction. Wall 14000 is a basalt wall approximately 3.0m wide. The interior face of this wall was not examined because it directly abuts the monumental brick wall 14033.

Wall 14033
Wall 14033 directly abuts the interior face of Wall 14000. Wall 14033 is approximately 3.0m wide and built of red bricks, measuring 0.45-0.50m x 0.12-0.15m. The brick wall is as large as the basalt city wall, and must have been part of a monumental structure of administrative or religious function or integral to the fortification system. The material culture in this area does not indicate an association with a religious or administrative function. Instead, the area to the south of Wall 14033 was either open to the elements (Phase II.1 and II.2.ii) or served a domestic function (Phase II.2.iii). Wall 14033 most likely served a defensive function as an inner buttress to the main fortification trace where the trace joins the gate structure (figure 45a-b). Such a feature would have strengthened the most vulnerable section of the defensive circuit and explains why the brick wall is not found in other parts of the site.

Wall 14030
Wall 14030 directly abuts the interior face of Wall 14033. It is a one course wide basalt wall that survives to a height of 2.5m (six courses). Wall 14030 is of particular importance to the phasing of the fortifications in Area 14, because it separates the relevant stratigraphy in the area from Walls 14000 and 14033. The interpretation of this wall's function and it's relationship to Walls 14000 and 14033 is thus of great significance.

Following Phase II.1, a foundation trench for Wall 14030 was cut in the northern part of Area 14. The foundation trench was cut from surface 14038, through the preceding deposits of Phase II.1, to the level of bedrock. The foundation trench extends 1.0m south of the interior face of Wall 14030 and runs parallel to the wall line over the exposed area. Following the construction of Wall 14030, the foundation trench was filled with debris (levels 17019, and17024).

Wall 14030 served either a domestic or defensive function. It is unlikely that it served a domestic function, because the area to the south was open to the elements when the wall was built. It was only later (Phase II.2.iii) that Wall 14030 was incorporated into a domestic structure. If this area was open to the elements at the time of construction, the function of Wall 14030 may have been to protect the interior face of brick Wall 14033 from the elements. Wall 14030 likely served a function integral to Wall 14033. Both of these walls appear to be integral to basalt city wall 14000.

Although the diagnostic pottery from the Wall 14030 foundation trench is limited, it is included in this discussion because the pottery from the foundation trench represents the construction phase of the fortifications (figure 42a). The results of the comparative study for this phase are similar to the results for the same phase in Areas 5 and 8, with close parallels found only with MBI and MBII sequences and none with MBIII period sequences.

Structural Relationship
Without detailed stratigraphy, it is impossible to confirm the true structural sequence for walls 14030 and 14033. It is most likely that the basalt city wall (14000), the large brick wall (14033) and the smaller basalt wall (14030) were built as a single architectural unit. If so, the Wall 14030 foundation trench represents the foundation of the fortifications in Area 14 (Phase II.2.i). This interpretation fits well with the evidence from Areas 5 and 8, where a brief period of MBA settlement (Phase II.1) was followed by the fortification construction (Phase II.2.i) and two intensive phases of urban settlement (Phase II.2.ii-iii). The same sequence is evident in Area 14, implying that the construction of walls 14000, 14033 and 14030 should be associated with Phase II.2.i.

Phase II.2.ii-iii

In Area 14, the stratigraphy for Phase II.2.ii-iii abuts the interior face of Wall 14030. Phase II.2.ii immediately follows the construction of the fortifications and is represented by a 0.5m deep deposit of alternating surfaces and earthen layers. During this phase, the area to the south of the fortification circuit was open to the elements.

Phase II.2.ii is superseded by a distinct architectural phase during Phase II.2.iii, a pattern also evident in Areas 5 and 8. During this phase, a pillared building is established to the south of the fortification wall. The new building incorporates Wall 14030, which functions as the northern wall for the building. To the south of Wall 14030, a new wall (14025) was built parallel to the fortification circuit. These two walls are integral to the new pillared building and form a 4.0m wide room, characterised by occupational surfaces and stone pillars to support the roof.

Once the building was in use, a series of inhumation burials were placed beneath the floor surface in the corners of the room. A series of surfaces, slurries and ash then accumulated to a depth of 1.0m above the original floor surface. The addition of Wall 14031, that connects Wall 14025 with one of the pillars within the building, is evident at this time.

The comparative ceramic analysis for Phase II.2.ii (figure 42b) finds over 50% of parallels with MBII assemblages, 25% of parallels with MBI assemblages, and 20% with MBIII assemblages. There are no close parallels with the MBIII period. A similar picture emerges for Phase II.2.iii (figure 43) with most close parallels with the MBII period.

Area 16

Area 16 is located to the north of the site, west of Area 14, and was established to examine the city gate (figures 28-29). When excavation began in 1995-96, the trench area measured 17m x 13m. A 3m x 5m trench was placed against the south-east corner of the gate during 1999 (Area 17). The plans for the Area 16 gate are illustrated in figures 44-45.

The Gate

The gate construction is the earliest excavated phase in Area 16. Part of the gate plan is visible on the surface due to erosion. Most of the excavated material within the gate area consists of contaminated fill and collapse; however, some occupation levels are intact and provide datable material.

The gate is based on a rectangular plan with the long axis running east-west parallel to the *wadi*. The gate is structurally linked to the main line of the fortification trace. The exterior face of the gate does not extend beyond the line of the wall trace, presenting a flat façade on the exterior with the bulk of the structure behind the line of the wall trace. The gate is comprised of two large towers flanking a central passage. The two towers have similar dimensions and ground-plan, except for a few minor variations. Both towers are approximately 5.0m wide at the south and 6.5m wide at the north, are buttressed with imposing bastions at the gate exterior, and contain guardrooms within the gate interior.

The gate is placed at the most defensible position. The *wadi* to the north of the site limits the line of approach. Any attackers would have to approach the gate from the west, parallel to the fortification wall. Although an approach ramp was never identified, evidence of an approach from the west may be seen in a possible attempted breach of the city wall immediately to the west of the gate (figure 45b).

The exterior face of the gate measures 15.0m; the interior face measures 13.5m; the east flank measures 11.0m; and, the west flank measures 10.5m. Both flanks curve inwards toward the gate structure. The entire structure is built with large basalt boulders, some of which are cut to shape. The largest stones are located along the flanks and exposed faces of the structure and measure 1.5m x 1.5m x 0.5m. The average stone size is approximately 1.0m x 0.75m x 0.5m. Smaller stones are used behind the wall faces and flanks. As with the main curtain wall, the entire gate is built using dry-stone construction.

Walls 1-4

The flanking and interior walls form the south, east and west sides of the gate. Wall 1, the east flank, is 11.0m in length and is 2.0m wide. From south to north, the wall curves slightly inwards before curving out towards the north-east corner. Only the southern 7.0m of Wall 1 is a true wall; the northern 4.0m are integrated into the north-east bastion. A sondage in the east guardroom exposed Wall 1 to a depth of 1.5m, but the foundation courses were never reached. Excavations abutting the south-east corner of Wall 1 revealed the top of the foundation trench at a depth of 1.5m.

Wall 2 runs from the south-east corner of the gate to the interior gate passage, and forms the southern extent of the east guardroom. The wall is 5.0m long and 2.0-2.5m wide. Worked boulders form the lining of the gate passage. Wall 3 has a parallel function to Wall 2, running from the south-west corner of the gate to the interior gate passage. Wall 3 forms the southern extent of the west guardroom and is 5.0m long and 2.0m wide. The interior face of Wall 3 is lined with a basalt stone revetment, one course wide and 1.3m deep. The stratigraphy within the gate abuts this revetment rather than Wall 3. The revetment was possibly added to Wall 3 during a later phase of construction.

Wall 4, the west flank, measures 10.5m long by 2.0m wide. The wall curves in towards the centre and out towards the north-west and south-west gate corners. The wall is clearly defined for the southern 7.5m of its extent. The northern 3.0m of the wall is integrated with the north-west bastion. Wall 4 is exposed to a depth of 0.5m for its entire length and to a depth of 1.6m in the sondage in the west guardroom. The lowest wall courses were reached at a depth of 1.5m below the existing surface.

East Tower

The east tower is the more solidly constructed of the two bastions. The exterior 6.5m of the tower consists of an almost completely solid block of basalt boulders, except for a small chamber leading off from the guardroom. The chamber measures 2.0m by 0.5m in width and was probably used for storage. The chamber was cleared to a depth of 0.75m.

A room measuring 4.0m x 3.5m is located on the interior of the east bastion. This room was exposed to a depth of 1.0m across its entire extent and to a depth of 1.5m in a sondage in the corner of the room. Within the room, two basalt pillars are preserved to a height of 1.0m. These pillars originally supported the ceiling of the chamber. A semi-circular basalt bench, one stone course wide and 3.0m long, is partly preserved in the south-west of the room. The room probably functioned as a guardroom

The earliest levels reached in the east guardroom are in the 1.0m square sondage located in the corner of the room abutting Walls 1 and 2. The lowest levels of Walls 1 and 2 were not reached in this area. The lowest excavated deposit was a hard, compacted level of orange slurry, approximately 0.6m deep. This level formed a compact surface (16012) that functioned as the original surface associated with the gate use. In some areas, this surface consisted of a patchwork of plaster and small, flat stones. Layers of ash, stone collapse, and fill layers covered surface 16012.

West Tower

Like the east tower, the west tower has a heavily strengthened bastion towards the exterior and a guardroom towards the interior. The bastion measures 7.0m x 6.0m. A chamber measuring 2.5m x 2.5m was constructed within the bastion core, possibly to be used for storage; although, it appears to have been deliberately filled with rubble.

The major feature of the west tower guardroom is a well preserved stone bench located against the wall of the bastion. The bench is 3.0m long and one course wide. There is a difference in the configuration of the guardrooms. The east guardroom is open to the passage and contains pillars. The west guardroom is entered through a 1.0m wide door, and is separated from the passage by a wall. No pillars have survived in the west guardroom; although, they were probably employed in the original structure.

The lowest level reached in the guardroom sondage is a deposit of brick collapse. Above this level is a 0.5m deposit of compact, hard orange material (levels 16006,011,and 013) that equates to level 16012, the original gate floor, in the east guardroom. In the west tower guardroom this surface was covered with layers of ash and collapse, representing either a destruction layer or a localized fire on the chamber floor.

The Passage

The gate passage is remarkably well preserved considering its proximity to the present surface. The original passage floor, exterior door socket and covered drain are still in place. The passage floor is a patchwork of hard, compact earth and small, flat stones. Fallen lintels from the entrance to the west guardroom are also found on the passage floor. No good stratigraphic contexts were excavated above the passage floor.

The entry passage into the gate is 2.0m wide at the northern extremity. The passage is narrowed to 1.5m at the exterior entrance by the construction of a wall measuring 1.0m x 2.5m. A stone door socket with two bored holes is located across the passage, against the west line of the gate passage and marks the position of the original door. To the south of the door, the passage widens to 3.0m, a width that continues for the remaining extent of the passage to the south.

Dating the Gate

Due to the gate's proximity to the existing surface, the pottery sample associated with the gate construction is limited. Most of the stratigraphy that post-dates the original gate surface is contaminated, mixed debris. The only areas of excavation that went below the gate surface (the small trenches in the corner of each guardroom) were almost completely devoid of diagnostic pottery. To address this problem, a further area of excavation was opened abutting the south-east exterior corner of the gate (Area 17). Although pottery finds in this area are limited, they provide the most useful dating evidence for the gate.

The new excavation area revealed the foundation trench (17015) of the gate flank, but only produced a limited amount of pottery. An earthen layer (17009) seals the foundation trench and is associated with an architectural structure (Wall 17007) to the east gate flank exterior. Occupational levels abut both the gate flank and wall 17007.

Because the levels associated with this structure are stratigraphically later than the gate foundation trench, they clearly post-date the construction of the gate. The ceramic evidence from this secondary structure provides a date for the gate construction (figure 46). Although the data is limited, the ceramic assemblage representative of this phase finds over 64% of parallels with MBII assemblages. Over 90% of close parallels are with the MBII period. None are found with MBIII period sequences elsewhere, indicating a construction date no later than the MBII period.

Phasing the Gate

Phasing the gate construction with the rest of the site is problematic because only the upper phases of Area 16 were excavated. The foundational stratigraphy of the gate was not examined nor was the relationship between the gate and the circuit wall. The central question is whether the gate is contemporary with the wall circuit (II.2.i) or whether it was built during a later phase (II.2.ii-iii or II.5), possibly as a rebuild of an earlier gate. The starting point when addressing the gate phasing is the MBII period pottery that post-dates the gate. This ceramic evidence implies that the gate should either be associated with Phase II.2.i, Phase II.2.ii or Phase II.2.iii.

Structurally, the gate is almost certainly integral to the basalt city wall (Phase II.2.i). The gate foundations, however, are relatively shallow (1.5-2.0m) in comparison to the city wall, which may suggest that the gate was constructed during a later phase of development (Phase II.2.ii-iii). The difference in levels between the gate and wall foundations, however, may be explained by the fact that the foundations were only encountered towards the gate interior. Further excavation could well demonstrate that the foundations are more substantial towards the gate exterior. A combination of substantial foundations towards the structure exterior and shallow foundations towards the structure interior is also evident at Pella (see figure 22b).

Given the available evidence, a gate construction date in Phase II.2.i seems most likely. Therefore, Phase II.2.i includes the building of the basalt city wall (Areas 5, 8, 14 and 16), the gate (Area 16) and the monumental brick wall and stone wall internal buttresses (Area 14). Combined, these features functioned as a single defensive unit.

Summary

• The fortification system at Rukeis is the result of a single construction effort during Phase II.2.i. The pottery associated with the fortifications indicates that construction was undertaken during the MBII. One limitation to this conclusion is the regional nature of the pottery at Rukeis. The MBA pottery from Rukeis finds limited parallels with MBA sequences elsewhere in the Levant, indicative of a strong degree of regionalism in the Hauran (see p.46).

• The major feature of the fortification system is the basalt city wall. The city wall is preserved around the entire perimeter of the site and was studied in detail in Areas 5 and 8.

• In Area 14, the basalt city wall is abutted on the interior by a monumental brick wall (14033) and a smaller stone wall (14030). Walls 14033 and 14030 appear to be contemporary with the construction of the main basalt circuit wall and probably function as internal revetments for that wall. These internal revetments likely strengthened the gate flank to the west. It is possible that a similar revetment exists on the other gate flank.

• The most prominent feature of the fortification circuit is the city gate in Area 16. The gate is structurally linked to the city all circuit and was almost certainly built during the same construction phase (Phase II.2.i). The Rukeis gate is the only known MBA gate in Jordan.

Chapter 4: Comparative Study

Walls and Related Features

Solid Walls

Single trace, solid walls were first used in fortification systems in Jordan during the EBA, as is evident at Khirbet Zeraqun (Ibrahim and Mittman 1997:388), Dibon (Winnett 1952:12), Tell Abu al-Kharaz (Fischer 2000:448), Tell Jawa (Geraty *et al.* 1989:144), Tell Jamid (Leonard 1992:24), Tell el-Handaquq North (Mabry 1989:65; 1996:123), Jawa (Helms 1975:27; 1977b:29), Khirbet Iskander (Richard 2001:441; Richard and Boraas 1984:68) and possibly Jerash (Glueck 1939b:26). Solid walls were also commonly used in MBA fortification systems, as is evident at Pella (pp.15,18), Rukeis (pp.21-23), Amman, Tell Irbid, Sahab, Tell Deir Alla and Tell Nimrin (pp.8-9)[16]. A survey of solid defensive walls of the Bronze Age Levant reveals that walls built of stone, brick and a combination of stone and brick were commonly employed.

Solid walls constructed of stone have been excavated at numerous EBA sites, including: Yanik tepe (Burney 1961:144) and Demirchihuyuk (Korfmann 1977:37) in Anatolia; Tell Ashara (Buccellati and Kelly-Buccellati 1978:75), Tell Mu'azzer, Tell Malhat adh-Dhra and Tell Mabtuh ash-Sharqi (McClellan and Porter 1995:51) in Syria; and, Arad (Amiran 1978), Tell Ta'anach (Lapp 1964:10), Ai (Callaway 1980:41), Beth Yerah (Esse 1991:39), Tell Marjamah (Zohar 1980:219), Tell Kinrot (Yakar 1984b:191) and Megiddo (Loud 1948:66) in the Southern Levant. Stone walls constructed during the MBA have been revealed at: Carchemish (Woolley 1952:209), el-Qitar (McCleland 1986:423) and Tell Hadidi (Dornemann 1979:141) in Syria; and Tell Beit Mirsim (Albright 1938a:18), Bethel (Albright 1934:5; Kelso 1958:4), Jerusalem (Kenyon 1966a:75),Tell Marjamah (Zohar 1980:219), Tell el-Far'ah North (Mallet 1987 Plan 14) and Beth Zur (Funk 1958:10; Sellers *et al.* 1968:10) in the Southern Levant.

Solid walls constructed of brick have been revealed at: EBA Hacilar (Mellaart 1960:97) and Arslan tepe (Palmieri 1984:208) in Anatolia; Al-Ubaid (Delougaz 1938:3), Grai Resh (Lloyd 1940:14) and Abu Salabikh (Postgate 1990:95) in Mesopotamia; Tell Abu Danne (Tefnin 1983:141-152), Tell Ashara (Chavalas 1996:93), Tell Bderi (Pfalzner 1987:294), Tell Selenkahiye (van Loon 1973:146), Tell Jidle (Mallowan 1946:134) and Hammam et-Turkman (van Loon *et al.* 1987:311) in Syria; and Tell Poran (Gophna 1992:267), Beth Yerah (Maisler *et al.* 1952:172; Esse 1991:38), Tell Gath (Yeivin 1960:201) and Tell el-Hesi (Petrie 1891:21; Rose and Toombs 1978:142) in the Southern Levant. Sites defended by brick walls were also common during the MBA, as seen at: Boghazkoy (Bittel 1957:21) in Anatolia; Carchemish (Woolley 1921:59) in Syria; Kabri (Prausnitz and Kempinski 1977:166), Tell Akko (Dothan and Conrad 1983:113), Tell Nagila (Amiran and Eitan 1963b:334), Tell el-Milh (Kochavi 1967:272), Tell Batash (Mazar 1997:38), Tell el-Hammam (Cahill *et al.* 1988:194), Yavneh Yam (Kaplan 1967:269), Megiddo (Loud 1948:8,105) and Tell Poleg and Tell Zeror (Kochavi *et al.* 1979:133, 155) in the Southern Levant; and, Buhen (Emery 1958; 1959; 1960) in Egypt.

Solid walls constructed with a stone substructure and a brick superstructure are also found throughout the Bronze Age Levant. EBA sites include: Demirchihuyuk (Korfmann 1978:16) in Anatolia; Nineveh (Stronach and Lumsden 1992:230) in northern Mesopotamia; Tell Taya (Reade 1968:241), Tell Selenkahiye (van Loon 1968:25) and Idlib (Egami 1983:76) in Syria; and, Tell el-Far'ah North (Mallet 1973:29), Tell Yarmouth (de Miroschedji 1990:52-53), Beth Yerah (Esse 1991:53), Jericho (Kenyon 1953:88-90); Mithiam Leviah (Kochavi 1994:141), Khirbet el-Mahruq (Yeivin 1974:259) and Tell Halif (Seger 1983:1) in the Southern Levant. MBA examples include: Korucutepe (van Loon 1970:10) and Tilmen Huyuk (Alkim 1964:24) in Anatolia; Mumbaqat (Orthmann 1975:131), Carchemish (Woolley 1921:45), Tell Kannas (Finet 1977:83), Tell Hadidi (Dornemann 1979:141), Tell Tuqan (Baffi Guardata 1996:66) and Qadesh (Pezard 1931:4-12; Parr 1983:106) in Syria; and Aphek (Kochavi 1975:30-321; 1997:149), Ain Zurekiyeh (Gophna and Ayalon 1982:69), Megiddo (Loud 1948:6), Jericho (Kenyon and Holland 1981:110), Hazor (Yadin 1989:51,217) and Shechem (Toombs and Wright 1963:7) in the Southern Levant.

Single trace, solid walls were commonly employed throughout the various regions of the Levant during the EBA and the MBA. The dimensions and building materials were random and not generally indicative of regionalism. The brick on stone walls that were employed at Pella, Tell Deir Alla, Tell Nimrin and the stone walls at Rukeis, Tell Irbid and Amman have too many random parallels elsewhere in the Levant to be indicative of a pattern of urban development. The common use of brick in the Jordan Valley and the common use of stone on the Jordanian Plateau reflects availability of building materials, rather than any pattern of regionalism. As a result, the MBI-II period wall at Pella finds it's closest parallels with other sites in the Jordan Valley, and the MBII period wall at Rukeis finds it's closest parallels with other sites in the Hauran, such as Khirbet Umbashi (Dubertret and Dunand 1955:64; Braemer 1994:124-125) and possibly Tell Debbeh (Braemer 1984:242-246).

[16] This trend continued in Jordan into the LBA and subsequent periods. Solid walls dating to the LBA have been excavated at Sahab (Ibrahim 1974:56; 1987:76), Tell Safut (Wimmer 1997:449), Tell es-Sa'idiyeh (Tubb 1988:41; 1990:26) and, possibly, Tell Irbid (Lenzen *et al.*1985). Later period solid walls have been excavated at Dibon (Winnett 1952:8-12), Tell el-Umeiri (Geraty *et al.* 1989:244-247; Herr *et al.* 1991a:53-57; 1991b:159), and Petra (Hammond 1960:30). Defensive systems of the LBA are largely unknown in the southern Levant. In Palestine, LBA walls are only attested at a few sites, such as Gezer (Dever *et al.* 1970:44). The possible existence of LBA defensive walls at four sites in Jordan emphasizes the unique role of Jordan during this period.

Casemate Walls

The casemate fortification wall is less common than the solid wall and thus is more indicative of patterns of architectural and urban development. In Jordan, casemate fortifications have been revealed in EBII-III contexts at Bab edh-Dhra (Rast and Schaub 1980:25; 1981:19) and Numeira (Rast and Schaub 1980:41; 1981:37). A possible EBII-IV casemate fortification has been revealed at Khirbet Iskander (Richard 1990:36; 2001:441). MBA casemate fortification walls in Jordan are rare compared with EBA examples; the only possible examples are at Tell Deir Alla (p.8) and Pella (p.18). Evidence for a casemate wall at each of these sites is limited. The most extensive evidence for a casemate fortification is at Pella[17].

Excavation in Area XXVIIIC, Locus 20 at Pella has revealed a possible casemate fortification to the east of Tower 1 (p.18; figure 25a). Evidence for a casemate is implied by Wall 15, which was built at a right angle, and was structurally integral to, the city wall (Wall 10). Evidence for stone packing within the area enclosed by Wall 10, Wall 15 and the east tower flank also indicates that this area was originally a solid casemate. Provided that Wall 15 is a casemate wall, a second, interior fortification trace must have existed to the north, running off the east tower flank. This entire casemate system would be approximately 10m wide. Excavation of the west flank of Tower 1 (Locus 10) did not reveal any evidence for a casemate fortification; however, a combination of single trace walls and casemate walls has been revealed at MBII-III Carchemish (Woolley 1921:95-104) and Hazor (Yadin 1989:283). Therefore, the possible combination of a single trace wall and a casemate wall at Pella would not be unusual.

There has been considerable debate regarding the origin of the casemate wall (see p.2). The earliest possible casemate fortifications are at Chalcolithic period Kurucay Huyuk in Anatolia (Duru 1981:196). In Syria, casemate walls have been excavated at EBII-III Tell Ashara (Buccellati and Kelly-Buccellati 1983:47; Chavalas 1996:93), Tell Taya (Reade 1968:241), Tell Abu Danne (Tefnin 1983:141-152) and Tell Selenkahiye (van Loon 1977:165); and at EBIV Hammam et-Turkman (van Loon *et al.* 1987:311). In the Southern Levant, EBII-III casemate walls have been revealed at Megiddo (Loud 1948:70) and Tell Halif (Cole 1978:119). It is difficult to establish a specific point of origin for the casemate fortification. The current available evidence indicates that the earliest common use of casemate walls was during the EBII-III in Anatolia, Syria, Mesopotamia and the Southern Levant.

The casemate was commonly employed during the MBA and later periods[18]. During the MBA, the casemate fortification wall emerges as a standardised architectural feature throughout the Levant. Casemate fortifications have been revealed at MBI Beycesultan (Lloyd 1958:96), Karahuyuk (Alp 1959:30), and Qadesh (Parr 1983:106; 1997:114); MBII-III Alalakh (Woolley 1955:133), Carchemish (Woolley 1921:102), Boghazkoy (Bittel 1957:21), Hazor (Yadin 1989:279) and Tell el-Far'a South (Petrie 1930:17); and MBIII Shechem (Bull *et al.* 1965:30; Bull and Campbell 1968:3; Dever *et al.* 1974:39; Sellin 1927:268-272; Toombs and Wright 1963:49; Wright 1957:14), Tell Ta'anach (Lapp 1969:19-21) and possibly Beth Zur (Sellers *et al.* 1968:6). Possible defensive casemates have also been revealed in the Hauran at MBA Tell Zheir (Braemer 1991:143-152; 1993:128) and Umm el-Alak (Braemer 1984:246).

During the MBA, there is only limited evidence for the use of the defensive casemate in Jordan. In this regard, Jordan falls outside of the sphere of this important architectural feature of MBA architectural development. The only realistic possibility for a casemate wall is at Pella; however, the evidence is limited. Assuming that the Pella wall is part of a true casemate fortification, the closest parallel is at MBII-III Hazor (Area K). At Hazor (figure 47a), the exterior trace and cross-walls are of similar dimensions to Walls 10 and 15 at Pella; and both sets of walls were constructed using brick on a stone substructure.

Sawtooth Walls

Excavation in Area XXVIIIC, Locus 20 at Pella revealed a sawtooth wall. Approximately 7.0m to the east of the Tower 1 flank, the line of the Wall 10 trace steps back 0.5m to the north (p.18; figure 25a). The wall line then continues to the east, sloping slightly north to south. The effect of this construction method is a wall line in a sawtooth pattern; the step-back forming a 'tooth'. The sawtooth wall is a defensive feature rarely employed in Bronze Age fortifications elsewhere in the Levant and is unknown elsewhere in Jordan. The sawtooth wall is found at EBIII Demirchihuyuk (Korfmann 1979:192) and MBIII Tilmen Huyuk in Anatolia (Alkim 1964:24) and at EBIII-IV Tell Jidle in Syria (Mallowan 1946:134, figure 5).

The closest parallel to the construction and date of the sawtooth wall at Pella is at MBIII Tell Shiloh in the Central Highlands (Finkelstein *et al.* 1993:17,37,49,57). Finkelstein suggests that the Shiloh wall was not necessarily intended as a fortification trace, but as support for earthen fills abutting it on either side. This suggestion is based on the lack of evidence for a brick superstructure; the surviving wall is no higher than the surrounding earthen fill (Finkelstein *et al.* 1993:20,374). Evidence that refutes this suggestion is noted in Area M, where

[17] Other evidence for casemate walls in Jordan is confined to LBA Tell Abu al-Kharaz (Fischer 2000:457) and Iron Age Tell el-Umeiri (Herr *et al.* 1991a:57; 1991b:159)

[18] Iron Age defensive casemate walls have been uncovered at numerous Levantine sites (Lapp 1976) such as Tell Yoqne'am (Ben-Tor *et al.* 1983:34), Tell el-Ful (Albright 1924:79; Sinclair 1960 plate 35; Lapp 1976:25-42), Hazor (Yadin 1960:1), Tell Beit Mirsim (Albright 1930:8; 1932a:14), Ein Gev (Mazar *et al.* 1964:32) and Kadesh Barnea (Dothan 1965:142). Casemate walls have been commonly found in Hellenistic period contexts such as Tell Dor (Stern 1988:6-14) and Roman period contexts such as Beitar (Carroll 1925:103).

a layer of brick collapse (L2008) potentially represents the remains of the original brick superstructure for the wall. Finkelstein admits that an original brick superstructure may have eroded away (Finkelstein *pers. comm.*). Providing the wall at Shiloh served a defensive function, it provides the closest parallel to the sawtooth wall at Pella.

Wall Buttresses

During the Bronze Age, buttresses were commonly used to add structural support to fortification walls. Buttresses have been revealed in the Northern Levant at sites such as EBI Tell Sheikh Hassan (Boese 1987:68); EBIII-IV Tell Selenkahiye (van Loon 1977:171) and Tell es-Sweyhat (Holland 1976:51); MBI Qadesh (Parr 1983:106); MBII Tell Umm el-Marra (Schwartz *et al.* 2003:341); and MBII-III Mumbaqat (Orthmann 1975:129-131). Buttresses were employed in the Southern Levant at sites such as EBII-III Tell Ta'anach (Lapp 1969:5), Ai (Callaway 1980:73,185-186), Tell Poran (Gophna 1992:269-271), Jericho (Kenyon 1952:68) and Tell el-Far'ah North (de Vaux 1955 fig 15); MBI Aphek (Kochavi 1975:30-32) and Tell Beit Mirsim (Albright 1938a:17); MBII-III Megiddo (Kenyon 1969:56; Loud 1948:6,15,87,113); and MBIII Jerusalem (Shiloh 1984b:58), Tell Shiloh (Finkelstein *et al.* 1993:49) and Gezer (Dever *et al.* 1970:57).

The only known MBA fortification wall buttresses in Jordan are at Pella. These buttresses include internal Buttress 16 and external Buttress 12 in Area XXVIIIC (p.18; figure 24a); external Buttress 100 and Buttress 37 in Area IIID (pp.13-14; figure 8a); external Buttress Wall 7 in Area IIIF (p.15; figure 8a); and, internal Buttress 83 in Area IIIC (p.14). EBI Jawa is the only other Jordanian site where buttresses have been revealed (Betts 1991:48). Direct parallels with the Pella buttresses are difficult to establish. Because most known buttresses were built of brick on a stone substructure, comparing building materials is of limited use. Buttress dimensions also vary randomly depending on the required support and offer little information. To illustrate this point, the MBI-II Buttress 16 in Area XXVIIIC at Pella has similar dimensions to buttresses at EBIII-IV Tell Selenkahiye (van Loon 1977:171), MBI Aphek (Kochavi 1975:30-32) and MBIII Tell Shiloh (Finkelstein *et al.* 1993:49).

Buttress 12 in Area XXVIIIC at Pella (p.18; figure 24a)is unusual because it forms a triangular shaped buttress that abuts the exterior face of the city wall (Wall 9) and the west tower flank (Tower 1). The single possible parallel for a triangular buttress is in Area AA at MBI-II Megiddo, abutting what was a gate or tower (Loud 1948: figure 7). The connection between the Pella and Megiddo buttresses (figure 47b) is fragmentary, and the limited nature of the evidence must be emphasized. Nonetheless, it should be noted that the closest potential parallel to the triangular shaped buttress at Pella is with a site of similar date (MBI-II) located in the near vicinity (Esdraelon Plain).

Regular spacing of buttresses is another aspect of buttress construction that adheres to a degree of standardization. Regularly spaced buttresses are evident in the buttress-niche wall systems (figure 47c) at EBI Tell Sheikh Hassan (Boese 1987:68); EBIV Tell es-Sweyhat (Holland 1976:49); and MBII-III Megiddo (Loud 1948:15) and Tell el-Far'ah North (Mallet 1973:45; 1987 Plans 5 and 14). There is no evidence in Jordan for buttress-niche wall systems, which implies that Jordan fell outside of this sphere of MBA architectural development.

Foundational Construction

The foundations of the fortification systems at Pella and Rukeis are the most comprehensively studied of any MBA defences in Jordan. Foundation trenches for the fortifications were revealed in Areas IIIC (p.13; figure 8b), IIIF (p.15; figure 9) and XXVIIIC (p.17; figures 18,24a),at Pella; and in Areas 5 (p.21; figures 30,32), 14 (p.23; figure 41) and 16 (p.24), at Rukeis. Because the foundational construction of military architecture is rarely studied in detail, there is limited comparative material.

Comparative Bronze Age material from Jordan is only found at EBI Jawa and LBA Tell Safut. Comparative material from elsewhere in the Levant is also limited. In all probability, the methods of foundation construction observed at Pella and Rukeis were commonly employed elsewhere but have not been examined and discussed in detail.

Some features of foundational construction were too commonly and randomly employed to be indicative of regional development patterns. For example, the use of a foundation trench on only one side of a fortification wall, as at Rukeis, finds parallels with EBII-III Bab edh-Dhra (Rast and Schaub 1980:25) and MBII-III Hazor (Yadin 1989:54). The choice of foundational base also varies randomly depending on whether there had been previous occupation at a given site. At Pella and Rukeis, fortifications were constructed directly on bedrock, on occupational debris from earlier periods, or on a combination of the two. Fortifications constructed directly on bedrock are known at EBII-III Tell Yarmouth (de Miroshedji 1990:58); EBIII Beth Yerah (Maisler *et al.* 1952:172; Esse 1991:38); and MBIII Tell Ashdod (Dothan 1973:10). Fortifications constructed on earlier occupational debris are known at EBIII Beth Yerah (Maisler *et al.* 1952:172); MBI Tell Beit Mirsim (Albright 1938a:18); MBII-III Jericho (Kenyon and Holland 1981:347); and, LBA Gezer (Dever *et al.* 1970:44). Fortifications constructed above a combination of the two are known at EBI Jawa (Betts 1991:33); EBI-II Arad (Amiran and Ilan 1996:22); EBII-III Ai (Callaway 1980:41); MBII-III Hazor (Yadin 1989:51); and, MBIII Tell Jerishe (Geva 1983:14).

A less common foundation feature involves stepping the substructure of a fortification feature up a slope to provide a solid, horizontal platform for the superstructure. This construction technique is evident at Pella (pp.15,17; figures 9,22b) and finds parallels with EBIII Bab edh-Dhra (Rast 1995:126); MBI Aphek (Kochavi *et al.* 2000:71); MBII-III

Carchemish (Woolley 1921:58) and Hazor (Yadin 1989:51); and possibly MBIII Gezer (Dever *et al.* 1970:19).

The most unusual foundation trait is the construction of a foundation platform for the city wall at Pella (p.18). In Area XXVIIIC at Pella, the base of the foundation trench is filled with a brick platform, upon which the true city wall (Wall 9) is constructed. In effect, the wall is stepped in from the line of the foundation course, making it narrower than the foundations (figure 47d). This construction technique is highly unusual, but does find some interesting parallels at MBII Bethel (Kelso 1958:4) and MBII-III Hazor (Yadin 1989:51). At both of these sites, the wall was stepped in from the line of the substructure. This building technique is peculiar to MBA sites in the Southern Levant.

Brick Sizes

According to Wright, 'it would seem axiomatic that the dimensions of moulded bricks should be standardised in terms of a unit of measure and that variations should have some chronological or regional significance' (Wright 1985a:354). Because information on brick sizes is sporadic and limited, a detailed survey of brick types should provide information on issues such as, chronology, metrology and foreign connections (Wright 1985a:358). Table 4 reviews brick sizes employed in fortifications and demonstrates that brick sizes became increasingly standardized during the MBA. The most commonly employed MBA brick size is 0.40m x 0.40m x 0.10m[19].

The bricks used in the fortifications at Pella were of the same standardized size, usually measuring 0.40m x 0.40m x 0.12m. The closest parallels to the Pella bricks are found at MBI-II Megiddo and Tell Kannas, and MBII-III Tell Deir Alla and Tell Nimrin. Significantly, three of the four closest parallels to Pella are at sites in close proximity to Pella. Megiddo is located to the west of Pella and uses identical sized bricks in fortifications of the same period. Tell Deir Alla and Tell Nimrin are a short distance to the south of Pella, on the eastern side of the Jordan Valley, and use the same size bricks; although, the defences are of a slightly later date. The brick sizes employed in the fortifications at these sites are indicative of a degree of regionalism.

Table 4: Comparison of Brick Sizes

Period	Site	Brick Size	Reference
EBII-III	Tell Bderi	0.20m x 0.20m	Pfalzner 1987:294
EBIII	Tell Taya	0.36m x 0.32 x 0.10m	Reade 1968:241
EBIII	Beth Yerah	0.40 x 0.30m x 0.10m	Maisler *et al*. 1952:172
EBIII	Tell el-Hesi	0.57m x 0.31m x 0.11m	Petrie 1891:35
		0.48m x 0.21-0.31m x 0.12m	Petrie 1891:35
EBIII-IV	Tell Jidle	0.44m x 0.44m x 0.14m	Mallowan 1946:134
		0.32m x 0.16m x 0.17m	Mallowan 1946:134
EBIII-IV	Tell es-Sweyhat	0.50m x 0.40m x 0.10m	Holland 1976:49
EBIV	Khirbet Iskander	0.35m x 0.35m	Richard 1983:50
MBI-III?	Carchemish	0.40m x 0.38m x 0.13m	Woolley 1921:144
MBI-II	Pella	0.40m x 0.40m x 0.12m	
MBI-II	Tell Kannas	0.40m x 0.40m x 0.10m	Finet 1977:83
MBI-II	Megiddo	0.35m x 0.35m x 0.10m	Loud 1948:87
MBII	Tell Hadidi	0.74m x 0.36m x 0.13m	Dornemann 1979:144
MBIII	Tell Nimrin	0.40m x 0.40m x 0.16m	Flangan *et al*. 1992:90
MBIII	Tell Deir Alla	0.60m x 0.40m	Franken and Ibrahim 1978:75
MBIII	Shechem	0.36m x 0.36m x 0.16m	Wright 1957:15
MBIII	Tell Jerishe	0.55m x 0.40m	Geva 1983:15
		0.50m x 0.40m	Geva 1983:15
		0.35m x 0.35m	Geva 1983:15

[19]The same sized brick was employed at a number of LBA sites, such as Tell Hamidi (Wafler 1990:222).

Brick Type and Combination in Construction

The composition of bricks used in Bronze Age defences varies between sites because soil types vary between regions. Occasionally, different brick types were employed in the construction of a single defensive work, as seen in Wall 41 and Wall 7 at Pella (Area III). Wall 41 (pp.13,15) is comprised of three different brick bands. The lowest and uppermost are built with grey, porous bricks bonded with orange mortar, and the central band is built with hard green bricks bonded with brown mortar. Wall 7 (p.15) uses orange bricks bonded with brown mortar in the lower section and dark red bricks bonded with white mortar in the upper section. In each case, the change in brick types occurs across a single horizontal course.

A fortification system encompassing a city the size of Pella would require an enormous number of bricks and a considerable construction time[20]. The defences were likely built in stages, using bricks from various clay sources. As one section was completed using bricks from a specific clay source, the following horizontal section would be constructed using bricks from a different clay source. This method would account for the change in brick types along the length of a single horizontal brick course.

The different brick bands employed in the construction of Wall 41 and Wall 7 probably facilitated drainage. Bricks of different texture and permeability were placed in bands to allow water to drain through the structure, enhancing stability. The use of this method implies that engineers had a good understanding of porous and less porous soil types and knew how to effectively use them in monumental building. This knowledge can be directly compared to a similar use of varying soil types employed in the construction of MBA earthen ramparts (see p.38).

Other sites that use a combination of brick types in fortifications include: EBII-III Aphek (Kochavi *et al.* 2000:63) and Bab edh-Dhra (Rast and Schaub 1978:13); MBI-II Megiddo (Loud 1948:87); and, MBII-III Gezer (MacAlister 1911:240) and Tell Batash (Mazar 1997:39). The most direct parallel to Wall 41 and Wall 7 at Pella is at MBII-III Hazor. The early city wall (W375) in Trench 500 at Hazor uses two types of alternately laid brick, soft, black earthen bricks and lighter, stronger lime-like bricks. The core of the wall consists of the black bricks, and the wall faces are lined with the lighter coloured brick. According to Yadin, this construction method was intended 'to provide a strong outer face which would withstand weathering' (Yadin 1989:51). The use of different brick types at Pella and Hazor probably served a similar function related to drainage.

The method of brick laying is a further feature of brick fortification construction. The use of headers and stretchers that is employed at Pella was commonly employed elsewhere in the Levant, as found at EBA Jericho (Garstang 1932) and MBA Carchemish (Woolley 1952 fig 28). At these sites the bricks were laid with the horizontal joints in the centre of the bricks on the previous level. Available material from other sites is too limited to allow for a detailed analysis of brick laying techniques.

Plaster Facing

Plaster facing is found on a number of architectural features at Pella, including the exterior face of Wall 9, the west flank of Tower 1 and Buttress 12 in Area XXVIIIC (p.18), and the interior face of Wall 41 in Area IIIF (p.13). During the Bronze Age, plaster was commonly applied to the exposed surface of *glacis* and defensive walls to facilitate runoff and reduce erosion. Plaster facing on fortifications has been noted at: EBII-III Tell Ta'anach (Lapp 1969:7) and Tell Yarmouth (de Miroschedji 1990:57); EBIII Bab edh-Dhra (Rast and Schaub 1980:26) and Tell Mozan (Kelly-Buccellati 1990:123); EBIII-IV Tell Jidle (Mallowan 1946:134); MBI Tell Akko (Dothan and Conrad 1978:265; 1983:113); MBII Tell Beit Mirsim (Albright 1938a:19); and, MBII-III Hazor (Yadin 1989:52) and Mumbaqat (Orthmann 1975:131). This feature is too commonly employed to indicate patterns of regional development.

Towers and Related Features

The projecting tower is one of the most standardised features of EBA military architecture (see p.1). EBA towers have been revealed at a number of sites in Jordan, including: EBI Jawa (Helms 1975:27; 1976:7; 1977:30); EBII-III Tell el-Handaquq North (Mabry 1989:65; 1996:123), Khirbet Zeraqun (Ibrahim and Mittman 1997:388), Numeira (Rast and Schaub 1980:42; Coogan 1984:80) and Bab edh-Dhra[21] (Rast and Schaub 1978; 1980; 1981); and EBII-IV Khirbet Iskander (Richard 1983;

[20]The construction of the Pella defences would have taken a period of years. Ancient textual material relating to the construction of defences is rare, and modern estimates are conjectural. Late 3rd millennium sources from southern Mesopotamia state that each worker in the construction of earthen embankments moved three cubic metres per day (Civil 1994:128). Assurnasirpal II records that the fortifications at Nimrud took five years to build (Mallowan 1950:159). Mazar estimates that the ramparts at Tell Batash would have required 300 labourers for a period of two years (Mazar 1997:250). Biran estimates that the construction of the Tell Dan ramparts required 1000 labourers, each moving one cubic metre of earth per day, over a total of three years (Biran 1994:71-72). Finkelstein estimates that, at Tell Shiloh, a population of 3000 people would require five years to construct the ramparts around the site (Finkelstein *et al.* 1993:379; see Finkelstein 1992:208-210). Although these studies refer to earthen ramparts, the construction of monumental brick walls was most likely as labour intensive.

[21]This tower possibly guarded an entryway into the city. The excavators originally considered it to be freestanding (Rast and Schaub 1978:14); however, in later reports, suggested that the tower was a 'postern tower' (Rast and Schaub 1981:21). This interpretation would indicate that the Field XI tower is part of the defensive *enceinte*, similar to the tower in Field XIII (Schaub and Rast 1984:43-46). The Field XI tower consists of two rectangular segments (each 4.0m X 10.0m) flanking an open passage (Rast and Schaub 1980:26). These segments include a stone substructure with a brick superstructure, that is covered with a brick platform at a later date (Rast and Schaub 1978:13-14; 1981:19).

1987; 1990; 2001:441). Pella has the only concrete evidence for a MBA tower in Jordan (p.17). Fragmentary evidence for MBA towers in Jordan is found at Amman, Tell el-Umeiri and Tell Nimrin (see pp.8-9)[22].

Tower Shape

During the MBA, the semi-circular tower that had characterized the EBA (see p.1) was almost completely abandoned in favour of the square or rectangular shaped tower[23]. The only MBA examples of semi-circular towers are at MBI Tell Kannas in Syria (Finet 1977:83) and MBA Buhen in Egypt (Emery 1958; 1959; 1960). However, the semi-circular towers at these sites were used in conjunction with square towers[24]. The MBA tower at Pella is a rectangular structure, measuring 8.0m x 12.0m.

Because square and rectangular towers have been revealed in fortification systems at a wide range of Bronze Age Levantine sites, it is difficult to associate the Pella tower with a specific pattern of architectural development based purely on shape. For example, EBA square and rectangular towers have been revealed at: Tell Sheikh Hassan (Boese 1987:89), Halawa (Meyer and Orthmann 1983:100; Meyer 1988), Habuba Kabira (Srommenger 1980:33-36), Abu Salabikh (Postgate 1990:98), Tell Bderi (Pfalzner 1987:294), Tell Taya (Reade 1968 Pl LXXVIII), Tell es-Sweyhat (Holland 1976:49) and Tell Selenkahiye (van Loon 1977:166; 1979:99-104) in Syria; and, Tell Dothan (Helms 1977a:103), Tell Ta'anach (Lapp 1969:12), Tell Yarmouth (de Miroschedji 1990), Ai (Callaway 1980:11), Jericho (Kenyon 1953:88-90) and Tell Halif (Seger 1977:46) in the Southern Levant.

MBA square and rectangular towers have been revealed at: at MBII-III Boghazkoy (Bittel 1957:21) in Anatolia; MBI Tell Kannas (Finet 1977:83) and Ebla (Peyronel 2000:1353); MBII Tell Tuqan (Baffi Guardata 1990:66); MBII-III Carchemish (Woolley 1921:45,50,54); and MBIII el-Qitar (Culican and McClellan 1983:292; McClellan 1986:423) in Syria; and, at MBI Tell Akko (Dothan and Conrad 1978:265-266), Tell Beit Mirsim (Albright 1938a:18), Tell Poleg, Tell Burga and Tell Zeror (Kochavi *et al.* 1979:133,142,155); MBII-III Tell el-Far'ah North (Mallet 1987 Plan 11); MBIII Beth Zur (Sellers and Albright 1931:6), Tell el-Milh (Kochavi 1967:272), Tell Shiloh (Finkelstein *et al.* 1993:73) and Gezer (Dever *et al.* 1970:19; MacAlister 1911:241); and possibly MBII Jericho (Marchetti *et al.* 2000:877) in the Southern Levant.

Tower Spacing

Tower spacing varied randomly between Bronze Age sites. For example, towers are 20m apart at EBI Tell Sheikh Hassan (Boese 1987:73), 25-40m apart at EBII Arad (Amiran 1978:12), 15m apart at EBII-III Habuba Kabira (Strommenger 1980:35), 23m apart at MBI Tell Beit Mirsim (Albright 1938a), 40m apart at MBII-III Carchemish (Woolley 1921:50), 12m apart at MBII-III Korucutepe (van Loon 1970:10), and 30-35m apart at MBIII Gezer (MacAlister 1911:238)[25]. The irregular tower spacing in a defensive circuit is due to topography. As a result, tower spacing is not a particularly useful indicator of urban or regional patterns of development.

Tower Projection

An interesting feature of MBA towers is an apparent movement away from the externally projecting tower systems that characterised the EBA. As Wright explains, projecting towers are 'a prominent feature of the EB cities and virtually unknown later' (Wright 1985a:188). Externally projecting towers were commonly employed during the EBA at sites such as Demirchihuyuk (Korfmann 1977:37) in Anatolia; Halawa (Meyer and Orthmann 1983:100), Abu Salabikh (Postgate 1990:98), Habuba Kabira (Strommenger 1979:64), Tell Taya (Reade 1968), Tell Selenkahiye (van Loon 1979:99-104;1977:166) and Tell Brak (Oates 1983:86) in Syria; and Ai (Amiran 1978:13) and Arad (Amiran 1978:12) in the Southern Levant. During the MBA, however, externally projecting towers have only been revealed at Karahoyuk (Alp 1959:30) and Boghazkoy (Bittel 1957:21) in Anatolia; Carchemish (Woolley 1921:54) in Syria; Tell el-Far'ah North (Mallet 1987 Plan 11), Tell Burga and Tell Zeror (Kochavi *et al.* 1979:142,155) in the Southern Levant; and Semna (Lawrence 1965 Pl XVI) and Buhen (Emery 1958; 1959; 1960) in Egypt[26].

---

[22]Towers that are later than the MBA have been excavated at numerous Jordanian sites including Tell Deir Alla (Franken 1960:388), Dibon (Winnett 1952:12), Jebel Sartaba (Smith *et al.* 1983:73) and Lejjun (Parker 1982:6; 1995:259).

[23]The square shaped tower remained the standard tower shape throughout the LBA, Iron Age and Classical periods. Square and rectangular towers have been excavated at: LBA Miletus (Weickert 1958:31); Iron Age Sacke gozu (Plat Taylor *et al.* 1950:81), Panemoteichos (Aydal *et al.* 1997:150), Karmir-blur (Barnett 1959:2), Tell Yoqne'am (Ben-Tor *et al.* 1983:34), Tell Batash (Kelm and Mazar 1982:21), Tell Beit Mirsim (Albright 1943:16), Tell el-Ful (Sinclair 1960 Pl 35) and Tell Seh Hamad (Kuhne 1983:241); Persian period Beth Yerah (Esse 1991:40) and Pasargadae (Stronach 1963:41;1965:18); Hellenistic period Tell Dor (Stern 1980:210; 1982:110;1988) and Tell el-Hajj (Stucky 1975:166); Roman period Beitar (Carroll 1925:103), Jerusalem (Ben-Arieh and Netzer 1974:99) and Yotvata (Meshel 1989); and, Byzantine period Dibsi Faraj (Harper 1974:34) and Caesarea (Lehmann 1995:123). During the Classical and Islamic periods, the semi-circular tower was more commonly employed, as is evident at: Persian period Beth Yerah (Esse 1991:40); Hellenistic period Caesarea (Raban 1987:73) and Tell el-Hajj (Stucky 1975:166); Roman period Beitar (Carroll 1925:103); Parthian Kish (Langdon 1935:121); and, Islamic period Samarra (Northedge *et al.* 1990:fig 7), Qasr al-Hayr al-Sharqi (Graber 1965) and Djebel Seis (Brisch 1963).

[24]A combination of both tower types during later periods was equally unusual. Two rare examples are Hellenistic period Tell el-Hajj (Stucky 1975:166) and Roman period Beitar (Carroll 1925:103).

[25] Evenly spaced towers were uncommon until the Classical periods, as seen at sites such as Palmyra (see Gerkan 1935 Pl XII).

[26] Following the MBA, externally projecting towers were commonly employed, as found at LBA Miletus (Weickert 1958:31), Iron Age Tell Yoqne'am (Ben-Tor *et al.* 1983:34); Hellenistic period Tell Dor (Stern 1980:210; 1982:110; 1988:6-14) and Tell el-Hajj (Stucky 1975:166); Roman period Jerusalem (Ben-Arieh and Netzer 1974:99), Yotvata (Meshel 1989:228-238) and Dibsi Faraj (Harper 1974:34); and Islamic period Samarra (Northedge *et al.* 1990:137).

Wright notes that towers are commonly replaced by a system of salients and recesses during the MBA. Whereas the EBA towers served a truly defensive function, the salients and recesses of the MBA incorporated an aesthetic aspect, providing 'an image with overtones of the sacred place' (Wright 1985a:191). With this argument, Wright proposes a Mesopotamian influence as is evident in the walls of the sacred enclosure at Ur. Wright suggests that the system of salients and recesses is 'invested with a sacred significance as enclosing the holy city - the city invincible' (Wright 1985a:516). A possible variation of this salient-recess tower system may be seen in the buttress-niche wall system that characterizes the fortifications at a range of MBA sites in the Levant (see p.29). Wright considers the salient-recess fortification to be 'the only possible approach to any old Middle Eastern building order' (Wright 1985a:516). A Syro-Mesopotamian origin for the salient-recess tower system, as suggested by Wright, is reinforced by the evidence from EBII-MBI period Mari in Syria (see Margueron 2000:fig2).

The tower at Pella projects very slightly, approximately 1m, from the wall line, as is common in a salient-recess system (p.17). The closest parallels to the Pella tower in terms of tower projection are at MBI Tell Beit Mirsim (Albright 1938a:18) and MBIII Gezer (MacAlister 1911:241). The tower at Tell Beit Mirsim provides a particularly close parallel to the Pella tower, because it only projects 1.5m from the exterior wall face and is of a similar date and size. Further excavation is needed to determine if the method of tower construction at Pella is consistent across the entire site. Providing the method of slight tower project is used elsewhere in the defensive circuit, Wright's theory of religious significance could apply to the Pella fortifications.

Tower Chambers

The internal chamber found in Locus 30 at Pella is the most unusual feature of Tower 1 (figures 22a,23b). Built as part of the brick superstructure, the chamber is an integral feature of the tower construction. The available evidence implies that the chamber extends throughout much of the tower's width. Comparable chambers probably exist in other parts of the tower superstructure. The chamber's function is unclear; although, it was almost certainly designed to be an open space. The closest comparable examples are found in casemate walls at EBII Tell Yarmouth (de Miroschedji 1990:58); EBIV Hammam et Turkman (van Loon *et al.* 1987:311); and MBI Qadesh (Parr 1983:106; 1997:114).

Evidence of tower chambers at other sites is limited. Towers with chambers have been revealed at EBII-III Tell Ta'anach (Lapp 1967:3) and Tell Halif (Jacobs 1984:199); EBIII Tell Selenkahiye (van Loon 1977:166) and Tell el-Hesi (Ross 1979:13); MBI Ebla (Peyronel 2000) and Carchemish (Woolley 1921:95); and, MBII-III Tell el-Far'ah North (Mallet 1987 Plan 3) and Gezer (MacAlister 1911:238). The morphology of the towers and chambers at each of these sites is different from Tower 1 at Pella. Gezer is the closest parallel in terms of date and the oblong shape of the chambers. Dever suggests that the chambers in the Gezer towers may originally have been hollow, before being filled with stone debris (Dever *et al.* 1970:19). The chamber face at Pella is plastered. There are no direct parallels for plastered chambers within tower systems,

Tower Dimensions

Because the dimensions of MBA towers varied randomly, the Pella tower cannot be associated with a pattern of architectural development based purely on tower dimensions. For example, towers of similar dimensions may be noted at MBI Tell Beit Mirsim (Albright 1938a:18) and Tell Poleg (Kochavi *et al.* 1979:133); MBII Tell Tuqan (Baffi Guardata 1996:67); MBII-III Beth-Zur (Sellers and Albright 1931:6); and MBIII Tell el-Milh (Kochavi 1967:272).

Tower Materials

The choice of building materials used in tower was largely dependent on available building resources. As a result, the use of stone and brick varied from region to region throughout the Bronze Age. Towers constructed purely of brick have been revealed at: EBI Tell Gath (Yeivin 1960:201); EBII-III Tell Bderi (Pfalzner1987:294); EBIII Tell Selenkahiye (van Loon 1979:104); MBI Tell Kannas (Finet 1977:83) and MBII-III Carchemish (Woolley 1921:45) in Syria; and, at EBIII Tell el-Hesi (Ross 1979:13); MBI Tell Akko (Dothan and Conrad 1978:265-266); MBII-III Jericho (Kenyon and Holland 1981:353); and, MBIII Gezer (MacAlister 1911:241) and Tell el-Milh (Kochavi 1967:272) in the Southern Levant. Towers constructed purely of stone have been revealed at: EBI Arslan tepe (Palmieri 1984:208); MBII Tell Hadidi (Dornemann 1979:141) and MBIII el-Qitar (McClellan 1986:423) in the Northern Levant; and, at EBII-III Tell Dothan (Free 1958:14), Tell Ta'anach (Lapp 1967:3), Tell Yarmouth (de Miroschedji 1990:59) and Ai (Callaway 1980:152); MBI Tell Beit Mirsim (Albright 1938a:18); MBII-III Tell el-Far'ah North (Mallet Plan 11) and MBIII Gezer (MacAlister 1911:238), Bethel (Kelso 1958:7) and possibly Tell Marjamah (Zohar 1980:219) in the Southern Levant[27].

Towers were most commonly constructed using a stone foundation and a brick superstructure. Evidence of this building technique appears as early as the Neolithic period at Hacilar in Anatolia (Mellaart 1960:97) and has been commonly noted at EBA and MBA Levantine sites, including: EBIII Tell Selenkahiye (van Loon 1977:166); EBIII-IV Tell es-Sweyhat (Holland 1976:49); MBI Tell Kannas (Finet 1977:83); and, MBII-III Carchemish (Woolley 1921:54) in Syria; and, EBII Tell Yarmouth (de Miroschedji 1990:59) and Tell Halif (Seger 1977:46; 1983:2) and MBIII Gezer (Dever *et al.* 1970:41-43) in the Southern Levant.

[27] Stone towers were commonly constructed in later periods at numerous sites, including: Iron Age Panemoteichos (Aydal *et al.* 1997:150) and Tell Yoqne'am (Ben-Tor *et al.* 1983:34); Hellenistic and Byzantine period Casearea (Lehmann 1995:123; Raban 1987:73).

Wooden superstructures were probably used at numerous Bronze Age sites in the Levant, but evidence is extremely limited. Rare examples where evidence of wooden superstructure beams have been preserved include EBII-III Ai (Callaway 1980:65-66,72,80) and Jericho (Kenyon 1957:137). There is no evidence of wood in the construction of the towers at Pella.

Gates and Related Features

EBA gates have been excavated at a number of sites in Jordan, including Jawa (Helms 1973:41; 1975:27; 1976:7; 1977a:29; 1989:149: Betts 1991:34), Khirbet Zeraqun (Ibrahim and Mittman 1997:388), Tell el-Handaquq North (Mabry 1989:65; 1996:123), Bab edh-Dhra (Schaub and Rast 1984:43-44) and Khirbet Iskander (Richard 1983:50; 1990:44; 2001:441; Richard and Boraas 1984:79). By comparison, the only confirmed MBA gate is at Rukeis (pp.24-25). Fragmentary evidence of a MBA gate is also found at Amman and Tell Deir Alla (see p.9). Future excavation would certainly reveal gate systems at other sites[28].

The Rukeis Gate Plan

The triple-entry gate is a defining architectural feature of the MBA (figure 4). These gates have been revealed at numerous MBA Levantine sites, such as Tell Tuqan (Guardata 1996:64-72), Ebla (Matthiae 1968:8; 1997a:181; 2000:1034), Carchemish (Woolley 1921:104), Alalakh (Woolley 1955:150), Beth Shemesh (Mackenzie 1912-1913:21-30; Dever 1997b:311), Yavneh Yam (Kaplan 1969:120), Shechem (Seger 1974:117-130), Tell el-Far'a South (Weinstein 1997a:305), Hazor (Yadin 1989:277-282; Ben-Tor *et al.* 1997:353-363) and possibly Megiddo (Loud 1948:16-22; Kempinski 1989:58; Plan 4; Herzog 1997:150-1). During the mature phases of the MBA, the triple-entry gate became increasingly standardised in plan (see p.1). It is possible to chart the evolution of this gate type from solid bastions flanking a central passage to guardrooms and stairwells located within the bastions (Wright 1985a fig 96). The triple-entry gate provides the prototype for the multi-chambered gates of the LBA and Iron Age.

The Rukeis gate is comprised of two towers flanking a direct entry passage, without piers, but with well-developed guardrooms, storerooms and external bastions. The Rukeis gate plan falls outside of the sphere of the standard MBA triple-entry gate. The Rukeis gate finds parallels with other sites that use a less common category of MBA gate, the double-entry gate. Examples of MBA double-entry gates come from Tell el-Farah North (Mallet Plan 9), Tell Ashdod (Dothan 1969;1971;1973b) and Shechem (Wright 1985b:8) and date to the later phases of the MBA (figures 48a-d). A variation on this gate type may be noted in the interior part of the gate at Tell Akko (figure 48e). Double-entry gates consist of a square plan with an inner and outer entrance and two guardrooms flanking a central passage. The main difference between the double-entry gate and the Rukeis gate is the absence of piers in the passage and the use of exterior bastions at Rukeis. Without the exterior bastions, the Rukeis gate essentially follows a double-entry gate plan (see figure 48g).

EBII-IV Khirbet Iskander (figure 48f) provides a further interesting comparison to the Rukeis gate (figure 48g). There are two phases of gate construction in Area C at Khirbet Iskander, an earlier gate and a reconstructed 'bench lined thoroughfare with two guardrooms' (Richard 1990:48). The Khirbet Iskander gate lacks the robust exterior bastions evident at Rukeis, but the two flanking rectangular towers contain guardrooms that protect a direct entry passage, as at Rukeis.

Another possible parallel is Tell Debbeh, to the north of Rukeis in the Syrian Hauran. Although the gate has not been excavated, the site was surveyed by Braemer, who suggests that the MBA is the major period of occupation at the site (Braemer 1984:246). Like Rukeis, it is possible to identify much of the Tell Debbeh gate plan on the existing surface. The Tell Debbeh gate is built of basalt, has similar dimensions to the Rukeis gate (a 4m x 14m passage) and has a large bastion flanking the passage to the west (Braemer 1984:244). Any comparison based on observation without excavation is dubious, however, the results of Braemer's survey, combined with a visit to the site during 1996, allowed the author to note the remarkable similarity in gate plans between the two sites[29]. The similarity between the two gates potentially hints at a degree of regionalism in the Hauran.

A final important aspect of the Rukeis gate plan is the possible use of an internal buttress to strengthen the gate flanks. Wall 14033 (Area 14) is a massive brick wall that likely buttressed the city wall adjacent to the gate flank (figure 45a-b) . A good parallel for heavily reinforced walls flanking a gate is found at MBII-III Hazor (Yadin 1989:283). The main difference between Hazor and Rukeis is the construction of the reinforcing walls. The walls flanking the gate at Hazor are reinforced with casemates, which become single, solid walls at a distance from the gate (figure 47a). At Rukeis, the walls flanking the gate consist of two solid walls built directly beside one another.

The Rukeis Gate: Individual Features

Many of the architectural features of the Rukeis gate find parallels with gates of various periods throughout the Levant. These architectural features were fundamental to most gates and cannot be considered indicative of a specific period or region. For example, the plaster-covered cobbled floor in the guardrooms and passage at Rukeis is well-attested at: EBI Arslan tepe (Palmieri 1981:104); EBIII Tell Taya (Reade

[28] Iron Age gates have been revealed at Tell Jawa (Daviau 1996:93) and Khirbet al-Mudayna (Daviau 2000:284) on the Jordanian Plateau.

[29] Other variations on the Rukeis gate plan can be observed at Iron Age period sites such as Tell Jawa on the Jordanian Plateau (Daviau 1996:93).

1968:247); MBI Ebla (Matthiae 1968:13); MBII Bethel (Kelso 1961:7); MBII-III Hazor (Yadin 1989:281-2; Ben-Tor *et al.* 1997:353-363); and, MBIII Shechem (Wright 1957:17)[30]. Similarly, the door socket in the passageway at Rukeis finds parallels with EBI Jawa (Helms 1977:29); EBIII Tell Selenkahiye (van Loon 1977:166); and MBII Bethel (Kelso 1961:7)[31]. The stone benches in the guardrooms at Rukeis find parallels with EBII-III Tell el-Far'ah North (Mallet 1973:35); MBI Tell Akko (Dothan 1978:265); and MBII-III Yavneh Yam (Kaplan 1969:120). The passage in the EBII-IV gate at Khirbet Iskander is lined by benches, though none are preserved in the guardrooms (Richard 1990:48).

Building materials and dimensions also find parallels with sites of different periods. Gate materials primarily depended upon the available building resources. Basalt, used in the Rukeis gate, was the most readily available building material in the Hauran. Gate dimensions also varied randomly between sites. For example, similar sized passages to that of the Rukeis gate are found at EBII Tell Dalit (Cresson and Gophna 1979:122) and MBII Tell Tuqan (Guardata 1996:65). Similar sized gate walls are found at EBII-IV Khirbet Iskander (Richard 1983:50) and MBIII Tell Ashdod (Dothan 1971:175).

Some aspects of the Rukeis gate are more unique. The covered drain finds one rare parallel at MBII-III Boghazkoy (Bittel 1957:21)[32]. The pillars in the guardrooms are also unusual, with the only known parallel at EBII-IV Khirbet Iskander. The main difference between the Rukeis and Khirbet Iskander gates is the placement of the pillars on each side of the passage at Khirbet Iskander (Richard 1990:48-50) and in the guardrooms at Rukeis. No pillars were required in the Rukeis passage, because the walls provide support for the ceiling of the gate along the centre of the passage.

The Rukeis Gate: Superstructure

Due to the almost complete erosion of the gate superstructure, reconstructing the vertical appearance of the gate is problematic. Comparative material from other Bronze Age gates is limited, because only the ground plans usually survive. The best-preserved gate superstructures are at MBI period Tell Dan and Ashkelon. At Tell Dan (figure 49a), the original arched entrance has survived intact, offering physical proof that the piers of the standard triple-entry gate provided support for the arched superstructure that spanned the main gate body (Biran 1984; 1994:75-90; Biran *et al.* 1996:58-61). The Ashkelon gate is also a monumental arched gate with a brick superstructure (Stager 1999:237).

The plans and sizes of these gates have little in common with the Rukeis gate. The available evidence indicates that the Rukeis gate had a flat ceiling rather than an arched one, as stone lintels over 2m long are found in the gate passage. There is no evidence that a brick superstructure was employed. The Rukeis gate probably had a horizontal ceiling, similar to that seen at EBII-III Bab edh-Dhra (figure 49b) or in the reconstruction of the MBII gate at Alalakh (figure 49c).

Earthen Ramparts and Related Features

A defining feature of the MBA is the widespread use of the earthen rampart (see p.1). During the EBA, earthen ramparts were primarily constructed to protect the base of a city wall[33]. During the MBA, the earthen rampart develops into a range of forms, including: exterior-wall ramparts; double-wall ramparts; and freestanding earthen ramparts that were either constructed around *tell* summits, or to form circular, rectangular or square enclosures. The relationship between the different ramparts types is unclear and is 'a question needing much precise investigation and analysis' (Wright 1985a:54).

There has been considerable debate regarding the function of these monumental architectural features. Earthen ramparts may have provided a purely defensive function to deter chariots (Kenyon 1952:71), battering rams (Yadin 1955:24) or tunnelling (Stager 1991:8). Alternately, earthen ramparts may have protected the city wall from erosion (Parr 1968:43-4; Kaplan 1975:2; Pennels 1983); provided a platform for a palace or temple (Finkelstein *et al.* 1993:374); or functioned as a temenos wall for a religious structure (Weinstein 1997b:368). It is also possible that earthen ramparts embodied a symbolic demonstration of power (Bunimovitz 1992).

The majority of known earthen ramparts in Jordan were constructed during the MBA[34]. Most known MBA ramparts are located on the Jordanian Plateau at Amman, Tell el-Umeiri, Sahab and Tell Irbid. There is fragmentary evidence for earthen ramparts at Pella, Tell Deir Alla and Tell Nimrin in the Jordan Valley (pp.8-9). All of the Jordanian ramparts date to the mature phases of the MBA, except for the MBI period *glacis* at Tell Irbid and the possible *glacis* at Pella that dates to the MBI-II period (see p.15).

Exterior-Wall Ramparts

The most common form of rampart in Jordan consists of an earthen deposit against the exterior face of a defensive wall (see figure 50a), as seen at MBII-III Amman (p.9) and possibly at MBI-II Pella (p.15) and MBII-III Tell Deir Alla, Tell Nimrin and Sahab (pp.8-9). The rampart at Amman is employed as part of an elaborate system involving retaining-walls and

[30] This construction feature was commonly employed in gates that post-date the MBA, such as the 'Musalu' gate in the Nebi Yunus armoury at Nineveh and Gate 7 at Khorsabad, both of the Neo-Assyrian period (Scott and MacGinnis 1990:65).

[31] Door sockets have been revealed in numerous post-MBA gates, such as the Neo-Assyrian 'Musalu' gate in the Nebi Yunus armoury at Nineveh (Scott and MacGinnis 1990:64).

[32] The lack of parallels may well be the result of a lack of excavation or preservation at other sites.

[33] The evolution of this process is evident at Tell Ta'anach (Parr 1968:42).

[34] Ramparts in Jordan have also been revealed at EBII Bab edh-Dhra (Rast and Schaub 1981:15), Iron age Tell el-Umeiri (Herr *et al.* 1991a:54-57) and Nabataean Petra (Hammond 1960:30).

possible outworks. At most MBA sites in Jordan, however, the earthen rampart is confined to a scarped apron at the base of a wall. At Pella the rampart is more like a plastered surface laid over an existing slope than a large earthen rampart. Although the evidence for a plastered rampart at Pella is limited, an important parallel may be noted in Area B at MBI period Aphek, where the excavators noted a mini *glacis* consisting of a plastered surface that had been deposited over a thin layer of packed soil (Kochavi *et al.* 2000:71). The Aphek 'mini *glacis*' provides the closest parallel for the possible Pella *glacis*. A similar *glacis* rampart may be noted at MBI Tell Poleg, also located on the coast (Kochavi *et al.* 1979:13).

Exterior-wall ramparts were commonly employed at Levantine sites during the EBA and MBA and are thus not a useful indicator of urban or regional development[35]. Exterior-wall ramparts dating to the EBA have been revealed in Syria at: Tell Ashara (Chavalas 1996:93) and Tell Bderi (Pfalzner 1987:294); EBIII Tell Mozan (Kelly-Buccellati 1990:123) and Tell Selenkahiye (van Loon 1968:25); MBI Tell Kannas (Finet 1979:83); MBII Tell Umm el-Marra (Schwartz *et al.* 2003:341) and Tell Hadidi (Dornemann 1979:141); and, MBIII el-Qitar (McClellan 1986:424). They have been revealed in the Southern Levant at: EBII Tell Yarmouth (de Miroschedji 1990:50), Ai (Callaway 1980:152) and Khirbet el-Mahruq (Yeivin 1974:259); EBI Tell Gath (Yeivin 1960:201); EBII-III Tell Halif (Cole 1978:119; Seger 1977:46; 1987:193; Seger *et al.* 1990:10; Jacobs 1984:199); MBI Aphek (Kochavi *et al.* 2000:71), Kabri (Kempinski and Miron 1987:177), Tell Poleg, Tell Burga and Tell Zeror (Kochavi *et al.* 1979:133,142,155); MBI-II Tell Kisan (Seton-Williams 1980:390); MBII Tell el-Hammam (Cahill *et al.* 1988:194) and possibly Tell el-Ajjul (Stewart 1974:11); MBII-III Megiddo (Loud 1948:6), Tell Nagila (Amiran and Eitan 1963a:144; 1963b:334), Shechem (Toombs and Wright 1961:30), Bethel (Kelso 1958:5; 1961:15), Tell Beit Mirsim (Albright 1932a:8; 1938a:28), Tell el-Far'ah North (Mallet 1987 Plans 3 and 4) and Tell el-Hammah (Cahill *et al.* 1988:194); and, MBIII period Gezer (Dever *et al.* 1970:42; see MacAlistar 1911:236).

Two-Wall Ramparts

The two-wall rampart was commonly used throughout the Levant during the EBA and MBA and is a limited indicator of urban and regional patterns of development. The two-wall rampart is an exterior-wall rampart with a retaining-wall at the foot of the earthen embankment (see figure 50b). Evidence for EBA two-wall ramparts has been revealed throughout the Levant at: EBII-III Tell Ta'anach (Lapp 1969:10), Tell Yarmouth (de Miroschedji 1990:56), Tell Halif (Jacobs 1984:199; Seger *et al.* 1990:10) and Tell Dothan (Helms 1977a:110); MBI Kabri (Kempinski and Miron 1987:177); MBII Bethel (Kelso 1958:5) and Tell Hadidi (Dornemann 1979:141); MBII-III Tell Ta'anach (Lapp 1964:15), Shechem (Toombs 1961:28-30) and Tell el-Far'ah North (Mallet 1987 Plan 4); and, MBIII Gezer (Dever *et al.* 1970:42) and Tell Shiloh (Finkelstein *et al.* 1993:15). The only known example in Jordan is at MBII-III Amman (p.9). At Amman, there is no evidence for a *fosse* beyond the line of the retaining-wall as seen at MBI-II Tell Kisan (Seton-Williams 1980:390), MBII Tell Masos (Fritz *et al.* 1983:187) and MBIII el-Qitar (McClellan 1986:424).

Freestanding Ramparts

The widespread use of the freestanding earthen rampart is one of the hallmarks of the MBA (see figure 50c). This rampart type consists of mounds of earth, 'heaped up to form defensive ramparts in themselves, with crests standing high above the internal ground level' (Wright 1985a:54). The origin of the freestanding rampart has been the focus of considerable debate, particularly in relation to the movement of peoples (see p.2). Kaplan argues that the freestanding earthen rampart is a by-product of digging irrigation canals in Mesopotamia (Kaplan 1975:2). Excavations at EBA Mari indicate that the earliest phase of freestanding ramparts served as protective dikes against flooding (Margueron 1997:414). During later phases at Mari, the rampart developed into a truly defensive feature (figure 51).

The earliest known examples of freestanding earthen ramparts are in the Northern Levant at: EBI Tell Mu'azzar, Tell Malhat adh-Dhari, Tell Mabtuh ash-Sharqi (McClellan and Porter 1995:51); EBII Mari (Margueron 1983:284) and Karatas-Semahuyuk (Mellink 1966:36); and, EBIII-IV Tell es-Sweyhat (Holland 1976:36). Freestanding ramparts dating to the EBA are rarely found in the Southern Levant, with a possible example at EBIII Tell el-Hesi (Petrie 1891:31; Dahlberg 1989:47 for section). The freestanding rampart was thus of Syrian origin.

Freestanding earthen ramparts become more commonly employed during the MBA, as is evident at: MBI-III Qatna (du Mesnil du Buisson 1935:40-41); MBI Alalakh (Woolley 1955:133) and Ebla (Matthiae 1970:61); MBI-II Tell Rimah (Oates 1967:91-93); and, MBII-III Carchemish (Woolley 1921:69-71) and Tell Umm el-Marra (Schwartz *et al.* 2003:341) in Syria. They are found in the Southern Levant at: MBI Ashkelon (Stager 1999) and Achzib (Prausnitz 1963:337; 1975:209); MBI-II Megiddo (Finkelstein *et al.* 2000:78); MBII Jaffa (Kaplan 1964:286; 1967:269), Tell Mevorakh (Stern 1973:256; 1984:49), Tell Batash (Mazar 1997:3) and Tell Masos (Fritz *et al.* 1983:186-187); and, MBII-III Tell Jemmah (van Beek 1992:5; 1997:214), Tell el-Milh (Fritz *et al.* 1983:186), Jericho (Kenyon and Holland 1981:108), Yavneh Yam (Kaplan 1967:269), Tell el-Far'a South (Petrie 1930:16), Lachish (Tufnell 1958:33-34), Shechem (Wright 1956; 1957; Toombs and Wright 1961; 1963; Bull *et al.* 1965; Bull and Callaway 1968), Hazor (Yadin 1958b:71; 1960:76; 1989:213) and Tell Haror (Oren and Yekutieli 1996:87).

[35] Following the MBA, this type of rampart was employed at Iron Age Tell el-Kheleifeh (Glueck 1940:6), Tell el-Ful (Albright 1924:17), Ein Gev (Mazar *et al.* 1964:32), and Kadesh Barnea (Dothan 1965:142); and possibly at LBA Megiddo (Loud 1948:16) and Gezer (Dever 1976:35-36), although Finkelstein has argued that the rampart at Gezer dates to the Iron Age (Finkelstein 1981:143).

The freestanding rampart is built in three ways: around the summit of a *tell*, as at Jericho; in a circular plan around a large city forming a lower town, as at Mari, Carchemish, Ebla and Ashkelon; or, in a large rectangular plan, as seen at Tell es-Sweyhat, Qatna, Hazor, Yavneh Yam, Tell Mevorakh, Tell Batash, Tell Haror, Tell Masos, Tell el-Milh, Tell el-Yehudiyeh and Heliopolis. The area enclosed by these ramparts ranges in size from 1ha at Tell Mevorakh to 50ha at Hazor. The smaller, square shaped fortresses, found in south-western Palestine at sites such as Tell Masos, Tell Malhata and Tell Mevorakh, have been variously interpreted as Middle Kingdom Egyptian fortresses (Stern 1984:49) or *Hyksos* fortresses (Fritz and Kempinski 1983:194-7).

The only confirmed example of a freestanding earthen rampart in Jordan is at MBIII Tell el-Umeiri (see p.9). The existence of a freestanding rampart at MBII-III Sahab is possible, although is yet to be confirmed (see p.9). In size, these sites best compare to the small square-shaped rampart sites of south-western Palestine, such as Tell Masos (6ha). Whether the ramparts at Tell el-Umeiri or Sahab were also built according to a square plan is yet to be confirmed. The existence of a possible corner tower at Tell el-Umeiri (Clark *pers. comm.*) implies that the ramparts at that site followed a square plan. At Sahab, the LBA defences followed a square plan (Ibrahim 1989:518; 1997:451) which implies that the earlier MBA defences may also have followed a similar plan.

Core-Walls and Crowning-Walls

Core-wall ramparts consist of earthen fill abutting both the inner and outer faces of a central wall (figure 50d). In section, this type of rampart is similar to the freestanding earthen rampart (figure 50c). Although the core-wall rampart is less common than the exterior-wall rampart and the freestanding earthen rampart, it was employed at EBII Mari (Margueron 2000:910); EBII-III Tell Ta'anach (Lapp 1969:10); MBI Tell Dan (Biran 1994:57-73; Biran *et al.* 1996:10-19); MBII Tell Batash (Mazar 1997:250); and, MBIII Tell Shiloh (Finkelstein *et al.* 1993). A variation of the core-wall rampart incorporates earlier city walls into the rampart to enhance stability. This feature is found at MBI Tell Akko (Dothan 1981:111; Dothan and Conrad 1983:113); MBI-II Tell Poran (Gophna 1992:270); MBII Tell Mevorakh (Stern 1984:49); and, MBIII Tell Malhata (Kochavi 1967:272).

Occasionally, core-walls were completely sealed within the ramparts, as at MBI Tell Akko (Dothan and Conrad 1979:227), Achzib (Prausnitz 1975:209) and Tell Dan (Biran 1994:57-73; Biran *et al.* 1996:10-19). The lack of a crowning wall above these ramparts indicates that the ramparts possibly served a non-defensive function (Finkelstein 1992:205-206). It is more likely, however, that the core-wall usually provided a foundation for a crowning-wall that has since eroded. This assumption is based on the extensive evidence for crowning-walls built above the rampart summit and suggests that the earthen rampart primarily served a defensive function (Mazar 1997:250).

Crowning-walls (figure 50e) have been revealed at: EBI Tell Mu'azzer, Tell Malhat adh-Dhari, and Tell Mabtuh ash-Sharqi (McClellan and Porter 1995:51); EBIII-MBII Mari (Margueron 2000:910-913); MBI Ebla (Matthiae 1970:61); MBI-III Qatna (du Mesnil du Buisson 1935:41); MBII Tell Tuqan (Baffi Guardata 1996:64), Tell Umm el-Marra (Schwartz *et al.* 2003:341); and, MBII-III Alalakh (Woolley 1955:133-137) and Carchemish (Woolley 1921:71) in Syria; and, MBII-III Tell Nagila (Amiran and Eitan 1965:118), Hazor (Yadin 1989:214), Lachish (Tufnell 1958:34), Jericho (Kenyon and Holland 1981:372) and Tell Jemmah (van Beek 1992:7); and, MBIII Yavneh Yam (Kaplan 1967:269), Tell Malhata (Kaplan 1967:272) and Tell Jerishe (Geva 1983:14) in the Southern Levant. At MBI Ebla (Peyronel 2000:1353-1358) and Ashkelon (Stager 1991:4; 1999:237-238), crowning-walls as well as towers have been identified on the summit of the earthen rampart.

There is no evidence in Jordan for core-wall ramparts or freestanding ramparts with crowning walls. The only evidence for a freestanding rampart with a defensive structure on the rampart crest is at MBIII Tell el-Umeiri, where a possible corner tower has been partially excavated on the north-west corner of the rampart (see p.9). The limited evidence for core-wall ramparts or freestanding ramparts with crowning-walls in Jordan contrasts with other regions of the Levant where these rampart types were commonly employed throughout the MBA. In this regard Jordan fell outside of the sphere of one of the major urban developments that characterised the MBA Levant.

Retaining-Walls and Stabiliser-Walls

During the MBA, retaining-walls were commonly used at the base of earthen embankments to provide extra rampart support and a further obstacle to any attacker. Retaining-walls were either employed in Two-Wall ramparts or in Freestanding Ramparts. Because retaining-walls were commonly employed throughout the MBA Levant, they have limited use as an indicator of urban or regional development. Retaining-walls associated with earthen ramparts are rare in Jordan[36]. The only MBA example is the two-wall rampart at Amman (p.9), where Wall 2015 retains an earthen rampart. There is no evidence for a retaining-wall used in association with a freestanding earthen embankment in Jordan. It is difficult to establish parallels between the Amman retaining-wall (2015) and other retaining-walls, because the walls vary randomly in dimensions and building materials and are not built to a standard technique.

Retaining-walls for earthen embankments have been revealed throughout the Bronze Age Levant at: EBII-III Tell Yarmouth (de Miroschedji 1990:56), Tell Ta'anach (Lapp 1964:10; 1969:10), Tell Dothan (Helms 1977a:110,113) and Tell Halif

[36] A possible retaining-wall is associated with the Field XIII *glacis* at EBIII Bab edh-Dhra (Rast and Schaub 1981:15). A possible Iron Age retaining-wall (7K90:10) was excavated in Field B at Tell el-Umeiri (Geraty *et al.* 1989:247).

(Jacobs 1984:199; Seger 1987:193; Seger *et al.* 1990:12); MBI Ebla (Matthiae 1970:61), Tell Umm el-Marra (Schwartz *et al.* 2003:341), Kabri (Prausnitz and Kempinski 1977:166; Kempinski and Miron 1987:177), Tell Burga (Kochavi *et al.* 1979:143) and Tell Akko (Dothan 1976:7); MBI-II Megiddo (Finkelstein *et al.* 2000:78) and Tell Kisan (Seton-Williams 1980:390); MBII Tell Mevorakh (Stern 1973:256; 1984:47,49) and Tell Masos (Fritz and Kempinski 1978:188); MBII-III Tell el-Far'ah North (Mallet 1973:35), Hazor (Yadin 1958b:74; 1960:81; 1989:214), Bethel (Kelso 1958:5), Tell Beit Mirsim (Albright 1928:29; 1932a:8; 1938a:28) and Shechem (Wright 1956:17); and, MBIII Yavneh Yam (Kaplan 1967:269), Tell Malhata (Kochavi 1967:272), Tell Shiloh (Finkelstein *et al.* 1993:20) and Gezer (Dever *et al.* 1970:42,57).

The stabilizer-wall is another structure used to support earthen ramparts is by strategically locating walls at different points within the rampart body. The use of stabiliser walls was relatively common, as found at MBII-III Jericho (Kenyon and Holland 1981:109,168), Hazor (Yadin 1960:76,81) and Tell Batash (Mazar 1997:22;36); and, at MBIII Tell Shiloh (Finkelstein *et al.* 1993:43,76). In some instances, earlier period walls were simply incorporated into the earthen rampart, effectively functioning as stabiliser walls. Examples of this technique are found at MBII Tell Mevorakh (Stern 1984:49) and MBIII Tell Ta'anach (Lapp 1969:19). Stabilizer walls occasionally ran at a right angle from a gate flank into an earthen embankment, bonding the two features. This method is seen at MBI Tell Akko (Dothan and Conrad 1978:266) and MBIII Tell Ashdod (Dothan 1971:175; 1973:10).

There are no known examples of MBA stabiliser walls in Jordan. Wall 2015 at Amman is the only possible exception, if the wall was originally covered by the earthen rampart, which has since eroded away. Ussishkin (1989:29-54; 1992:274-276) argues that stabiliser walls have occasionally been misinterpreted as retaining-walls; however, Wall 2015 at Amman is almost certainly a true retaining-wall at the base of the rampart (p.9).

Rampart Composition

During the EBA, most ramparts had a simple composition, like those at EBII Khirbet el-Mahruq (Yeivin 1974:259), EBII-III Tell Halif (Seger 1977:46; 1983:2; Cole 1978:119) and EBIII Beth Yerah (Maisler *et al.* 1952:173). EBA experimentation with different soil types in rampart construction is only rarely observed at sites such as EBII Tell Yarmouth (de Miroschedji 1990:55-56). During the MBA, engineering knowledge of rampart composition becomes increasingly sophisticated (Pennels 1983:61). The use of different soil types to create rampart stability is seen at MBI Ebla (Matthiae 1970:61; 1997b), Kabri (Prausnitz and Kempinski 1977:166), Achzib (Prausnitz 1975:208), Tell Poran (Gophna 1992:270), el-Qitar (McClellan 1986:424) and Tell Akko (Dothan 1973:257; 1974:277; 1976:5-7). The trend towards a more complicated, deliberate use of soil types continued into the mature phases of the MBA, as seen at: MBII Bethel (Kelso 1961:15), Tell Batash (Kelm and Mazar 1979:242), Tell Jemmah (van Beek 1992:5), Tell Mevorakh (Stern 1984:49) and Tell Masos (Fritz *et al.* 1983:187); MBII-III Hazor (Yadin 1958b:74-75; 1989:214), Tell Nagila (Amiran and Eitan 1963a:144) and Tell Ta'anach (Lapp 1964:15); and, MBIII Tell el-Milh (Kochavi 1967:272), Tell Shiloh (Finkelstein *et al.* 1993:299-300) and Tell Jerishe (Geva 1983:14,18)[37].

The use of superimposed layers at Tell Akko indicates that the rampart builders had used 'careful consideration of their physical qualities' to 'stop penetration of water' (Dothan 1976:5-7). The composition of the rampart at Tell Jerishe also 'prevented drainage problems' (Geva 1983:14,18). A similar use of soil types was employed at Tell Shiloh, where granulometric studies revealed important differences in the clay content and permeability of various layers. The clay layers give weight to the structure, lend stability to the rampart and seal the *glacis* surface, while the porous layers facilitate drainage (Finkelstein *et al.* 1993:299-300). The earthen rampart was stabilised by 'the crushed limestone reducing the plasticity of the clay, and the ash exerting a binding action similar to cement' (Wright 2000:72). These examples demonstrate that the builders of MBA ramparts had developed an advanced understanding of the properties of soil types, how those soil types facilitate or impede drainage, and how various soil types affect the stability of the rampart.

Engineering ability developed to such a level during the MBA that a number of ramparts were constructed using a standard technique known as the 'sandwich method' (see p.1; figure 4). The sandwich method utilised evenly spaced plaster layers to separate the different earthen deposits and act as anchors between the rampart fill and the plaster *glacis* surface. Evidence of this technique has been revealed at: MBI Kabri (Prausnitz and Kempinski 1977:166) and Tell Dan (Biran 1994:59; Biran *et al.* 1996:10-19,53-57); MBII Tell Jemmah (van Beek 1992:5); MBII-III Jericho (Kenyon and Holland 1981:108,168,218,374) and Lachish (Tufnell 1958:45); and, MBIII Gezer (Dever *et al.* 1970:43). A variation of the 'sandwich method' is noted at MBIII Tell Shiloh, where the surface layer of the rampart was anchored into the lower layers by means of 'interfingering' (Finkelstein *et al.* 1993:299-300).

The MBA ramparts of Jordan were of simpler construction than the complex ramparts noted elsewhere in the Levant. There is no evidence for the standardised sandwich method of construction. The most complex rampart composition in Jordan is at Amman (p.9), where a combination of earthen deposits was used. The ramparts at Pella (p.15), Tell Deir Alla (p.8) and Tell Nimrin (p.8) are also simple, but the evidence of their construction is limited.

[37] Granulometric studies of ramparts at Tell Beersheba indicate that engineering knowledge used in rampart construction evolved to an even higher level during the Iron Age. The use of different soil types to increase drainage and rampart stability continued and was combined with the construction of earthen arches within the core of the slope to give even greater stability (Aharoni 1973:11).

Rampart Surface

MBA ramparts were frequently sealed with a plastered *glacis* to protect the earthen embankment from water and enhance stability. The plastered *glacis* is a common feature of earthen ramparts at: EBI Tell Gath (Parr 1968:41) and Tell el-Far'ah North (Joffe 1997:303); EBII-III Tell Ta'anach (Lapp 1969:12) and Tell Yarmouth (de Miroschedji 1990:56); EBIII Tell el-Hesi (Dahlberg 1989:46) and Tell Mozan (Kelly-Buccellati 1990:123); MBI Tell Akko (Dothan 1981:111; Dothan and Conrad 1983:113), Aphek (Kochavi 1975:31) and Tell Poleg (Kochavi *et al.* 1979:133); MBI-II Megiddo (Loud 1948:6); MBII Tell Mevorakh (Stern 1973:256) and Tell Batash (Mazar 1997:22); MBII-III Alalakh (Woolley 1995:133), Jericho (Kenyon and Holland 1981) and Lachish (Tufnell 1958:46); and MBIII Shechem (Toombs 1964:9). Earthen ramparts were occasionally protected by a layer of tightly packed stones or crushed brick rather than plaster, as found at EBII-III Tell Yarmouth (de Miroschedji 1990:56); EBIII Beth Yerah (Maisler *et al.* 1952:173), MBI Tell Akko (Dothan 1974:277), MBII Tell Masos (Fritz *et al.* 1983:188) and MBII-III Carchemish (Woolley 1921:47).

Evidence for a *glacis* in Jordan is possibly found at MBI Tell Irbid (p.8), MBII-III Sahab and Amman (p.9) and MBI-II Pella (p.15). No details are given regarding the *glacis* at Tell Irbid or Sahab. The Amman *glacis* consisted of 'rubble with a grayish clay' that sealed the rampart body (Zayadine *et al.* 1989:359). The *glacis* at Amman finds parallels with MBII-III Hazor (Yadin 1989:214). There is fragmentary evidence for a *glacis* at Pella in the form of a plaster layer mixed with fist-sized stones. The Pella *glacis* is comprised of a simple plastered level formed out of the existing slope. The possible *glacis* at Pella finds a close parallel at MBI Aphek (Kochavi *et al.* 2000:71). In general, the *glacis* is too common to be indicative of any pattern of urban or regional development.

Chapter 5: Conclusions

Summary of Comparative Study

Two constraints emerge from the comparative study. The first constraint is the disparity of data between MBA sites in Jordan. Most of the available data for fortification walls, buttresses, towers and gates comes from excavation at Pella and Rukeis. Data from Amman, Tell Irbid, Sahab, Tell el-Umeiri, Tell Deir Alla and Tell Nimrin supplements the more comprehensive data from Pella and Rukeis. Secondly, many architectural features have limited use as indicators of urban and regional patterns of development, because they were commonly and randomly employed at numerous Levantine sites during the Bronze Age. Features of limited use are solid walls, tower shape typology, spacing of towers along the line of a defensive circuit, exterior-wall ramparts, two-wall ramparts, retaining-walls and stabilizer walls for earthen ramparts, the plastered *glacis*, and the building materials and dimensions of walls, towers and buttresses. Despite data disparity and the limitations of some architectural features, a number of salient points emerge from the comparative study.

Walls and Related Features

• During the EBA, casemate walls were rare in the Levant but were common in Jordan at Bab edh-Dhra, Numeira and Khirbet Iskander. During the MBA, casemate walls were a standard architectural feature in the Levant, but were rare in Jordan. The only possible MBA casemates in Jordan are found at Pella and Tell Deir Alla. Evidence for a casemate wall at Pella is extremely limited. If the Pella wall *is* a casemate, MBII-III Hazor is the closest parallel in terms of materials, dimensions and the combined use of single trace and casemate walls.

• The only known example of the sawtooth wall in Jordan is found at MBI-II Pella. The only other South Levantine example is at MBIII Tell Shiloh.

• The only known wall buttresses in MBA Jordan are found at MBI-II Pella. The triangular-shaped Buttress 12 is an unusual feature found there. Although the evidence is thin, a contemporary parallel possibly exists at Megiddo, which indicates a degree of regionalism in the North Jordan Valley and Esdraelon Plain during the MBI-II period. There is no evidence in Jordan for the buttress-niche method of wall construction, indicating that Jordan falls outside the sphere of this architectural development.

• Three types of distinctive MBA foundation construction are found in Jordan. Stepping in a wall from the line of foundations is a technique only evident at MBI-II Pella and MBII-III Hazor and Bethel. Stepping the substructure up the slope and using deeper foundations at the fortification exterior than interior are two other techniques, most clearly evident at MBI-II Pella, MBII-III Hazor and MBIII Gezer. These parallels are indicative of a degree of regionalism in the North Jordan Valley and the Huleh Valley.

• The study of brick sizes and types indicates a degree of regionalism in the Jordan Valley. Good parallels exist between the brick sizes used in the fortifications at MBI-II Pella, MBI-II Megiddo, MBII-III Tell Deir Alla and MBII-III Tell Nimrin. The use of different brick types to facilitate drainage and enhance wall stability is most clearly evident at MBI-II Pella, MBII Megiddo and MBII-III Hazor.These parallels are indicative of a degree of regionalism in the North Jordan Valley, Huleh Valley and Esdraelon Plain.

Towers and Related Features

• MBI-II Pella has the only concrete evidence for a MBA tower in Jordan. The tower at EBII-IV Khirbet Iskander is the closest parallel in Jordan by date (Richard 1983:50; 1987:243; 1990:36; 2001:441). Both the Pella tower and the Khirbet Iskander tower straddle the curtain walls.

• A shift from externally protruding towers to a salient-recess system is one of the defining features of MBA military architecture. Tower 1 at Pella protrudes slightly from the defensive line, indicative of a salient-recess system. The tower at MBI Tell Beit Mirsim is the closest parallel in terms of tower projection, dimensions and date.

• The most distinctive feature of Tower 1 at Pella is the chamber within the tower superstructure. The function of this chamber is debatable. The closest South Levantine parallel in date and construction is MBII-III Gezer.

Gates and Related Features

• The only known MBA gate in Jordan is found at MBII Rukeis. The Rukeis gate plan differs considerably from the triple-entry gate plan that is typical throughout the MBA Levant. Some tentative parallels may be drawn between the Rukeis gate and double-entry gate plans at MBII-III Tell el-Far'ah North, Shechem and Tell Ashdod; and at EBII-IV Khirbet Iskander. The well-developed guardrooms are typical of the later MBA. The Rukeis gate is unique, consisting of a combination of elements found in gates elsewhere.

• The use of pillars in the guardrooms is the most unusual individual feature of the gate plan. Close parallels for this feature are only found with EBII-IV Khirbet Iskander.

Earthen Ramparts and Related Features

• In Jordan, the greatest number of rampart sites and the most complex ramparts are found on the Jordanian Plateau. These ramparts date to the MBII-III period. MBA ramparts in Jordan are less complex than ramparts noted elsewhere in the Southern Levant. The most complex rampart in Jordan is the two-wall rampart at MBII-III Amman. Information about the ramparts at Pella, Tell Deir Alla, Tell Nimrin, Tell Irbid, Sahab and Tell el-Umeiri is limited.

• The most commonly employed rampart in MBA Jordan is the exterior-wall rampart. The preference for exterior-wall ramparts, instead of the freestanding rampart, indicates a degree of architectural continuity between the EBA and the MBA in Jordan.

• The only confirmed example of a MBA freestanding rampart in Jordan is at MBIII Tell el-Umeiri. A similar rampart may have existed at Sahab. The limited evidence for freestanding ramparts, either with or without core-walls or crowning- walls, indicates that Jordan was on the periphery of this development that characterised the MBA Levant. There is a possibility that the freestanding rampart at Tell el-Umeiri on the Jordanian Plateau may be compared to the small rampart sites found in the Central Highlands and in south-western Palestine.

• There is limited evidence for a MBI-II period *glacis* surface at Pella (Area IIIF). If this surface is a *glacis*, it is unusual because it is a simple plastered surface, rather than a plastered surface sealing an earthen rampart. The only close parallels for this feature are at MBI period Aphek and Tell Poleg.

Military Architecture and MBA Settlement in Jordan

Fortifications provide physical evidence of centralized control. The identification and dating of military architecture is central to understanding patterns of MBA urbanism in the Levant. The role of fortifications in reconstructing urban patterns is best illustrated by the debate between Yadin (1978) and Kochavi, Beck and Gophna (1979) regarding the date and location of the earliest MBA fortifications in the Southern Levant.

MBI Coastal Settlement (c.2000-1775 B.C.)

MBI urban settlement in the Southern Levant is most evident on the coast (figure 52). During this period, fortified sites were established at Ashkelon, Tell Akko, Kabri, Aphek, Tell Poleg, Tell Burga, Tell Zeror, Tell Hefer, Ain Zurekiyeh, Tell Poran and Achzib. Ashkelon and Tell Akko were important urban centres characterised by monumental, elaborate fortifications. There are multiple phases of MBI period fortifications at Ashkelon, Aphek and Tell Zeror. At Ashkelon, there are four phases of earthen ramparts dating to this period (Stager 2003 *pers. comm.*). By comparison, there is very little evidence for urban fortifications in other regions of the Southern Levant at this time. Regions such as the *Shephelah*, the Central Highlands, the Esdraelon Plain, the Jordan Valley and the Jordanian Plateau were largely devoid of urban settlement during the MBI.

Why MBA urbanism first developed on the coast is debatable. The access to sea-trade at the coast suggests a connection with Syria or Egypt. Stager attributes the foundation of Ashkelon to the *Coastal Amorites* and suggests that the characteristic defensive features, such as massive freestanding ramparts and arched gateways, find their origins in the rampart sites of the EBIII period in north-eastern Syria (Stager 1999:237). An alternative explanation to Syrian colonisation is the introduction of the *concept* of the Syrian freestanding earthen rampart followed by an indigenous development of rampart styles.

Less credence is given to a possible Egyptian influence, even though the beginning of the MBI in the Southern Levant coincides with the Middle Kingdom in Egypt, a period of expanded trade and prosperity. There is significant evidence of Egyptian trade objects from this period found at sites on the Syrian coast and mainland (Bietak 1987a:97; 1991:54; Matthiae 1984:23; 1987:113;166; Porada 1966; O'Connor 1983; 1985; Tufnell and Ward 1966; Ward 1978a). Sea-trade between Egypt and Syria followed the route along the South Levantine coast, and most likely resulted in interaction between Egyptian traders and the local population.

This interaction may have provided the initial stimulus for the development of urban centres along the South Levantine coast at this time. It is interesting to note that many of the MBI coastal sites, such as Tell Burga and Tell Zeror, are small settlements, characterised by simple fortifications. It is possible that these early, simple settlements preceded the construction of the monumental Syrian style ramparts at sites such as Ashkelon and represent a brief phase of development that was triggered by trade with Egypt.

Ultimately, current evidence supports a stronger degree of Syrian influence than Egyptian influence on the MBI period coastal settlement. The evidence for four MBI phases of massive freestanding ramparts at Ashkelon certainly implies that Syrian influence on the South Levantine coast was consolidated early in the MBI period. Even if an initial Egyptian inspiration triggered MBI settlement on the coast, this stimulus was soon superseded by a Syrian influence.

The MBI-II Transition (c.1800-1750 B.C.)

During the MBI-II transition, settlement, trade and prosperity continued to expand (figure 53). Fortified settlements were established in geographical regions directly linked to the Coastal Plain, such as Tell Beit Mirsim in the *Shephelah*, Megiddo in the Esdraelon Plain, and Tell Dan in the North Jordan Valley. Evidence for this settlement pattern has led to the development of a coastal theory of settlement, involving an initial phase of settlement on the coast and subsequent settlement inland (Gerstenblith 1983:118).

Magness-Gardiner argues that 'the presence of MBIIA (MBI) town walls at Pella and a thriving MBIIA (MBI) community at Tell el-Hayyat counters the coastal theory of Gerstenblith (Magness-Gardiner 1997:315). More recent excavation at Pella point to a date closer to the MBI-II period transition for the construction of the fortifications (see p.19). A date in the MBI-II transition lends a degree of support to Gerstenblith's theory, because it implies that Pella's fortifications were constructed slightly later than at the sites on the coast. A coastal influence at Pella may also be noted in the possible *glacis* rampart, which only finds close parallels at MBI period Aphek (Kochavi *et al.* 2000:71) and Tell Poleg (Kochavi *et al.* 1979:13) in the coastal region.

There are several reasons why Pella's defences were constructed early in the MBA. Pella is located in a good agricultural area in the North Jordan Valley, is strategically situated to trade routes and is directly connected to the coast by way of the Esdraelon Plain. An early MBA date for Pella's defences is supported by evidence from nearby Megiddo, which was also fortified during the MBI-II transition (Finkelstein *et al.* 2000:78,591). The fortifications at Megiddo share some similar features with those at Pella, including brick sizes and possibly the use of 'triangular' buttresses'. These parallels hint at a degree of regionalism in the North Jordan Valley and Esdraelon Plain around the MBI-II transition. Because Pella and Megiddo occupy important locations for agriculture and trade, settlement spreading from the coast would first be felt at sites such as these.

Tell Irbid on the North Jordanian Plateau is the only fortified site in Jordan that dates to the beginning of the MBI. The fortifications at Tell Irbid are contemporary with the earliest coastal settlement and pre-date the fortifications at MBI-II Pella and Megiddo, even though Pella and Megiddo were more directly connected to the coast. Tell Irbid is more closely linked geographically to southern Syria, the Hauran and the Damascene, than to the coast. The existence of a large, fortified urban centre on the Jordanian Plateau at the beginning of the MBI period does not fit neatly with the coastal theory. Instead, development at Tell Irbid suggests that Syrian urban influence moved south overland as well as along the coast. An 'overland theory' may also account for the early date of the fortifications at Tell Dan, which pre-date the defences at Pella, Megiddo and Hazor. Tell Dan is further from the coast than these sites and is geographically connected to Syria by way of the Beq'a Valley.

Lack of excavation in southern Syria is the main hindrance to testing a theory of Syrian overland influence. Currently, we are unable to correlate north and South Levantine pottery (Nigro 2000:1187) or establish patterns of architectural and urban development between the two regions. The Hauran and the Damascene are crucial areas for understanding the relationship between the Northern and Southern Levant in the early MBA. These regions are characterised by numerous large *tells*, that may reveal evidence for an urban MBI period culture.

The MBII-MBIII in the Jordan Valley (c.1775-1550 B.C.)

The trend towards increased trade and urban development accelerated during the MBII-MBIII periods (figure 54). Fortified urban centres emerged in the Central Highlands, the South Jordan Valley and the Jordanian Plateau south of Tell Irbid. These regions have limited agricultural potential compared to the fertile regions of the Coastal Plain, Esdraelon Plain, and North Jordan Valley. The Central Highlands and the Jordanian Plateau are hilly areas conducive to pastoralism. Nearly all of the MBA urban fortifications in Jordan date to the MBII-III periods. In the East Jordan Valley, fortified settlements developed at Tell Deir Alla and Tell Nimrin. On the Jordanian Plateau, fortified settlements emerged at Amman, Sahab, Tell el-Umeiri and Khirbet el-Mekhayyer. The fortifications at Rukeis in the Hauran also date to this period.

Regionalism

During the MBII-III periods, the architecture of the East Jordan Valley exhibits a degree of regionalism. Concrete evidence of regionalism includes: the use of standard brick sizes at Pella, Tell Deir Alla and Tell Nimrin; the use of a simple exterior-wall rampart *glacis* at Pella, Tell Deir Alla and possibly Tell Nimrin; and, the possible use of casemate defences at Pella and Tell Deir Alla. Considering that the sites of Tell Deir Alla and Tell Nimrin are immediately south of Pella and date to a later phase in the MBA, the military architecture of MBI-II Pella may have influenced the fortifications at MBII-III Tell Deir Alla and Tell Nimrin. Architectural characteristics that developed at MBI-II Pella thus spread south during the following phases of the MBA.

In the same way, urban development spread north from MBI-II Pella to MBII-III Hazor in the Huleh Valley. Common construction techniques used at Pella and Hazor indicate that the engineers who constructed the Hazor defences may have been influenced by the methods employed in the North Jordan Valley. The most distinct parallels between Pella and Hazor are the construction techniques used in the city walls. At both sites, evidence shows the wall superstructure constructed above a platform of brick or stone, using a deliberate combination of brick types to facilitate drainage and enhance stability. Both of these construction features are rare. Parallels may also be noted in the use of casemate walls and the method of stepping fortification foundations up a slope. These parallels hint at a degree of regionalism in the North Jordan Valley and Huleh Valley.

Shared Knowledge

The fortifications at Pella also find close parallels with other South Levantine sites. The platform method of foundation construction, noted at Pella and Hazor, is only evident elsewhere at MBII-III Bethel. The foundation technique of stepping the stone substructure up a slope that has been observed at Pella and Hazor is evident at MBIII Gezer. The use of a sawtooth wall at Pella finds the closest parallels at MBIII Tell Shiloh. Features of the Pella tower, including the date, size and tower projection from the circuit wall find the closest parallels at MBI Tell Beit Mirsim. The use of narrow chambers within the Pella tower superstructure best compare to the towers at MBIII Gezer. These various parallels are not particularly instructive for establishing patterns of regional development. The salient point is that many features employed in the defences at Pella are evident at other South Levantine sites that were established during a later phase of the MBA.

The similarity in construction features between sites is evidence for a degree of shared knowledge regarding the civil engineering of fortification systems. Parallels between sites indicate that once new and effective construction methods were implemented at a given site, knowledge of those techniques spread to neighbouring regions. Some degree of shared knowledge must have existed at the time, otherwise standard architectural features, such as the sandwich method of rampart construction and the use of different brick types in a wall, would have had to develop independently at a variety of sites.

One explanation is that single engineering corps were employed to construct the defences at a number of sites. This theory would account for the almost identical nature of the fortifications at some sites. Alternately, cities may have shared engineering knowledge, if not actual engineers. If this theory is true, distinct methods of military construction may be interpreted as reflections of political alliances. For example, the almost identical rampart composition at Lachish and Gezer may be indicative of a political relationship between those two sites. Similarly, the unusual features of brick wall construction evident at Pella and Hazor may be indicative of a political relationship between those two sites.

Without the confirmation of textual material, this theory remains at a hypothetical stage. The MBA Mari texts disclose the hiring of foreign artisans by various Syrian city-states (*ARM* XIII,42; *ARM* I,46; *ARM* I,83; and *ARM* XII,747). These texts generally refer to skilled workers such as carpenters and metalworkers. Although civil engineers are never specifically mentioned, the Mari texts provide evidence of skilled workers being traded between cities to fulfil specific functions. Mari text *ARM* XII,747 describes 'a group of men from various cities as far apart as Hazor, Yamkhad, Carchemish, Emar, Elahut, and Eshnunna who stay together in a special quarter of the palace at Mari'. These texts demonstrate that artisans with required skills were employed 'from foreign countries to perform specific tasks requiring special skills' (Gerstenblith 1983:13; Sasson 1968:46-54). It is possible that a similar situation existed in the Southern Levant.

The MBII-III on the Jordanian Plateau (c.1775-1550 B.C.)

Regionalism

The greatest number of MBA fortified sites in Jordan are found on the Jordanian Plateau and date to the MBII-III period. The military architecture of the Jordanian Plateau exhibits features that are unusual elsewhere in the Levant, indicating a degree of regionalism. During the EBA, for example, casemate walls were common in Jordan but rare elsewhere in the Levant. During the EBIV and LBA there is strong evidence for fortified urban centres on the Jordanian Plateau, while the Southern Levant is distinguished by a lack of military architecture. During the MBA, Jordanian sites used the exterior-wall rampart at a time when the freestanding rampart was commonly employed throughout the Levant.

The absence of military architectural features in Jordan also provides useful insights into the nature of urban development there. A range of important defensive features, commonly employed throughout the MBA Levant, were never utilized in Jordan. For example, core-wall ramparts, freestanding ramparts with crowning-walls, ramparts of elaborate composition, and triple-entry gates were rarely or never employed in MBA Jordan. This absence is either because excavation in Jordan has revealed only fragmentary evidence or because Jordan was outside of the sphere of influence of these important architectural developments.

Continuity

A study of gates and ramparts indicates greater architectural continuity on the Jordanian Plateau from the EBA to the MBA, than is evident elsewhere in the Levant. The exterior-wall rampart was commonly employed in Jordan during both the EBA and the MBA. Elsewhere in the Levant, the freestanding rampart typically superseded the exterior-wall rampart during the MBA. Further evidence for architectural continuity may be seen in the EBII-IV gate at Khirbet Iskander and the MBII gate at Rukeis, sites that are geographically linked by their location on the frontier of the Eastern Desert. Despite the difference in date, the gates share a number of similarities and are different to the standard MBA gates employed elsewhere in the Levant. Standard building practices possibly existed for longer periods of time in marginal zones than in the more rapidly evolving verdant regions.

One explanation for continuity may be found in the potential purpose for desert frontier settlement. Sites such as Khirbet Iskander and Rukeis fulfilled an intermediary function between the desert and the verdant regions of the Levant. Rukeis may have been established by a larger centre, such as Bosra or Tell Debbeh, to function as a fortress on the desert frontier. Sahab, to the south, may have served a similar function in relation to the large centre of Amman. Ibrahim suggests that Sahab functioned as a fortress on the desert frontier and 'must have been important in defending the highland from desert attack' (Ibrahim 1987:76).

Ibrahim calls for an archaeological survey 'with an eye to locating a *limes*, or series of fortifications', dating to the MBA, located along the desert frontier (Ibrahim 1987:76). Chains of fortresses along the desert frontier are known to have existed during the Middle Kingdom in southern Egypt (Lawrence 1965:71). Whether sites such as Sahab and Rukeis formed part of a *limes* system along the desert frontier in Jordan is debatable. The available evidence suggests a political landscape characterised by independent city-states, each centre primarily concerned with protecting individual interests. Although city-states possibly worked together to protect trade routes running through the region, they were more likely protecting their individual territory. Only textual evidence will clarify this issue.

Jordanian Plateau and Central Highlands

The Jordanian Plateau and the Central Highlands exhibit similar patterns of urban development, transitioning from pastoral to sedentary settlement during the MBII-III periods. These regions shared a similar topography and climate, on opposite sides of the Jordan Valley. Fortified urban settlements developed in these regions only during the MBII-III periods. Sites in both regions, such as Tell el-Umeiri, Sahab and Tell Shiloh, were of similar size (4-6ha), were characterised by rampart fortifications often with a defensive wall, and date to the same time period (the MBII-III). Finkelstein has proposed that sites such as Shechem and Jerusalem in the Central Highlands were surrounded by smaller centres at Tell Shiloh, Bethel and Hebron (Finkelstein *et al.* 1993:380). A similar pattern may be observed on the Jordanian Plateau, where the large centre of Amman is comparable to Shechem or Jerusalem and the small sites, such as Tell Safut and Sahab, are comparable to Tell Shiloh, Bethel and Hebron. This model may also be applied to Tell Jalul and Tell el-Umeiri. Future excavation may reveal that these sites served a similar function, whether as strongholds or as cultic or administrative centres.

Conclusion

Four main points arise from the study of the military architecture of MBA Jordan. First, settlement patterns and urban developments in MBA Jordan essentially parallel developments in other regions of the Southern Levant. Second, MBA regionalism is clearly identified in the North Jordan Valley/ Huleh Valley/Esdraelon Plain; and on the Jordanian Plateau. To a lesser extent, there are elements of regionalism in the East Jordan Valley. Third, the military architecture of the Jordanian Plateau and the Hauran displays a number of traits that are rare in MBA fortifications elsewhere in the Levant. These features indicate continuity in architectural traditions on the desert frontier, from the EBA to the MBA. Fourth, the Hauran, the Damascene and southern Syria have not been adequately explored in order to fully understand Syrian influence on MBA settlement patterns in the Southern Levant. Excavation in these regions may well confirm Syrian urban influences moving overland towards the Southern Levant at the beginning of the MBA. Excavation at Tell Dan and Tell Irbid and survey in the Irbid region of the Jordanian Plateau has already indicated this possibility. An overland urban influence would complement the previously established coastal theory of MBA settlement in the Southern Levant.

Appendices

Introduction

The following appendices present the comparative data for the ceramic material from Pella and Rukeis. Appendix A presents the data from Pella XXVIIIC. Appendix B presents the data from Pella IIF. Appendix C presents the data Rukeis Area 0705. Appendix D presents the data from Rukeis Area 0708. Appendix E presents the data from Rukeis Area 0714. Appendix F presents the data from Rukeis 0716.

The data is presented in a standard format, whereby each relevant sherd is presented consecutively according to locus level number (as is the system at Pella) or according to a running number system (as is the system at Rukeis). Within each level, sherds are listed consecutively according to catalogue number. The vessel form is listed below the catalogue number. Each of these details is highlighted.

The comparative data for each sherd presented using a series of abbreviations, listed below. Matches fall into 'Parallel' and 'Variable' categories, each followed by the relevant reference, and the associated period.

Abbreviations

P	Parallel
V	Variable

Alalakh
Heinz, M.
1992 *Tell Atchana / Alalakh. Die Schichten VII-XVII.* Wiesbaden

Aphek 75
Beck, P.
1975 The Pottery of the MBIIA at Tel Aphek. *Tel Aviv* 2, 45-85

Aphek 85
Beck, P.
1985 The MBIIA Pottery From Aphek 1972-1984 : First Summary. *Tel Aviv* 12.1, 181-203

Arqa 78
Thalmann, J-P.
1978 Tell Arqa : Campagnes 1972-74. *Syria* 55, 88-144

Arqa 79
Thalmann, J-P.
1979 Tell Arqa 1978-1979 : Rapport Provisoire. *BMB* 30, 61-80

Arqa 91
Thalmann, J-P.
1991 L'Age du Bronze 'a Tell 'Arqa. Bilan et Perspectives (1981-1991). *Berytus* XXX, 21-38

Bosra
Seeden, H.
1986 Bronze Age Village Occupation at Busra : AUB Excavations on the Northwest Tell, 1983-1984. *Berytus* XXXIV, 11-82

Dan
Biran, A., Ilan, D., Greenberg, R.
1996 *Dan I*, Jerusalem

Gezer i
Dever, W.G., Lance, H.D. and Wright. G.E.
1970 *Gezer I*, Jerusalem

Gezer ii
Dever, W.G., Bullard, R.G., Cole, D.P. and Seger, J.D.
1974 *Gezer II*, Jerusalem

Gezer iv
Dever W.G.,
1986, *Gezer IV*, Jerusalem

Ghassil
Doumet, C.
1986 *Les Fouilles de Tell el Ghassil de 1972 'a 1974*, Unpublished Doctoral Dissertation, Sorbonne, Paris

Hama
Fugmann, E.
1958 *Hama : Fouilles et Recherches, 1931-1938, II. L'Architecture des Periodes pre-Hellenistiques,* Copenhagen.

Hazor i
Yadin, Y.
1958b *Hazor I*, Jerusalem

Hazor iii-iv
Yadin, Y.
1989 *Hazor III-IV*, Jerusalem

Hazor v
Ben-Tor, A., Bonfil, R., Garfinkel, Y., Greenberg, R., Maeir, A.M., Mazar, A.
1997 *Hazor V*, Jerusalem

Jawa
Betts, A.V.G. (ed)
1991 *Excavations at Jawa 1972-1986*, Edinburgh

Jericho iv
Kenyon, K.M., Holland, T.A.
1982 *Excavations at Jericho IV. The Pottery Type Series and Other Finds*, London

Jericho v
Kenyon, K.M., Holland, T.A.
1983 *Excavations at Jericho V. The Pottery Phases of the Tell and Other Finds*, London

KEL
Marfoe, L.
1979b *Tell Kamid el-Loz, 1972-73. Preliminary Type Series Catalogue. North Slope and Palace Areas*, Unpublished manuscript, University of Chicago

Lachish
Tufnell, O.
1958 *Lachish IV*, London

Meg
Loud, G.
1948 *Megiddo II*, Chicago

Mev
Stern, E.
1984 *Tel Mevorakh 2*, Qedem 18, Jerusalem

Sali
von der Osten, H.H.
1956 *Svenska Syrien Expeditionen, 1952-53, Die Grabung von Tell Es-Salihiyeh*,
Lund

Sarepta
Anderson, W.P.
1988 *Sarepta I : The Late Bronze and Iron Age Strata of Area II.Y*,
Beirut

Shec
Cole, D.P.
1984 S*hechem I : The Middle Bronze IIB Pottery*, Winona Lake

TBM 32
Albright, W.F.
1932b The Excavation of Tell Beit Mirsim in Palestine, Volume 1 : The Pottery of the First Three Campaigns. *AASOR* XII

TBM 33
Albright, W.F.
1933 The Excavation of Tell Beit Mirsim, vol 1A : The Bronze Age Pottery of the Fourth Campaign. *AASOR* XIII, 55-128

TNM
Bourke, S.J.
1991 *The Transition From the Middle To the Late Bronze Age in the Northern Levant : The Evidence From Tell Nebi Mend*, Unpublished Doctoral Dissertation, University of London

Tyre
Bikai, P.
1978 *The Pottery of* Tyre, Warminster

Pottery Methodology

Shape Typology
This is an architectural study that employs a pottery methodology to date the architecture at Pella and Rukeis. The pottery methodology is based upon a comparison of rim shapes with pottery sequences at other MBA sites. The results of the comparison are then subjected to a quantitative analysis. Because rims are the most common and instructive diagnostic form, they are the focus of the study rather than handles or bases. The widespread application of the fast-turning wheel at the beginning of the MBA (Beck 1985:181) resulted in vessel rims being manufactured to standard types throughout the Levant. Specific rim types were produced during various stages of the MBA and changed over time according to fashion. McGovern has also demonstrated the advantages of focusing on rim profiles and Harbottle in their work on MBA storage jars from Tell ed-Dab'a (1997:143). Surface treatment is not given the same consideration in this study, as there tends to be greater regional variation in decoration than with rim profiles.

Comparative Ceramic Analysis
The pottery from Pella and Rukeis is compared to the ceramic sequences from Alalakh, Aphek, Arqa, Bosra, Tell Dan, Gezer, Tell Ghassil, Hama, Hazor, Jawa, Jericho, Kamid el-Loz, Lachish, Megiddo, Tell Mevorakh, Tell es-Salihiyeh, Sarepta, Shechem, Tell Beit Mirsim, Tell Nebi Mend and Tyre. These sequences represent a broad range of MBA assemblages from the greater Levant. Ceramic parallels with the Pella and Rukeis data are tabulated in Appendices A-E with each sherd listed numerically according to trench and level number. Close matches are categorized as 'Parallel', while variants are categorized as 'Variation'.

It is preferable to have as broad a range of comparative ceramic material as possible. Although the sample of pottery sequences used here is not comprehensive, it provides a solid platform for a comparative pottery study. There are inevitably some limitations with this method of enquiry, as stylistic analysis 'is an imprecise, subjective process' (Helms 1989a:27). The size of published ceramic sequences, for example, varies from site to site. There is also a disparity in the MBA phases that are represented in each sequence, with some sequences dating predominantly to the earlier phases of the MBA and some dating to the later phases. Particular note should be taken of the Rukeis assemblage, which differs to a significant degree from ceramic sequences elsewhere in the Levant. This is the result of Rukeis being located in the relatively unexcavated region of the Hauran where there are few comparable ceramic sequences (McLaren 2001a, 2001b).

Quantitative Analysis
The ceramic material may be subjected to a qualitative or a quantitative analysis. The qualitative approach focuses on the latest material derived from any given phase. To some extent the qualitative approach can be problematic, as some ceramic types have long histories and can appear in multiple phases of the MBA. The boundaries between the various MBA phases

are indistinct and some ceramic types straddle these phases. To focus solely on the latest material thus fails to account for the greater part of the ceramic data and can result in a misleading date based on a ceramic type that spans multiple phases.

A quantitative analysis accounts for this problem as it focuses on the largest group of parallels and gives a 'mean' average of where a given context fits into the MBA sequence. This is the main reason why a quantitative analysis is adhered to in this study. It should be noted, though, that the 'Parallel' matches provide more indicative results than when combined with 'Variation' matches (which often find matches with multiple periods). The results of the comparative ceramic analysis are presented according to trench phase. A selection of illustrated sherds from each trench phase accompanies the tabulated results.

Appendix A: Pella, Area IIIF, Comparative Pottery Analysis

5.4
17997
Bowl (globular)

P Aphek 75. Fig 6. (A IVb) MBI
V Gezer ii. Pl 17.4. (FI. Tr 4168) MBII
V Ghassil. Pl 42.21. Str IX MBIII
V Hama. Fig 109.3C79. (H5) MBI

5.4
17998
Bowl (almond)

P Alalakh. Pl 34.12. Str IX MBII
P Meg. Pl 37.17. Str XI MBII
V Bosra. Pl 13.56. (Str 28) MBI

5.4
17999
Bowl (almond)

P Meg. Pl 14.6. Str XIV MBI
P Meg. Pl 21.21. Str XIII MBI
V Gezer ii. Pl 11.26. (FI. Str 10) MBII
V Meg. Pl 9.2. Str XV MBI
V Meg. Pl 19.7. Str XIIIA MBI
V Mev. Fig 12.18. Str XIII MBII

5.4
18000
Bowl (almond)

V Meg. Pl 14.7. Str XIV MBI
V Meg. Pl 21.17. Str XIII MBI

5.4
18001
Jar (holemouth)

P Jericho iv. Fig 116.2. (H.IX.xxxii-iii) MBII

5.4
18002
Jug (fine)

P Meg. Pl 23.20. Str XII MBII
V Hazor i. Pl XCIV.14. (D1 Str 3) MBIII
V Lachish. Pl 74.669. L173. (Jugs class A) MBII
V Meg. Pl 7.13. Str XV-XIIIA MBI
V Meg. Pl 10.10. Str XV-XIIIA MBI

5.5
17077
Cpot

P Jericho iv. Fig 143.2 (H.VII.xxvi) MBII
P Jericho iv. Fig 144.1,3-6,9-10. (H.VII-IX.xxiva-xxxiva) MBII**(x7)**
P Jericho iv. Fig 144.2,7-8,13. (Tr II.XXII.lxx-lxxiii) MBII**(x4)**
P Jericho iv. Fig 145.4-9. (H.VIII-XI.xxvii-xlvii) MBII**(x6)**
P Jericho iv. Fig 146.6 (H.IX.xxxvb) MBII
P Jericho v. Fig 22.13 (Tr I.XLIII.lxi) EBIV
P Jericho v. Fig 165.8 (H.VIII.xxviiia) MBII
P Jericho v. Fig 168.6 (H.IX.xxxiia) MBII
P Jericho v. Fig 169.9 (H.IX.xxxiii) MBII
P Jericho v. Fig 108.35 (Tr III.XXI.lxxxiiia) MBI
P Shec. Pl 23.d. (Cf A.3). Str XIX MBII
V Aphek 85. Fig 2.10. (Str X20-X19) MBI
V Gezer iv. Pl 4.15,22. (FVI. Str 12/11) MBII**(x2)**
V Jericho iv. Fig 140.1, 3-7,9-11. (H.IX-XI.xxxa-xlv) MBII**(x9)**
V Jericho iv. Fig 140.2,8 (H.VI.xx) MBI**(x2)**
V Jericho iv. Fig 141.1,3-5,7. (H.IX.xxxi-xxxiva) MBII**(x5)**
V Jericho iv. Fig 141.2,6. (H.VI.xvii-xx) MBI**(x2)**
V Jericho iv. Fig 142.1,4,6-7,9. (H.VIII-IX.xxviiia-xxxiv) MBII**(x9)**
V Jericho iv. Fig 142.2. (Tr II.XXII.lxix) MBII
V Jericho iv. Fig 142.3,5,8,10. (H.VI.xvii-xx) MBI**(x4)**
V Jericho iv. Fig 143.1,3 (H.VII-VIII.xxiva-xxviia) MBII**(x2)**
V Jericho v. Fig 21.23 (Tr I.XLIII. lviii) EBIV
V Jericho v. Fig 163.11 (H.VII.xxiva) MBII
V Jericho v. Fig 165.7,10 (H.VIII.xxviia-xxviiia) MBII**(x2)**
V Jericho v. Fig 166.3,10-12 (H.VIII.xxix) MBII**(x4)**
V Jericho v. Fig 167.10 (H.IX.xxxia) MBII
V Jericho v. Fig 179.24 (H.XI.xlvii) MBII
V Jericho v. Fig 186.13 (H.XII.li) MBII
V Jericho v. Fig 106.11,22,24 (Tr III.XXI.lxxxii-vi) MBI**(x3)**
V Meg. Pl 7.9,10. Str XVI EBIV**(x2)**
V Meg. Pl 9.19. Str XV MBI
V Meg. Pl 30.5. Str XII MBII
V Shec. Pl 23.a,b,c. (Cf A.2-3). Str XX-XIX MBII**(x3)**

5.5
17079
Bowl

P Aphek 75. Fig 13.7. (Str A II) MBI
P Jericho iv. Fig 104.11. (H.IX.xxxii) MBII

5.5
17080
Juglet

V Meg. Pl 10.20. Str XIV MBI

5.6
17081
Jar (tnn)

V Gezer ii. Pl 14.10. (FI. Str 9A) MBIII
V Mev. Fig 11.12. Str XII MBIII
V Aphek 85. Fig 4.15. (Str X18) MBI
V Gezer iv. Pl 2.20. (FVI. Str 12) MBI
V Hazor v. Fig III.6.2. (L) MBII

5.7
17084
Sjar (holemouth)

V Aphek 75. Fig 4.18. (Str A IVb) MBI
V Jericho iv. Fig 116.1. (H.XI.xlvi) MBII

5.7
17085
Bowl

P Jericho v. Fig 177.7 (H.X.xxxixa) MBII
V Aphek 75. Fig 6.11. (Str A IVb) MBI
V Shec. Pl 2.i,k. (Bp.22). Str XX MBII**(x2)**
V Gezer iv. Pl 5.21. (FVI. Str 11/10) MBIII
V TNM. Fig 27.7. Phase G MBIII

5.7
17086
Bowl

V Bosra. Pl 11.8. (Str 28) MBI
V Gezer ii. Pl 17.21. (FI. Tr 4168) MBIII
V Jericho iv. Fig 106.7. (H.XI.xliva) MBII
V Jericho iv. Fig 108.7. (H.X.xxxvc) MBII
V Jericho v. Fig 169.1 (H.IX.xxxii) MBII
V Jericho v. Fig 106.27 (Tr III.XXI.lxxxvii) MBI

5.7
17097
Jug

P Aphek 75. Fig 5.16. (Str A IVb) MBI
P Hazor i. Pl CXVII.1. (D5. Str 5) MBI
V Aphek 85. Fig 2.16. (Str X20-X19) MBI
V Hazor i. Pl CXVII.4. (D5. Str 5) MBI
V Lachish iv. Pl 78.804. L1513 EBIV
V Meg. Pl 10.3. Str XIV MBI

5.9
17106
Bowl

P Sali. Pl 41.1. Str XII MBII
V Jericho iv. Fig 105.15. (H.XIII.liii) MBIII
V Shec. Pl 5.b. (Bp.53). Str XIX MBII

5.9
17107
Jar (tnn)

P Aphek 75. Fig 16.8. (Str B IV) MBI
P Gezer ii. Pl 12.17. (FI. Str 9B) MBII
P Gezer ii. Pl 13.2. (FI. Str 9B) MBII
P Hazor v. Fig III.6.2. (L) MBII
P Mev. Fig 14.24. Str XIII MBII
P Shec. Pl 41.g. (JJ.21). Str XX MBII
V Aphek 75. Fig 1.15. (Str A VII) MBI
V Aphek 75. Fig 5.9. (Str A IVb) MBI
V Aphek 85. Fig 4.13,14. (Str X18) MBI**(x2)**
V Bosra. Pl 15.96. (Str 20) MBIII
V Meg. Pl 18.1. Str XIIIA MBI
V Sali. Pl 40.30. Str XI MBIII

5.9
17994
Jar

V Meg. Pl 8.3. Str XV MBI

5.9
17995
Bowl (bars)

V Hazor iii-iv. Pl CCLIX.6. (H Str 3) MBII
V Meg. Pl 9.3. Str XV MBI

5.11
17985
Bowl (almond)

P Meg. Pl 14.6. Str XIV MBI
P Meg. Pl 21.17. Str XIII MBI
V Gezer ii. Pl 18.31. (FI. Tr 4168) MBII
V Gezer iv. Pl 2.6. (FVI. Str 12) MBI
V Meg. Pl 9.2. Str XV MBI
V Meg. Pl 19.7. Str XIIIA MBI
V Meg. Pl 22.5. Str XIII MBI
V Meg. Pl 29.11. Str XII MBII
V Mev. Fig 12.18. Str XIII MBII

5.11
17986
Bowl (bars)

P Aphek 75. Pl 6.13. (A IVb) MBI
V Hazor iii-iv. Pl CCLIX.6. (H Str 3) MBIII
V Meg. Pl 9.3. Str XV MBI
V Meg. Pl. 15.12. Str XIV MBI

5.11
17987
Bowl (rounded)

P Aphek 75. Fig 6.11. (Str A VIb) MBI
P Bosra. Pl 11.10. (Str 28) MBI
P Bosra. Pl 13.54. (Str 26) MBI
P Gezer ii. Pl 11.29. (FI. Str 10) MBII
P Gezer iv. Pl 6.12. (FVI. Str 11/10) MBII
P Ghassil. Pl 13.18. Str X MBII
P Hazor iii-iv. Pl CCLXXXVI.3. (K Str 4) MBII
P Hazor v. Fig II.6.6. (A Str 9D) MBI
P Hazor v. Fig II.7.9. (A Str 9D) MBI
P Hazor v. Fig III.10.4. (L) MBII
P Jericho iv. Fig 104.7. (H.X.xxxviia) MBII
P Meg. Pl 9.4. Str XV MBI
P Shec. Pl 1.h. (Bp.12). Str XIX MBII
P Shec. Pl 3.f .(Bp.32). Str XIX MBII
V TNM. Fig 28.8. Ph G-C MBIII
V Dan. Fig 4.77. (T1025.6). Str XII MBI
V KEL. Pl 3.T102. Str 8 MBIII
V Gezer ii. Pl 18.14. (FI. Tr 4168) MBIII
V Hazor i. Pl XCIII.1. (D1 Str 5) MBII
V Hazor iii-iv. Pl CCXXXVI.8. (BA Str 12) MBII
V Hazor v. Fig III.7.14. (L) MBII
V Jericho iv. Fig 104.23. (H.IX.xxx-i) MBII
V Jericho v. Fig 70.1. (Tr II.XXII.lxix) MBII
V Jericho v. Fig 106.26. (Tr III.XXI.lxxxvi) MBI
V Jericho v. Fig 165.1. (H.VIII.xxviia) MBII
V Jericho v. Fig 181.21. (H.XI.xlixa) MBII
V Jericho v. Fig 186.2. (H.XII.li) MBII
V Meg. Pl 14.11. Str XIV MBI
V Meg. Pl 19.10. Str XIIIA MBI
V Meg. Pl 21.15. Str XIII MBI
V Meg. Pl 29.20. Str XII MBII
V Meg. Pl 37.26. Str XI MBII
V Mev. Fig 12.26. Str XIII MBII
V Sali. Pl 42.65. Str XII MBII
V Shec. Pl 2.m. (Bp.22). Str XIX MBII
V Shec. Pl 46.c. (Bp.23). Str XX MBII

5.11
17988
Bowl (rounded)

P Ghassil. Pl 13.12. Str X MBII
P Hazor i. Pl XCIII. (D1 Str 5) MBII
P Hazor v. Fig III.7.14. (L) MBII
P Shec. Pl 1.h. (Bp.12). Str XIX MBII
V Aphek 75. Fig 2.4. (Str AVI) MBI
V Gezer ii. Pl 11.29. (FI. Str 10) MBII
V Hazor iii-iv. Pl CCLXXXVI.3. (K Str 4) MBII
V Hazor v. Fig II.6.8. (A Str 9D) MBI
V Hazor v. Fig II.7.9. (A Str 9D) MBI
V Jericho iv. Fig 111.3. (H.X.xxxii-iii) MBII
V Jericho v. Fig 70.1. (Tr II.XXII.lxix) MBII
V Jericho v. Fig 106.26. (Tr III.XXI.lxxxvi) MBI
V Jericho v. Fig 160.2. (H.VII.xxia) MBII
V Jericho v. Fig 179.12. (H.XI.xlvi) MBII
V Meg. Pl 14.9. Str XIV MBI
V Meg. Pl 21.15. Str XIII MBI
V Meg. Pl 29.20. Str XII MBII
V Mev. Fig 12.25. Str XIII MBII
V Sali. Pl 42.65. (Str XII) MBII
V Shec. Pl 3.f. (Bp.32). Str XIX MBII

5.11
17989
Jug

P Meg. Pl 10.11. Str XIV MBI
P Meg. Pl 11.15. Str XIV MBI

P	Meg. Pl 12.21. Str XIV	MBI
V	Shec. Pl 31.e. (Jg.O.O.11). Str XIX	MBII
V	Shec. Pl 46.y. (JJ.43). Str XX	MBII
V	Shec. Pl 46.j. (Bd C.23). Str XX	MBII

5.12
17984
Bowl (rounded)

P	Aphek 75. Fig 2.4. (Str AVI)	MBI
P	Gezer ii. Pl 11.29. (FI. Str 10)	MBII
P	Ghassil. Pl 13.18. Str X	MBII
P	Hazor i. Pl XCIII.1. (DI Str 5)	MBII
P	Hazor iii-iv. Pl CCXXXVI.8. (BA Str 12)	MBII
P	Hazor iii-iv. Pl CCLXXXVI.3. (K Str 4)	MBII
P	Hazor v. Fig II.6.6. (A Str 9D)	MBI
P	Hazor v. Fig II.7.9. (A Str 9D)	MBI
P	Hazor v. Fig III.7.14. (L)	MBII
P	Hazor v. Fig III.10.4. (L)	MBII
P	Jericho iv. Fig 104.7. (H.X.xxxviia)	MBII
P	KEL. Pl 3.T179. Str 7	MBII
P	Meg. Pl 9.5. Str XV	MBI
P	Shec. Pl 1.h. (Bp.12). Str XIX	MBII
P	Shec. Pl 3.f (Bp.32). Str XIX	MBII
V	Bosra. Pl 13.54. (Str 26)	MBI
V	Dan. Fig 4.77. (T1025.6). Str XII	MBI
V	Hama. Fig 124.2C926. (H2)	MBIII
V	Hazor i. Pl CXIX.2. (D5 Str 4)	MBII
V	Jericho iv. Fig 111.3. (H.IX.xxxii-iii)	MBII
V	Jericho v. Fig 71.1. (Tr II.XXII.lxix)	MBII
V	Jericho v. Fig 106.26. (Tr III.XXI.lxxxvi)	MBI
V	Jericho v. Fig 160.2. (H.VII.xxia)	MBII
V	Jericho v. Fig 181.21. (H.XI.xlixa)	MBII
V	Jericho v. Fig 183.1. (H.XI.l)	MBII
V	Meg. Pl 14.11. Str XIV	MBI
V	Meg. Pl 19.11. Str XIIIA	MBI
V	Meg. Pl 21.15. Str XIII	MBI
V	Meg. Pl 29.20. Str XII	MBII
V	Meg. Pl 37.26. Str XI	MBII
V	Mev. Fig 12.25. Str XIII	MBII
V	Sali. Pl 42.65. (Str XII)	MBII

5.13
17976
Bowl (rounded)

P	Bosra. Pl 11.10 (Str 28)	MBI
P	Jericho v. Fig 106.26. (Tr III.XXI.lxxxiii)	MBI
P	Meg. Pl 14.9. Str XIV	MBI
P	Meg. Pl 29.10. Str XII	MBII
V	Aphek 75. Fig 6.11. (Str A IVb)	MBI
V	Bosra. Pl 13.54. (Str 26)	MBI
V	Jericho v. Fig 165.1. (H.VIII.xxviia)	MBII
V	Jericho v. Fig 186.2. (H.XII.li)	MBII
V	Meg. Pl 9.4. Str XV	MBI
V	Meg. Pl 19.10. Str XIIIA	MBI
V	Shec. Pl 2.1. (Bp.22). Str XVIII	MBII

5.13
17977
Bowl (almond)

P	Jericho iv. Fig 104.3. (H.IX.xxxii)	MBII
P	Jericho iv. Fig 105.22. (H.IX.xxxiva)	MBII
P	Meg. Pl 22.3. Str XIII	MBI
P	Sali. Pl 41.65. (Str XII)	MBII
P	Shec. Pl 2.m. (Bp.22). Str XIX	MBII
V	Hazor iii-iv. Pl CCLXXXVI.3. (K Str 4)	MBII
V	Jericho v. Fig 71.1. (Tr II.XXII.lxix)	MBII
V	Jericho v. Fig 184.4. (H.XI.l)	MBII
V	Meg. Pl 9.5. Str XV	MBI
V	Meg. Pl 9.5. Str XV	MBI
V	Meg. Pl 19.10. Str XIIIA	MBI
V	Shec. Pl 3.f. (Bp. 32). Str XIX	MBII

5.13
17978
Bowl (rounded)

P	Hazor i. Pl CXIX.2. (D5 Str 4)	MBII
P	Meg. Pl 9.3. Str XV	MBI
P	Shec. Pl 3.f (Bp.23). Str XVIII	MBII
V	Gezer iv. Pl 8.15. (FVI. Str 10B)	MBIII
V	Gezer iv. Pl 5.21. (FVI. Str 11/10)	MBII
V	Hazor iii-iv. Pl CCXXXXVI.8. (BA Str 12)	MBII
V	Hazor v. Fig III.10.2. (L)	MBII
V	Jericho iv. Fig 107.26. (H.VI.xxia)	MBII
V	Meg. Pl 14.9. Str XIV	MBI
V	Meg. Pl 19.11. Str XIIIA	MBI
V	Meg. Pl 21.15. Str XIII	MBI
V	Meg. Pl 29.20. Str XII	MBII
V	Meg. Pl 37.21. Str XI	MBII
V	Mev. Fig 12.25. Str XIII	MBII

5.13
17979
Bowl (rounded)

P	Alalakh. Pl 55.6. Str X	MBII
P	Dan. Fig 4.77.1. (T.4663). Str X	MBII
P	Dan. Fig 4.77.3. (T.4663). Str X	MBII
P	Jericho iv. Fig 103.15. (H.XII.li)	MBII
P	Meg. Pl 29.24. Str XII	MBII
P	Shec. Pl 113.h. (Bp.41). Str XVIII	MBII
P	Shec. Pl 4.j. (Bp.41). Str XVIII	MBII
V	Gezer iv. Pl 8.14. (FVI. Str 10B)	MBIII
V	Hazor i. Pl CI.1. (D3 Str2)	MBII
V	Hazor v. Fig III.6.3. (L)	MBII
V	Hazor v. Fig III.12.7. (L)	MBIII
V	Hazor v. Fig III.12.9. (L)	MBIII
V	Lachish. Pl 70.583. L157. (Bwl class G)	MBII
V	Meg. Pl 14.13. Str XIV	MBI
V	Meg. Pl 37.12. Str XI	MBII
V	Sali. Pl 40.28. (Str XII)	MBIII

5.13
17980
Bowl (bars)

P	Aphek 75. Fig 4.16. (A IVb)	MBI
V	Aphek 75. Fig 6.13. (A IVb)	MBI
V	Hazor iii-iv. Pl CCLIX.6. (H Str 3)	MBIII

5.13
17981
Bowl (almond)

V	Meg. Pl 9.2. Str XV	MBI

5.13
17983
Bowl (bars)

P	Aphek 75. Pl 12.15. (Str XIV)	MBI
V	Hazor iii-iv. Pl CCLIX.6. (H Str 3)	MBIII
V	Meg. Pl 12.15. Str XIV	MBI

5.15
17971
Bowl (almond)

P	Bosra. Pl 11.10. (Str 28)	MBI
P	KEL. Pl 3.1102. Str 8	MBIII
V	Aphek 75. Fig 4.9. (Str AIVb)	MBI

V Aphek 75. Fig 6.11. (Str AIVb) MBI
V Bosra. Pl 13.54. (Str 26) MBI
V Gezer iv. Pl 8.12. (FVI Str 10B) MBIII
V Jericho iv. Fig 104.17. (H.IX.xxxia) MBII
V Meg. Pl 14.11. Str XIV MBI
V Meg. Pl 19.11. Str XIIIA MBI
V Meg. Pl 29.20. Str XII MBII
V Shec. Pl 2.1. (Bp.22). Str XVII MBII

5.15
17972
Jar (tnn)

P Gezer iv. Pl 2.20. (FVI. Str 12) MBI
P Jericho iv. Fig 127.27. (H.XI.xlvia) MBII
P Mev. Fig 14.24. StrXIII MBII
P Mev. Fig 16.8. Str XI MBI
V Aphek 85. Fig 4.11. (X18) MBI
V Bosra. Pl 12.30. (Str 26) MBI
V Gezer ii. Pl 12.17. (FI. Str 9B) MBII
V Ghassil. Pl 7.13. Str XI MBII
V Hazor v. Fig II.7.14. (A Str 9D) MBI
V Jericho v. Fig 171.10. (H.IX.xxxiva) MBII
V Meg. Pl 11.20. Str XIV MBI

5.15
17973
Bowl (fine)

P Gezer ii. Pl 14.13. (FI. Str 9A) MBII
P Gezer iv. Pl 2.2. (FVI. Str 12) MBI
P Mev. Fig 11.2. Str XII MBIII
V Gezer ii. Pl 17.10. (FI. Tr 4168) MBII
V Gezer iv. Pl 5.16. (FVI. Str 11/10) MBII
V Hazor i. Pl XCIII.13. (D1 Str 5) MBII
V Hazor v. Fig II.7.16. (A Str 9D) MBI
V Shec. Pl 13.a. (Bg B.11). Str XIX MBII

5.15
17974
Bowl (fine)

V Gezer iv. Pl 1.2. (FVI. Str 12) MBI
V Shec. Pl 13.c. (Bg B.13). Str XVIII MBII

5.16
17968
Sjar (tnn)

P Aphek 85. Fig 5.12. (X17) MBI
P Gezer iv. Pl 4.19. (FVI. Str 12/11) MBII
P Meg. Pl 11.20. Str XIV MBI
P Meg. Pl 18.1. Str XIIIA MBI
V Bosra. Pl 15.96. (Str 20) MBIII
V Shec. Pl 41.g. ((JJ.21). Str XX MBII

5.18
17962
Bowl (almond)

P Gezer iv. Pl 5.22. (FVI. Str 12/11) MBII
P Meg. Pl 29.11. Str XIII MBI
P Meg. Pl 29.11. Str XII MBII
V Hazor v. Fig II.6.5. (A Str 9D) MBI
V Meg. Pl 9.2. Str XV MBI
V Meg. Pl 14.6. Str XIV MBI

5.18
17963
Bowl (rounded)

P Ghassil. Pl 13.15. Str X MBII
P Hazor i. Pl CXIX.10. (D5 Str 4) MBII
P Hazor v. Fig III.6.3. (L) MBII
P Jericho iv. Fig 103.15. (H.XI.li) MBII
P Meg. Pl 29.24. Str XII MBII
V Alalakh. Pl 13.15. Str X MBII
V Bosra. Pl 13.52. (Str 26) MBI
V Dan. Fig 4.77. (T.4663.1). Str X MBII
V Hazor iii-iv. Pl CCLIX.5. (H Str 3) MBIII
V Meg. Pl 9.4. Str XV MBI

5.21
17961
Jug

P Ghassil. Pl 37.58. Str IX MBIII
P Ghassil. Pl 20.42. Str X MBII
P Jericho v. Fig 22.28. (Tr I.XLIV.lxvi) MBIII
P Meg. Pl 23.6. Str XII MBII
P Shec. Pl 31.c. (Jg.O.O.11). Str XVIII MBII
P Tyre. Pl LV.12. Str XIX MBII
V Jericho v. Fig 166.14. (H.IX.xxx) MBII
V Meg. Pl 31.9. Str XIII MBI
V Shec. Pl 30.c. (Jg B). Str XIX MBII
V Shec. Pl 31.b. (Jg.O.O.11). Str XIX MBII

5.29
17949
Bowl (simple)

V Shec. Pl 46.a. (Bp.11). Str XX MBII

5.29
17950
Bowl (simple)

P Alalakh. Pl 37.39. Str IX MBII
P Meg. Pl 14.24. Str XIV MBI
V Gezer iv. Pl 5.27. (FVI. Str 11/10) MBII

5.29
17951
Bowl

V Gezer iv. Pl 6.25. (FVI. Str 11/10) MBII
V Ghassil. Pl 36.23. Str IX MBIII
V Hazor I. Pl CXVII.4. (D5 Str 5) MBII
V Hazor v. Fig II.6.27. (A Ph 9D) MBI

5.29
17952
Bowl (almond)

5.29
17953
Bowl (bars)

V Meg. Pl 15.12. Str XIV MBI

5.30
17944
Jug

P Meg. Pl 10.3. Str XIV MBI
P Meg. Pl 10.11. Str XIV MBI
P Meg. Pl 16.1. Str XIV-XIIIB MBI
P Shec. Pl 31.e. (Jg.O.O.11). Str XIX MBII
V Aphek 85. Fig 2.16. (X 20-19) MBI
V Bosra. Pl 11.6. (Str 28) MBI
V Gezer i. Pl 31.11. (FI. Str 8) MBIII
V Gezer iv Pl 4.6. (FVI. Str 12/11) MBII
V Hazor v. Fig II.6.38. (A Str 9D) MBI
V Meg. Pl 12.16. Str XIV MBI
V Shec. Pl 46.y. (JJ.43). Str XX MBII

5.30
17945
Jug

P Hazor i. Pl CXX.8. (D5 Str 4) MBII
V KEL. 44.T141. Str 8 MBIII
V Meg. Pl 11.11. Str XIV MBI
V Shec. Pl 31.c. (Jg.O.O.11). Str XVIII MBII

7.4
15476
Cpot

P Mev. Fig 16.19. Str XIV MBI

7.5
15474
Sjar (holemouth)

V Bosra. Pl 11.13. (Str 28) MBI
V Bosra. Pl 12.40. (Str 26) MBI
V Jericho iv. Fig 117.1. (H.IX.xxxva) MBII

7.6
15473
Bowl

P Meg. Pl 15.3. Str XIV MBI
V Gezer ii. Pl 16.18. (FI. Tr 4168) MBIII

7.8
15466
Cpot

P Jericho iv. Fig 143.2 (H.VII.xxvi) MBII
P Jericho iv. Fig 144.1,3-6,9-10. (H.VII-IX.xxiva-xxxiva) MBII**(x7)**
P Jericho iv. Fig 144.2,7-8,13. (Tr II.XXII.lxx-lxxiii) MBII**(x4)**
P Jericho iv. Fig 145.4-9. (H.VIII-XI.xxvii-xlvii) MBII**(x6)**
P Jericho iv. Fig 146.6 (H.IX.xxxvb) MBII
P Jericho v. Fig 22.13 (Tr I.XLIII.lxi) MBI
P Jericho v. Fig 173.12 (H.IX.xxxvc) MBII
P Jericho v. Fig 165.8 (H.VIII.xxviiia) MBII
P Jericho v. Fig 168.6 (H.IX.xxxiia) MBII
P Jericho v. Fig 108.35 (Tr III.XXI.lxxxiiia) MBI
P Shec. Pl 23.c,d. (Cf A.3). Str XIX MBII**(x2)**
V Gezer iv. Pl 4.15,22. (FVI. Str 12/11) MBII**(x2)**
V KEL. Pl 24.CT16. Str 8 MBIII
V Jericho iv. Fig 140.1, 3-7,9-11. (H.IX-XI.xxxa-xlv) MBII**(x9)**
V Jericho iv. Fig 140.2,8 (H.VI.xx) MBI**(x2)**
V Jericho iv. Fig 141.1,3-5,7. (H.IX.xxxi-xxxiva) MBII**(x5)**
V Jericho iv. Fig 141.2,6. (H.VI.xvii-xx) MBI**(x2)**
V Jericho iv. Fig 142.1,4,6-7,9. (H.VIII-IX.xxviiia-xxxiv) MBII**(x5)**
V Jericho iv. Fig 142.2. (Tr II.XXII.lxix) MBII
V Jericho iv. Fig 142.3,5,8,10. (H.VI.xvii-xx) MBI**(x4)**
V Jericho iv. Fig 143.1,3 (H.VII-VIII.xxiva-xxviia) MBII**(x2)**
V Jericho v. Fig 21.23 (Tr I.XLIII. lviii) MBI
V Jericho v. Fig 163.11 (H.VII.xxiva) MBII
V Jericho v. Fig 165.7,10 (H.VIII.xxviia-xxviiia) MBII**(x2)**
V Jericho v. Fig 166.3, 10-12 (H.VIII.xxix) MBII**(x3)**
V Jericho v. Fig 167.10 (H.IX.xxxia) MBII
V Jericho v. Fig 179.24 (H.XI.xlvii) MBII
V Jericho v. Fig 186.13 (H.XII.li) MBII
V Jericho v. Fig 106.11,22,24 (Tr III.XXI.lxxxii-vi) MBI**(x3)**
V Meg. Pl 7.9,10. Str XVI EBIV
V Meg. Pl 9.19. Str XV MBI
V Meg. Pl 30.5. Str XII MBII
V Shec. Pl 23.a,b,. (Cf A.2-3). Str XX MBII**(x2)**

7.8
15467
Sjar (holemouth)

P Sali. Pl 42.101. Str XII MBII
V Bosra. Pl 14.88. (Str 22) MBII
V Gezer ii. Pl 14.10. (FI. Str 9A) MBIII
V Sali. Pl 41.10. Str XII MBII

7.8
15468
Sjar (holemouth)

V Jericho v. Fig 73.7 (Tr II.XXII.lxxi) MBII

7.8
15470
Cpot

V Jericho v. Fig 70.1. (Tr II.XXI.lxviii) EBIV
V Jericho v. Fig 106.11,22. (Tr III.XXI.lxxxii-lxxxvi) MBI**(x2)**
V Jericho v. Fig 108.35. (Tr III.XXI.lxxxiiia) MBI
V KEL. Pl 24.CT16. Str 8 MBIII

7.8
15471
Jar (tnn)

P Bosra. Pl 12.30. (Str 26) MBI
P Jericho iv. Fig 128.5. (H.X.xxxvii) MBII
V Jericho v. Fig 171.10. (H.IX.xxxiva) MBII
V Sali. Pl 40.30. Str XI MBIII

7.8
17074
Jug

V Mev. Fig 11.12. Str XII MBIII

7.10
15460
Sjar

V Gezer ii. Pl 11.33. (FI. Str 10) MBII

7.10
15461
Jug

V Jericho v. Fig 73.1 (Tr II.XXI.lxix) EBIV

7.10
15462
Jar (fine)

7.12
15459
Jug

V Gezer iv. Pl 8.4. (FVI. Str 10B) MBIII
V Hazor i. Pl CXVII.2,4. (D5. Str 5) MBII**(x2)**
V Hazor v. Fig II.6.38. (A. Str 9D) MBI

7.13
15463
Bowl

V Gezer ii. Pl 16.12. (F1. Tr 4168) MBIII
V Hazor iii-iv. Pl CCLXXXVI.3. (K. Str 4) MBII

7.13
15464
Sjar

7.13
15482
Cpot

P	Jericho iv. Fig 143.2 (H.VII.xxvi)	MBII
P	Jericho iv. Fig 144.1,3-6,9-10. (H.VII-IX.xxiva-xxxiva)	MBII**(x7)**
P	Jericho iv. Fig 144.2,7-8,13. (Tr II.XXII.lxx-lxxiii)	MBII**(x4)**
P	Jericho iv. Fig 145.4-9. (H.VIII-XI.xxvii-xlvii)	MBII**(x6)**
P	Jericho iv. Fig 146.6 (H.IX.xxxvb)	MBII
P	Jericho v. Fig 22.13 (Tr I.XLIII.lxi)	MBI
P	Jericho v. Fig 173.12 (H.IX.xxxvc)	MBII
P	Jericho v. Fig 165.8 (H.VIII.xxviiia)	MBII
P	Jericho v. Fig 168.6 (H.IX.xxxiia)	MBII
P	Jericho v. Fig 108.35 (Tr III.XXI.lxxxiiia)	MBI
P	KEL. Pl 24.CT16. Str 8	MBIII
P	Shec. Pl 23.c,d. (Cf A.3). Str XIX	MBII
V	Gezer iv. Pl 4.15,22. (FVI. Str 12/11)	MBII**(x2)**
V	Jericho iv. Fig 140.1, 3-7,9-11. (H.IX-XI.xxxa-xlv)	MBII**(x9)**
V	Jericho iv. Fig 140.2,8 (H.VI.xx)	MBI**(x2)**
V	Jericho iv. Fig 141.1,3-5,7. (H.IX.xxxi-xxxiva)	MBII**(x5)**
V	Jericho iv. Fig 141.2,6. (H.VI.xvii-xx)	MBI**(x2)**
V	Jericho iv. Fig 142.1,4,6-7,9. (H.VIII-IX.xxviiia-xxxiv)	MBII**(x5)**
V	Jericho iv. Fig 142.2. (Tr II.XXII.lxix)	MBII
V	Jericho iv. Fig 142.3,5,8,10. (H.VI.xvii-xx)	MBI**(x4)**
V	Jericho iv. Fig 143.1,3 (H.VII-VIII.xxiva-xxviia)	MBII**(x2)**
V	Jericho v. Fig 21.23 (Tr I.XLIII. lviii)	MBI
V	Jericho v. Fig 163.11 (H.VII.xxiva)	MBII
V	Jericho v. Fig 165.7,10 (H.VIII.xxviia-xxviiia)	MBII**(x2)**
V	Jericho v. Fig 166.3, 10-12 (H.VIII.xxix)	MBII**(x3)**
V	Jericho v. Fig 167.10 (H.IX.xxxia)	MBII
V	Jericho v. Fig 179.24 (H.XI.xlvii)	MBII
V	Jericho v. Fig 186.13 (H.XII.li)	MBII
V	Jericho v. Fig 106.11,22,24 (Tr III.XXI.lxxxii-vi)	MBI
V	Meg. Pl 7.9,10. Str XVI	EBIV
V	Meg. Pl 9.19. Str XV	MBI
V	Meg. Pl 30.5. Str XII	MBII
V	Shec. Pl 23.a,b. (Cf A.2-3). Str XX	MBII

7.13
15483
Sjar

P	Jericho iv. Fig 137.18. (H.X.xxxvia)	MBII
V	Hazor i. Pl C.24. (D3. Str 3)	MBII

7.15
15479
Sjar

7.15
15480
Sjar

V	KEL. Pl 59.T246. Str 7	MBII

7.15
15481
Sjar (holemouth)

P	Jericho v. Fig 21.29. (Tr I.XLIII.lx)	MBI
P	Jericho v. Fig 22.3. (Tr I.XLIII.lxi)	MBI
P	Jericho v. Fig 22.15. (Tr I. XLIII.lxi)	MBI

Appendix B: Pella, Area XXVIIIC, Comparative Pottery Analysis

10.3
17144
Bowl

P	Hazor i. Pl C.19. (D3. Str 3)	MBII
P	Mev. Fig 11.2. Str XII	MBIII
V	Gezer iv. Pl 4.14. (FVI. Str 12/11)	MBII
V	Hama Fig 109.3C126. (H5)	MBI
V	Jericho v. Fig 191.10 (H.XII.lii)	MBII
V	Jericho v. Fig 163.2 (H.VII.xxiv)	MBII

10.3
17145
Bowl

P	Aphek 75. Fig 4.10. (Str A IVb)	MBI
V	Aphek 75. Fig 2.4. (Str A VI)	MBI
V	Aphek 75. Fig 4.11. (Str A Ivb)	MBI
V	Hazor i. Pl CI.1. (D3. Str 2)	MBIII
V	KEL. Pl 4.T125. Str 8	MBIII

10.3
17146
Sbowl

V	Aphek 75. Fig 13.7. (Str A II)	MBI
V	Jericho iv. Fig 104.11 (H.IX.xxxii)	MBII

10.15
17188
Bowl

P	Gezer ii. Pl 16.25. (FI. Tr 4168)	MBIII
P	Gezer iv. Pl 2.24. (FVI. Str 12)	MBI
P	Hazor i. Pl C.19. (D3. Str 3)	MBII
P	Hazor v. Fig II.7.12. (A. Str 9D)	MBI
P	Jawa. Fig 11.14	MBI
P	Mev. Fig 11.1. Str XII	MBIII
P	TBM 33. Pl 4.5. Str G-F	MBI
V	Aphek 75. Fig 8.2. (Str A IVa)	MBI
V	Gezer ii. Pl 15.22. (FI. Tr 4168)	MBIII
V	Hama Fig 109.3C75. (H5)	MBI
V	Hazor v. Fig II.8.3. (A. Str 9D)	MBI
V	Jericho v. Fig 160.1 (H.VI.xx)	MBII
V	Meg. Pl 15.2. Str XIV	MBI
V	Sali. Pl 42.9. Str XII	MBII

10.15
17189
Bowl

P	Aphek 75. Fig 8.2. (Str A IVa)	MBI
P	Hazor v. Fig II.8.3. (A. Str 9D)	MBI
P	Jawa. Fig 11.14	MBI
P	Jericho v. Fig 160.1 (H.VI.xx)	MBII
P	Jericho v. Fig 106.14 (Tr III.XXI.lxxxiii)	MBI
P	Sali. Pl 42.9. Str XII	MBII
V	Gezer ii. Pl 16.25. (FI. Tr 4168)	MBIII
V	Gezer ii. Pl 19.22. (Tr.4168)	MBIII
V	Hama Fig 109.3C75. (H5)	MBI
V	Hazor i. Pl C.19. (D3. Str 3)	MBII
V	Hazor v. Fig II.7.12. (A. Str 9D)	MBI
V	Jericho v. Fig 163.2 (H.VII.xxiv)	MBII
V	Meg. Pl 15.2. Str XIV	MBI
V	Mev. Fig 11.1. Str XII	MBIII

10.15
17190
Cpot

P	Jericho iv. Fig 144.11. (H.XI.xliva)	MBII
P	Jericho iv. Fig 146.1 (H.XII.li)	MBII
V	Jericho iv. Fig 140.9. (H.IX.xxxii)	MBII
V	Jericho iv. Fig 140.4. (H.VII.xxiva)	MBII
V	Jericho iv. Fig 146.5,7,12. (H.X-XI.xxxvii-li)	MBII**(x3)**
V	Jericho iv. Fig 146.9. (Tr II.XXII.lxxib)	MBII

10.15
17191
Bowl

P	Bosra. Pl 11.10. (Str 28)	MBI
P	Bosra. Pl 13.57. (Str 26)	MBI
P	Gezer iv. Pl 8.12. (FVI. Str 10B)	MBIII
P	Hama. Fig 124.2C926. (H2)	MBIII
P	Jericho iv. Fig 104.15. (Tr II.XXII.lxix)	MBII
P	Jericho v. Fig 186.2 (H.XII.li)	MBII
V	Bosra. Pl 13.54. (Str 26)	MBI
V	Gezer ii. Pl 11.29. (FI. Str 10)	MBII
V	Gezer ii. Pl 13.7. (FI. Str 9B)	MBII
V	Ghassil. Pl 13.12. Str X	MBII
V	Hazor v. Fig II.6.8. (A. Str 9D)	MBI
V	Jericho iv. Fig 104.9. (H.XII.lii)	MBII
V	Jericho iv. Fig 104.13. (H.XIII.liii)	MBIII
V	Jericho iv. Fig 105.8. (H.IX.xxxiva)	MBII
V	Jericho v. Fig 181.21. (H.XI.xlixa)	MBII
V	KEL. Pl 3.T179. Str 7	MBII
V	Meg. Pl 14.9. Str XIV	MBI
V	Mev. Fig 12.24. Str XIII	MBII

10.15
17192
Bowl

P	Dan. Fig 4.94.9 (T.187a). Str IX	MBIII
P	Gezer ii. Pl 16.11. (FI. Tr 4168)	MBIII
P	Gezer ii. Pl 17.27. (Tr. 4168)	MBIII
P	Gezer iv. Pl 8.15. (FVI. Str 10B)	MBIII
P	Ghassil. Pl 14.4. Str X	MBII
P	Hazor i. Pl CXIX.2. (D5. Str 4)	MBII
P	Hazor iii-iv. Pl CCLXXXVI.3. (K. Str 4)	MBIII
P	Hazor v. Fig II.6.8. (A. Str 9D)	MBI
P	Hazor v. Fig III.6.3. (L)	MBII
P	Jericho iv. Fig 104.6. (H.X.xxxviia)	MBII
P	Jericho v. Fig 189.4. (H.XII.lii)	MBII
P	Meg. Pl 9.5. Str XV	MBI
V	Dan. Fig 4.104.1 (T.1025). Str XII	MBI
V	Gezer ii. Pl 11.29. (FI. Str 10)	MBII
V	Gezer ii. Pl 13.7. (FI. Str 9B)	MBII
V	Ghassil. Pl 13.12. Str X	MBII
V	Hazor v. Fig II.6.6. (A. Str 9D)	MBI
V	Hazor v. Fig II.7.17. (A. Str 9D)	MBI
V	Jericho iv. Fig 111.2. (H.XI.xlvi)	MBII
V	Jericho iv. Fig 111.3. (H.IX.xxxii)	MBII
V	Jericho v. Fig 183.14. (H.XI.la)	MBII
V	Lachish iv. Pl 71.612. (L1539). Class J	MBII
V	Meg. Pl 14.14. Str XIV	MBI
V	Meg. Pl 19.10. Str XIIIA	MBI
V	Meg. Pl 21.15. Str XIIIA	MBI
V	Meg. Pl 37.21. Str XI	MBII
V	Mev. Fig 12.25. Str XIII	MBII
V	Sali. Pl 40.1. Str XII	MBII

10.15
17195
Bowl

10.15
17196
Bowl

P	Hazor v. Fig II.7.2. (A. Str 9D)	MBI
P	Hazor v. Fig III.12.7. (L)	MBIII
P	Jericho iv. Fig 105.2. (H.XI.xlviia)	MBII
V	Hazor v. Fig III.12.9. (L)	MBIII
V	Jericho v. Fig 172.3. (H.IX.xxxva)	MBII
V	Jericho v. Fig 71.6. (Tr II.XXII.lxx)	MBII
V	Jericho v. Fig 71.7. (Tr II.XXII.lxx)	MBII
V	Tyre. Pl LV.20. Str XIX	MBII

10.15
17197
Bowl

P	Jericho iv. Fig 105.4 (H.IX.xxxiia)	MBII
V	Alalakh. Pl 37.37. Str IX	MBII
V	Gezer i. Pl 30.8. (FI. Str 7)	MBIII
V	Jericho v. Fig 177.16 (H.X.xlia)	MBII
V	Sali. Pl 42.96. Str XII	MBII

10.15
17198
Bowl

P	Hazor i. Pl CI.1. (D3. Str 2)	MBIII
V	Aphek 75. Fig 4.10. (Str A IVb)	MBI
V	Jericho v. Fig 178.4. (H.X.xliia)	MBII

10.15
17200
Jar (tnn)

P	Aphek 75. Fig 5.6. (Str A IVb)	MBI
P	Aphek 75. Fig 7.8. (Str A IVb)	MBI
P	Aphek 75. Fig 13.11. (Str A II)	MBI
V	Gezer ii. Pl 12.23. (FI. Str 9B)	MBII
V	Gezer ii. Pl 13.9. (FI. Str 9B)	MBII
V	Ghassil. Pl 37.47. Str IX	MBIII
V	Jericho v. Fig 167.12. (H.IX.xxxia)	MBI

10.18
17209
Bowl

V	Hazor v. Fig II.6.14. (A. Str 9D)	MBI

10.18
17210
Sjar

P	Jericho iv. Fig 133.1. (H.VI.xx)	MBI

10.18
17211
Bowl

P	Dan. Fig 4.94.11 (T.187a). Str IX	MBIII
P	Hazor i. Pl XCIV.2. (D1.Str 3)	MBIII
P	Jericho v. Fig 165.1 (H.VIII.xxviia)	MBII
V	Dan. Fig 4.92.4 (T.4663). Str X	MBII
V	Hazor i. Pl CXIX.3. (D5. Str 4)	MBII
V	Hazor v. Fig II.7.9. (A. Str 9D)	MBI
V	Hazor v. Fig III.10.1,4. (L)	MBII
V	Jericho v. Fig 164.1 (H.VIII.xxvi)	MBII
V	Jericho v. Fig 179.11 (H.XI.xlvi)	MBII
V	Mev. Fig 12.25. Str XIII	MBII
V	Tyre. Pl LV.20. Str XIX	MBII

10.18
17212
Jug

10.18
17213
Sjar (holemouth)

V	Aphek 75. Fig 4.18. (Str A IVb)	MBI
V	Aphek 75. Fig 6.15. (Str A Ivb)	MBI
V	Hazor v. Fig II.6.24. (A. Str 9D)	MBI
V	Hazor v. Fig III.13.1. (l)	MBIII

10.18
17214
Sjar

P	Jericho iv. Fig 139.11. (H.XIII.liii)	MBIII

10.18
17215
Sjar

P	Hazor v. Fig III.13.8. (L)	MBIII
P	Jericho iv. Fig 135.28. (H.XI.l)	MBII
V	Bosra. Pl 12.29. (Str 26)	MBI
V	Alalakh. Pl 94.93. Str IX	MBII
V	Hazor iii-iv. Pl CCXCVI.13. (210/A. Str 3)	MBIII
V	Hazor v. Fig III.8.15. (L)	MBII
V	Sali. Pl 40.43. Str XI	MBIII

10.18
17216
Jar (tnn)

P	Jericho iv. Fig 133.37. (H.XII.l)	MBII
P	Jericho v. Fig 167.13. (H.IX.xxxia)	MBII
P	Jericho v. Fig 173.14. (H.IX.xxxvc)	MBII
P	Jericho v. Fig 178.24. (H.X.xliva)	MBII
V	Aphek 75. Fig 1.6. (Str A VII)	MBI

10.18
17217
Jug

P	Jericho iv. Fig 135.1. (H.VI.xx)	MBI
P	Jericho iv. Fig 135.7. (H.XII.lii)	MBII
P	Jericho v. Fig 167.2. (H.IX.xxxi)	MBII
V	Sali. Pl 42.78. Str XII	MBII

10.18
17218
Jug

P	Jericho iv. Fig 135.1. (H.VI.xx)	MBI
P	Jericho iv. Fig 135.7. (H.XII.lii)	MBII
V	Jericho v. Fig 185.9 (H.XII.li)	MBII

10.29
17701
Bowl (almond)

P	Gezer iv. Pl 5.21. (FVI. Str 11/10)	MBII
P	Meg. Pl 14.11. Str XIV	MBI
P	Shec. Pl 46.c. (Bp.23). Str XX	MBII
V	Hazor iii-iv. Pl CLVI.10. (A. Str 18)	EBIV
V	Meg. Pl 19.11. Str XIIIA	MBI
V	Meg. Pl 21.21. Str XIII	MBI
V	Shec. Pl 2.k. (Bp.22). Str XIX	MBII
V	Shec. Pl 3.a. (Bp.23). Str XVIII	MBII
V	TNM. Fig 27.7. Ph G	MBIII

10.30
17704
Jug

P Meg. Pl 17.2. Str XV-XIV MBI
V Meg. Pl 10.2. Str XV-XIV MBI
V Meg. Pl 25.14. Str XII MBII
V Shec. Pl 31.e. (Jg.O.O.11). Str XIX MBII

10.30
17705
Jug

P Meg. Pl 7.13. Str XV-XIIIA MBI
V Hazor i. Pl XCIV.14. (D1 Str 3) MBIII
V Lachish. Pl 74.669. (Jug Class A) MBII
V Meg. Pl 10.10. Str XV-XIIIA MBI
V Sali. Pl 40.12. Str XI MBIII

10.30
17706
Jug

V Gezer iv. Pl 4.9. (FVI. Str 12/11) MBII
V Hazor i. Pl CXVII.2. (D5 Str 5) MBII
V Hazor v. Fig II.6.42. (A Str 9D) MBI
V Lachish. Pl 74.675. (Jugs class B) MBI
V Meg. Pl 17.25. Str XIIIA MBI
V Meg. Pl 31.2. Str XI MBII
V Shec. Pl 32.d. (Jl.13). Str XIX. (Large Jar) MBII

10.30
17707
Bowl (simple)

P Shec. Pl 2.a. (Bp.21). Str XX MBII
V Meg. Pl 9.5. Str XV MBI
V Meg. Pl 21.15. Str XIII MBI

10.30
17708
Bowl (simple)

P Jericho iv. Fig 106.4. (Tr II.XXII.lxxi) MBII
P Shec. Pl 2.e. (Bp.21). Str XVIII MBII
V Hazor i. Pl CXIX.8. (D5. Str 4) MBII

10.30
17709
Bowl (simple)

V Shec. Pl 2.e. (Bp.21). Str XVIII MBII

10.30
17710
Bowl (simple)

P Alalakh. Pl 37.39. Str IX MBII
P Hazor i. Pl XCIII.1. (D1. Str 5) MBII
P Meg. Pl 14.24. Str XIV MBI
P Shec. Pl 46.a. (Bp.11). Str XX MBII
V Gezer ii. Pl 17.24. (FI. Tr 4168) MBII
V Gezer iv. Pl 5.28. (FVI. Str 11/10) MBII

10.30
17711
Bowl (almond)

P Ghassil. Pl 13.6. Str X MBII
V KEL. Pl 3.T234. Str 7 MBII

10.30
17713
Cpot (upright)

P Jericho iv. Fig 144.1-14. (H.VII-IX.xxiv-xxxii) MBII**(x14)**
P Jericho iv. Fig 144.15. (Tr III.XXI.lxxxvi) MBI
P Jericho v. Fig 106.24. (Tr III.XXI.lxxxvi) MBI
P Jericho v. Fig 108.35. (Tr III.XXI.lxxxiiia) MBI
P Jericho v. Fig 167.1,10. (H.IX.xxxi) MBII**(x2)**
P Jericho v. Fig 168.6. (H.IX.xxxii) MBII
P Jericho v. Fig 173.10,12. (H.IX.xxxvc) MBII**(x2)**
V Aphek 85. Fig 2.10. (X20-19) MBI
V Gezer ii. Pl 11.25. (FI. Str 10) MBII
V Gezer iv. Pl 4.15. (FVI. Str 12/11) MBI
V Gezer iv. Pl 4.22. (FVI. Str 12/11) MBI
V Gezer iv. Pl 6.11. (FVI. Str 11/10) MBII
V Hazor iii-iv. Pl CLVI.24. (A. Str XVII) MBII
V Hazor iii-iv. Pl CLVI.25. (A. Str XVII) MBII
V Hazor v. Fig III.8.12. (L) MBII
V Jericho v. Fig 70.1. (Tr II.XXI-XXII) MBI
V Jericho v. Fig 73.3 (Tr II.XXI-XXII) MBI
V Jericho v. Fig 106.11. (Tr III.XXI.lxxxi) MBI
V Jericho v. Fig 106.12. (Tr III.XXI.lxxxvi) MBI
V Jericho v. Fig 160.3-4. (H.VI-VII.xxia) MBI**(x2)**
V Jericho v. Fig 163.4-11. (H.VII.xxiv-xxva) MBII**(x8)**
V Jericho v. Fig 164.7-10. (H.VIII.xxvi) MBII**(x4)**
V Jericho v. Fig 165.3,7,8,10. (H.VIII.xxviia-xxixa) MBII**(x4)**
V Jericho v. Fig 166.3,4,10-12,17. (H.IX) MBII**(x6)**
V Jericho v. Fig 179.24. (H.XI.xlvi) MBII
V Jericho v. Fig 186.13,14. (H.XII.li) MBII
V KEL. Pl 24. CT16. Str 8 MBIII
V Meg. Pl 7.10. Str XVI MBI
V Meg. Pl 7.11. Str XVI MBI
V Meg. Pl 9.19. Str XV MBI
V Meg. Pl 30.5. Str XII MBII
V Shec. Pl 23.c. (Cf A.3). Str XIX MBII
V TBM 32. Pl 41.6. (Str E) MBII
V TBM 32. Pl 43.9. (Str D) MBIII
V TBM 32. Pl 44.10-14. (Str D) MBIII**(x5)**
V TBM 33. Pl 13.3-6. (Str D) MBIII**(x4)**

10.32
17714
Sjar

P Jericho iv. Fig 127.2. (H.XI.xliva) MBII
P Meg. Pl 12.18. Str XIV MBI
P Shec. Pl 33.c. (Jl.23). Str XVIII MBII
V Gezer i. Pl 31.16. (FI. Str 8) MBII
V Jericho v. Fig 164.11. (H.VIII.xxvi) MBII
V Meg. Pl 28.1. Str XII MBII
V TNM Pl 43.1. Ph G MBIII

10.32
17715
Jar (shortnecked)

V Meg. Pl 6.4. Str XVI EBIV

10.32
17717
Sjar

P Ghassil. Pl 7.14. Str XI MBII
V Gezer i. Pl 32.13. (FI. Gl 8012) MBIII
V Hazor i. Pl CI.27. (D3 Str 2) MBIII
V TNM. Pl 39.6. Ph G-C MBIII

10.32
17718
Bowl (almond)

P	Aphek 75. Fig 8.14. (Str A IVa)	MBI
P	Ghassil. Pl 13.6. Str X	MBII
P	Jericho iv. Fig 104.8. (H.XII.liia)	MBII
P	Meg. Pl 21.21. Str XIII	MBI
P	Mev. Fig 12.18. Str XIII	MBII
V	Gezer iv. Pl 4.4. (FVI. Str 12/11)	MBI

10.32
17722
Sjar (holemouth)

10.33
17724
Sjar (holemouth)

10.33
17725
Jar

V	Lachish. Pl 74.673. L.1542. (Jug classB)	MBII
V	Meg. Pl 20.19. Str XIII	MBI

10.33
17726
Sjar

P	Gezer ii. Pl 11.18. (FI. Str 10)	MBII
P	Jericho v. Fig 108.38. (Tr III.XXI.lxxxv)	MBI

10.33
17727
Sjar

P	Hazor i. Pl CXVII.2. (D5 Str 5)	MBII
P	Meg. Pl 40.3. Str XII	MBII
P	Shec. Pl 32.f. (Jl.23). Str XX	MBII
V	Shec. Pl 33.o. (Jl.33). Str XVIII	MBII

10.33
17728
Bowl (rounded)

P	Hazor iii-iv. Pl CCXCVI.1. (210/A1. Str 4)	MBII
P	Meg. Pl 29.17. Str XII	MBII
P	Shec. Pl 3.g. (Bp.33). Str XVIII	MBII
V	Sali. Pl 41.4. Str XII	MBII
V	Shec. Pl 1.i. (Bp.13). Str XIX	MBII

10.33
17730
Bowl

10.35
17731
Jar (tnn)

P	Aphek 85. Fig 4.11. (X18)	MBI
P	Bosra. Pl 12.30. (Str 26)	MBI
P	Mev. Fig 16.8. Str XV	MBI
V	Aphek 75. Fig 7.8. (A IVb)	MBI
V	Gezer ii. Pl 13.9. (FI. Str 9B)	MBII
V	Ghassil. Pl 7.13. Str XI	MBII
V	Hazor iii-iv. Pl CCXXXV.19. (BA Str 15)	EBIV
V	Hazor v. Fig II.7.14. (A Str 9D)	MBI
V	Jericho iv. Fig 127.27. (H.XI.xlvia)	MBII
V	Jericho v. Fig 171.10. (H.IX. xxxiva)	MBII
V	Mev. Fig 14.24. StrXIII	MBII

10.35
17732
Jar (tnn)

P	Jericho iv. Fig 133.22. (H.IX.xxxii-iii)	MBII
P	Jericho v. Fig 170.2. (H.IX.xxxii)	MBII
V	Meg. Pl 10.4. Str XIV. (Jug)	MBI

10.35
17733
Bowl (simple)

P	Alalakh. Pl 55.10. Str X	MBII
P	Gezer ii. Pl 11.26. (FI. Str 10)	MBII
P	Meg. Pl 14.6. Str XIV	MBI
P	Shec. Pl 3.6. (Bp.23). Str XX	MBII
P	Shec. Pl 46.c. (Bp.23). Str XX	MBII
V	Gezer ii. Pl 19.29. (FI. Tr 4168)	MBIII
V	Hazor iii-iv. Pl CLVI.10. (A. Str 18)	EBIV
V	Meg. Pl 19.11. Str XIIIA	MBI
V	Meg. Pl 29.9. Str XII	MBII
V	Shec. Pl 2.f. (Bp.21). Str XVIII	MBII

10.35
17734
Jug

P	Bosra. Pl 12.29. (Str 26)	MBI
P	Hazor iii-iv. Pl CCXXXV.32. (BA Str 14)	MBII
P	Hazor v. Fig II.7.24. (A. Str 9D)	MBI
P	Jericho iv. Fig 135.28. (H.XII.l-li)	MBII
P	Meg. Pl 23.7. Str XII	MBII
P	Shec. Pl 32.b. (Jl.11). Str XVII. (Large Jar)	MBII
V	Arqa 91. Fig 7.2. Str 16	EBIV
V	Dan. Fig 4.82. (T1025.3). Str XII	MBI
V	Ghassil. Pl 37.37. Str IX	MBIII
V	Meg. Pl 31.2. Str XI	MBII

10.35
17739
Jug

V	Gezer iv. Pl 2.11. (FVI. Str 12)	MBI
V	Ghassil. Pl 36.23. Str IX	MBIII
V	Hazor v. Fig II.6.27. (A Str 9D)	MBI
V	Jericho iv. Fig 126.19. (H.XIII.liii)	MBIII

10.35
17740
Sjar (holemouth)

10.35
17741
Sjar (holemouth)

10.35
17742
Sjar (holemouth)

10.38
17744
Bowl (almond)

P	Dan. Fig 4.77. (T.1025.6). Str XII	MBI
P	Shec. Pl 2.m. (Bp.22). Str XIX	MBII
P	Shec. Pl 3.f. (Bp.32). Str XIX	MBII
V	Ghassil. Pl 13.12. Str X	MBII
V	Jericho iv. Fig 104.26. (H.XII.l)	MBII
V	Jericho iv. Fig 107.26. (H.VII.xxia)	MBII
V	Jericho v. Fig 106.26. (Tr III.XXI.lxxxvii)	MBI
V	Jericho v. Fig 184.4. (H.XI.l)	MBII
V	Jericho v. Fig 186.2. (H.XII.li)	MBII

V KEL. Pl 3.T102. Str 8 MBIII
V Meg. Pl 19.11. Str XIIIA MBI
V TNM. Fig 28.8. Ph G-C MBIII

10.38
17745
Bowl (almond)

V Lachish. Pl 70.584. L119. (Class G Bowl) MBII
V Sali. Pl 43.TS198. Str XII MBII

10.38
17749
Sjar (holemouth)

V Aphek 75. Fig 16.3. (B IV) MBI
V Hazor i. Pl XCIII.23. (D1 Str 4) MBII

10.38
17750
Sjar

P Meg. Pl 17.6. Str XIV-XII MBI
V Jericho iv. Fig 127.6. (H.XI.xliva) MBII
V Meg. Pl 28.1. Str XII MBII
V Shec. Pl 32.k. (Jl.22). Str XIX MBII

10.38
17751
Jar (tnn)

P Aphek 75. Fig 7.9. (A IVb) MBI
P Meg. Pl 11.20. Str XIV MBI
V Gezer iv. Pl 6.2. (FVI. Str 11/10) MBII

10.38
17752
Pithoi

V Mev. Fig 11.13. Str XII MBIII

10.38
17753
Bowl (globular)

P Gezer ii. Pl 14.12. (FI. Str 9A) MBII
P Gezer iv. Pl 4.13. (FVI. Str 12/11) MBI
V Aphek 75. Fig 6.4. (A IVb) MBI
V Bosra. Pl 12.21. (Str 26) MBI
V Gezer ii. Pl 16.24. (FI. Tr 4168) MBII
V Gezer iv. Pl 2.24. (FVI. Str 12) MBI
V Hazor i. Pl XCIII.13. (D1 Str 5) MBII
V Hazor iii-iv. Pl CLVI.6. (A Str 18) MBII
V Hazor iii-iv. Pl CCXXXV.24. (BA Str 14) MBII
V Hazor v. Fig II.7.16. (A Str 9D) MBI
V Hazor v. Fig II.8.3. (A Str 9D) MBI
V Meg. Pl 14.26,37,39. Str XIV MBI**(x3)**
V Meg. Pl 15.2. Str XIV MBI
V Meg. Pl 16.14. Str XIIIB MBI
V Meg. Pl 19.3. Str XIIIA MBI
V Mev. Fig 11.2. (Str XII) MBIII

10.38
17754
Jar (tnn)

P Ghassil 95/6. Pl X.14.2. Str X MBII

10.38
17755
Cpot (upright)

P Jericho iv. Fig 144.1-14. (H.VII-IX.xxiv-xxxii) MBII**(x14)**
P Jericho iv. Fig 144.15. (Tr III.XXI.lxxxvi) MBI
P Jericho v. Fig 106.24. (Tr III.XXI.lxxxvi) MBI
P Jericho v. Fig 108.35. (Tr III.XXI.lxxxiiia) MBI
P Jericho v. Fig 167.1,10. (H.IX.xxxi) MBII**(x2)**
P Jericho v. Fig 168.6. (H.IX.xxxii) MBII
P Jericho v. Fig 173.10,12. (H.IX.xxxvc) MBII**(x2)**
V Aphek 85. Fig 2.10. (X20-19) MBI
V Gezer ii. Pl 11.25. (FI. Str 10) MBII
V Gezer iv. Pl 4.15. (FVI. Str 12/11) MBI
V Gezer iv. Pl 4.22. (FVI. Str 12/11) MBI
V Gezer iv. Pl 6.11. (FVI. Str 11/10) MBII
V Hazor iii-iv. Pl CLVI.24. (A Str XVII) MBII
V Hazor iii-iv. Pl CLVI.25. (A Str XVII) MBII
V Hazor v. Fig III.8.12. (L) MBII
V Jericho v. Fig 70.1. (Tr II.XXI-XXII) MBI
V Jericho v. Fig 73.3 (Tr II.XXI-XXII) MBI
V Jericho v. Fig 106.11. (Tr III.XXI.lxxxi) MBI
V Jericho v. Fig 106.12. (Tr III.XXI.lxxxvi) MBI
V Jericho v. Fig 160.3-4. (H.VI-VII.xxia) MBI**(x2)**
V Jericho v. Fig 163.4-11. (H.VII.xxiv-xxva) MBII**(x8)**
V Jericho v. Fig 164.7-10. (H.VIII.xxvi) MBII**(x4)**
V Jericho v. Fig 165.3,7,8,10. (H.VIII.xxviia-xxixa) MBII**(x4)**
V Jericho v. Fig 166.3,4,10-12,17. (H.IX) MBII**(x6)**
V Jericho v. Fig 179.24. (H.XI.xlvi) MBII
V Jericho v. Fig 186.13,14. (H.XII.li) MBII**(x2)**
V KEL. Pl 24. CT16. Str 8 MBIII
V Meg. Pl 7.10. Str XVI MBI
V Meg. Pl 7.11. Str XVI MBI
V Meg. Pl 9.19. Str XV MBI
V Meg. Pl 30.5. Str XII MBII
V Shec. Pl 23.c. (Cf.A3). Str XIX MBII
V TBM 32. Pl 41.6. Str E MBII
V TBM 32. Pl 43.9. Str D MBIII
V TBM 32. Pl 44.10-14. Str D MBIII**(x5)**
V TBM 33. Pl 13.3-6. Str D MBIII**(x4)**

10.38
17756
Sjar

P Hazor v. Fig II.6.27. (A Str 9D) MBI
V Lachish. Pl 74.673. L1542. (Jug class B) MBII

10.40
17760
Cpot (upright)

P Jericho iv. Fig 144.1-15. (H.IX.xxxii) MBII**(x15)**
P Jericho v. Fig 106.24. (Tr III.XXI.lxxxvi) MBI
P Jericho v. Fig 108.35. (Tr III.XXI.lxxxiiia) MBI
P Jericho v. Fig 167.1,10. (H.IX.xxxi) MBII**(x2)**
P Jericho v. Fig 168.6. (H.IX.xxxii) MBII
P Jericho v. Fig 173.10,12. (H.IX.xxxvc) MBII**(x2)**
V Aphek 85. Fig 2.10. (X20-19) MBI
V Gezer ii. Pl 11.25. (FI. Str 10) MBII
V Gezer iv. Pl 4.15. (FVI. Str 12/11) MBI
V Gezer iv. Pl 4.22. (FVI. Str 12/11) MBI
V Gezer iv. Pl 6.11. (FVI. Str 11/10) MBII
V Hazor iii-iv. Pl CLVI.24. (A Str XVII) MBII
V Hazor iii-iv. Pl CLVI.25. (A Str XVII) MBII
V Hazor v. Fig III.8.12. (L) MBII
V Jericho v. Fig 70.1. (Tr II.XXI-XXII) MBI
V Jericho v. Fig 73.3 (Tr II.XXI-XXII) MBI
V Jericho v. Fig 106.11. (Tr III.XXI.lxxxi) MBI
V Jericho v. Fig 106.22. (Tr III.XXI.lxxxvi) MBI
V Jericho v. Fig 160.3-4. (H.VI-VII.xxia) MBI**(x2)**
V Jericho v. Fig 163.4-11. (H.VII.xxiv-xxva) MBII**(x8)**

V Jericho v. Fig 164.7-10. (H.VIII.xxvi) MBII**(x4)**
V Jericho v. Fig 165.3,7,8,10. (H.VIII.xxviia-xxixa) MBII**(x4)**
V Jericho v. Fig 166.3,4,10-12,17. (H.IX) MBII**(x6)**
V Jericho v. Fig 179.24. (H.XI.xlvi) MBII
V Jericho v. Fig 186.13,14. (H.XII.li) MBII**(x2)**
V KEL. Pl 24. CT16. Str 8 MBIII
V Meg. Pl 7.10. Str XVI MBI
V Meg. Pl 7.11. Str XVI MBI
V Meg. Pl 9.19. Str XV MBI
V Meg. Pl 30.5. Str XII MBII
V Shec. Pl 23.c. (Cf A.3). Str XIX. MBII
V TBM 32. Pl 41.6. Str E MBII
V TBM 32. Pl 43.9. Str D MBIII
V TBM 32. Pl 44.10-14. Str D MBIII**(x5)**
V TBM 33. Pl 13.3-6. Str D MBIII**(x4)**

10.40
17761
Jar (tnn)

P Hazor iii-iv. Pl CCXXXV.18. (BA Str 15) EBIV

10.40
17762
Bowl (rounded)

P Bosra. Pl 11.10. (Str 26) MBI
P Hazor v. Fig II.6.7. (A. Str 9D) MBI
P Jericho v. Fig 163.1. (H.VII.xxiv) MBII
P Jericho v. Fig 165.1. (H.VIII.xxviia) MBII
P Jericho v. Fig 167.14. (H.IX.xxxi) MBII
P Jericho v. Fig 187.14. (H.XII.liib) MBII
P KEL. Pl 3.T179. Str 7 MBIII
P Meg. Pl 14.19. Str XIV MBI
V Bosra. Pl 11.10. (Str 28) MBI
V Ghassil. Pl 13.18. Str X MBII
V Hazor i. Pl XCIV.2. (D1. Str 3) MBIII
V Hazor i. Pl CXIX.10. (D5. Str 4) MBII
V Hazor v. Fig II.7.9. (A. Str 9D) MBI
V Jericho v. Fig 106.27. (Tr III.XXI.lxxxvii) MBI
V Meg. Pl 9.4. Str XV MBI
V Meg. Pl 21.15. Str XIII MBI
V Mev. Fig 12.26. Str XIII MBII
V Sali. Pl 42.96. (Str XII) MBII
V Shec. Pl 2.f. (Bp.21). Str XVIII MBII
V Shec. Pl 3.f. (Bp.32). Str XIX MBII
V Shec. Pl 4.g. (Bp.52). Str XVIII MBII

10.40
17763
Krater

10.42
17766
Sjar (holemouth)

P Aphek 75. Fig 16.4. (B IV) MBI
V Meg. Pl 21.4. Str XIII-XII MBI

10.42
17767
Jar (shortnecked)

P Lachish. Pl 87.1006. L119. (Sjar class A) MBII
P Sali. Pl 41.8. (Str XII) MBII
V Ghassil. Pl 7.10. Str XI MBII
V Hazor i. PL C.24. (D3 Str 3) MBII
V Jericho iv. Fig 136.26. (H.XIII.liii) MBIII

10.42
17768
Jar (tnn)

P Meg. Pl 20.19. Str XIII MBI
V Jericho iv. Fig 127.5 (H.X.xxxvii-iii) MBII
V TBM 32. Pl 46.5. Str D MBIII
V TNM Pl 42.3. Ph G MBIII

10.42
17769
Jug

P Alalakh. Pl 44.95. Str IX MBII
P Ghassil. Pl 20.26. Str X MBII
P Hazor v. Fig III.8.17. (L) MBII
P Meg. Pl 23.1. Str XII MBII
V Hazor i. Pl CXVII.2. (D5 Str 5) MBII
V Meg. Pl 31.2. Str XI MBII
V TNM. PL 37.10. Ph G MBIII

10.42
17770
Bowl (fine)

P Gezer ii. Pl 14.6. (FI. Str 9A) MBII
P Ghassil. Pl 34.3. Str IX MBIII
P Ghassil 95/6. Pl 3.6. Str IX MBII
V Hama. Fig 109.3B312. (H5) MBI
V Meg. Pl 19.3. Str XIIIA MBI

10.42
17771
Bowl (globular)

P Gezer ii. Pl 16.25. (FI. Tr 4168) MBII
P Gezer ii. Pl 17.10. (FI. Tr 4168) MBII
P Gezer iv. Pl 2.24. (FVI. Str 12) MBI
P Hazor iii-iv. Pl CCXXXVI.6,7. (BA. Str 12) MBII**(x2)**
P Jericho iv. Fig 109.19. (H.VI.xx-xxi) MBI
P Meg. Pl 14.33. Str XIV MBI
P Meg. Pl 19.2. Str XIIIA MBI
P Shec. Pl 7.h. (Bd C.12). Str XVIII MBII
V Dan. Fig 4.104.13,14. (T1025). Str XII MBI**(x2)**
V Gezer ii. Pl 19.22. (FI. Tr 4168) MBII
V Hazor i. Pl C.1. (D3 Str 4) MBIII
V Hazor v. Fig II.7.12. (A Str 9D) MBI
V Hazor v. Fig III.10.7. (L) MBII
V Meg. Pl 21.8. Str XIII MBI
V Shec. Pl 16.g. (Bc B.11). Str XVII MBII

10.42
17772
Bowl (rounded)

P Ghassil. Pl 14.4. Str X MBII
P Hazor i. Pl CXIX.2. (D5. Str 4) MBII
P Meg. Pl 14.11. Str XIV MBI
V Aphek 75. Fig 2.4. (Str AVI) MBI
V Jericho v. Fig 160.2. (H.VII.xxia) MBII
V KEL. Pl 3.T102. Str 8 MBIII
V Meg. Pl 9.5. Str XV MBI
V Meg. Pl 29.21. Str XII MBII
V Shec. Pl 2.m. (Bp.22). Str XIX MBII

10.42
17773
Bowl (simple)

P Ghassil. Pl 29.1. Str X MBII
P Hazor i. Pl CXIX.8. (D5. Str 4) MBII
P Sali. Pl 40.26. (Str XI) MBIII

P	Sali. Pl 42.26. (Str XII)	MBII
P	Shec. Pl 2.c. (Bp.21). Str XX	MBII
V	KEL. Pl 6.T68. Str 9	MBIII
V	Shec. Pl 46.b. (Bp.21). Str XX	MBII

10.42
17774
Bowl (simple)

10.42
17780
Sjar (holemouth)

10.42
17781
Sjar (holemouth)

P	Aphek 75. Fig 2.8. (Str AVI)	MBI
V	Hazor i. Pl XCIII.23. (D1 Str 4)	MBII
V	Hazor v. Fig II.8.7. (A Str 9D)	MBI

10.42
17782
Sjar (holemouth)

V	Hazor i. Pl XCIII.23. (D1 Str 4)	MBII

10.42
17783
Bowl (rounded)

P	Aphek 75. Fig 2.4. (Strr AVI)	MBI
P	Gezer ii. Pl 11.29. (FI. Str 10)	MBII
P	Hazor i. Pl XCIII.1. (D1. Str 5)	MBII
P	Hazor iii-iv. Pl CCXXXVI.8. (BA. Str 12)	MBII
P	Hazor iii-iv. Pl CCLXXXVI.3. (K. Str 4)	MBII
P	Hazor v. Fig II.6.8. (A. Str 9D)	MBI
P	Hazor v. Fig II.7.9. (A. Str 9D)	MBI
P	Jericho iv. Fig 104.23. (H.IX.xxx-xxxia)	MBII
P	Jericho iv. Fig III.3. (H.IX.xxxii-iii)	MBII
P	Meg. Pl 9.5. Str XV	MBI
P	Meg. Pl 14.11. Str XIV	MBI
P	Meg. Pl 29.20. Str XII	MBII
P	Sali. Pl 42.65. (Str XII)	MBII
P	Shec. Pl 1.h. (Bp.12). Str XIX	MBII
V	Bosra. Pl 11.10. (Str 28)	MBI
V	Bosra. Pl 13.54. (Str 26)	MBI
V	Dan. Fig 4.77. (T.1025.6). Str XII	MBI
V	Hama. Fig 124.2C926. (H2)	MBIII
V	Hazor i. Pl CXIX.3. (D5. Str 4)	MBII
V	Hazor v. Fig III.7.14. (L)	MBII
V	Hazor v. Fig III.10.4. (L)	MBII
V	Jericho iv. Fig 104.7. (H.X.xxxviia)	MBII
V	Jericho v. Fig 70.1. (Tr II.XXII.lxix)	MBII
V	Jericho v. Fig 106.26. (Tr III.XXI.lxxxvi)	MBI
V	Jericho v. Fig 160.2. (H.VII.xxia)	MBII
V	Jericho v. Fig 165.1. (H.VIII.xxviia)	MBII
V	Jericho v. Fig 189.17. (H.XII.lii)	MBII
V	KEL. Pl 3.T177. Str 7	MBII
V	Meg. Pl 19.10. Str XIIIA	MBI
V	Meg. Pl 21.15. Str XIII	MBI
V	Mev. Fig 12.25. Str XIII	MBII
V	Shec. Pl 3.f. (Bp.32). Str XIX	MBII
V	Shec. Pl 46.c. (Bp.23). Str XX	MBII

10.42
17784
Bowl (almond)

P	Meg. Pl 9.2. Str XV	MBI
V	Meg. Pl 14.7. Str XIV	MBI
V	Meg. Pl 19.7. Str XIIIA	MBI
V	Meg. Pl 21.17. Str XIII	MBI

10.42
17785
Bowl (almond)

V	Meg. Pl 9.2. Str XV	MBI
V	Meg. Pl 14.6. Str XIV	MBI
V	Meg. Pl 21.17. Str XIII	MBI

10.42
17786
Jar (shortnecked)

P	Aphek 85. Fig 5.15. (X17)	MBI
P	Hazor v. Fig II.8.12. (A Str 9D)	MBI
V	Aphek 75. Fig 5.13. (A IVb)	MBI
V	Meg. Pl 35.5. Str XII-XI	MBII
V	Mev. Fig 15.20. Str XIV	MBI
V	Shec. Pl 33.f. (Jl.23). Str XVIII	MBII
V	TNM PL 39.1. Ph G	MBIII

10.42
17787
Small Jar

V	Ghassil. Pl 20.17. Str X	MBII
V	Meg. Pl 17.25. Str XIII	MBI
V	Meg. Pl 20.19. Str XIII	MBI
V	Meg. Pl 23.8. Str XIII	MBI

10.42
17788
Sjar (holemouth)

P	Jericho iv. Fig 125.2. (H.IX.xxxiiia)	MBII
V	Hazor v. Fig III.13.1. (L)	MBIII
V	Hazor v. Fig V.1.10. (P. Phc)	MBIII
V	Meg. Pl 21.4. Str XIII	MBI

10.44
17795
Jar (tnn)

P	Gezer iv. Pl 4.20. (FVI. Str 12/11)	MBII
V	Gezer iv. Pl 8.1. (FVI. Str 10B)	MBIII

10.44
17796
Sjar

V	Hazor iii-iv. Pl CLVI.26. (A Str XVII)	MBII
V	Meg. Pl 18.3. Str XIIIA	MBI
V	Shec. Pl 33.I. (Jl.33). Str XIX	MBII
V	TNM Pl 37.6. Ph G	MBIII

10.44
17797
Jug

10.44
17798
Cpot (upright)

P	Aphek 75. Fig 5.1. (A IVb)	MBI
P	Aphek 75. Fig 6.18. (A IVb)	MBI
P	Jericho v. Fig 173.9. (H.IX.xxxvc)	MBII
V	Aphek 75. Fig 5.2. (A IVb)	MBI
V	Aphek 85. Fig 2.10. (X20-19)	MBI
V	Gezer ii. Pl 11.25. (FI. Str 10)	MBII
V	Gezer iv. Pl 3.7. (FVI. Str 12)	MBI
V	Gezer iv. Pl 4.15. (FVI. Str 12/11)	MBI
V	Gezer iv. Pl 4.22. (FVI. Str 12/11)	MBI
V	Gezer iv. Pl 6.11. (FVI. Str 11/10)	MBII

V Hazor iii-iv. Pl CLVI.24. (A. Str XVII) MBII
V Hazor iii-iv. Pl CLVI.25. (A. Str XVII) MBII
V Hazor iii-iv. Pl CCXXXV.12-14. (BA. Str 15) EBIV**(x3)**
V Hazor v. Fig III.8.12. (L) MBII
V Jericho iv. Fig 140.1-11. (H.VII-XI.xxv-xlv) MBII**(x11)**
V Jericho iv. Fig 141.1,3-5,7-8. (H.VIII-IX.xxvi-xxxiv) MBII**(x6)**
V Jericho iv. Fig 141.2,6. (H.VI.xvii-xx) MBI**(x2)**
V Jericho iv. Fig 142.1,4,6-7,9. (H.VIII-IX.xxviiia-xxxiv) MBII**(x5)**
V Jericho iv. Fig 142.2. (Tr II.XXII.lxix) MBII
V Jericho iv. Fig 142.3,5,8,10. (H.VI.xvii-xx) MBI**(x4)**
V Jericho iv. Fig 143.1-5. (H.VII-XI.xxiv-xlvi) MBII**(x5)**
V Jericho v. Fig 21.23. (Tr.I.XLIII.lvii) MBI
V Jericho v. Fig 70.1. (Tr II.XXI-XXII) MBI
V Jericho v. Fig 73.3 (Tr II.XXI-XXII) MBI
V Jericho v. Fig 160.3-4. (H.VI-VII.xxia) MBI**(x2)**
V Jericho v. Fig 163.7,9. (H.VII.xxiv-xxiva) MBII**(x2)**
V Jericho v. Fig 164.4-8. (H.VIII.xxvi) MBII**(x5)**
V Jericho v. Fig 165.3,7,10. (H.VIII.xxviia-xxixa) MBII**(x3)**
V Jericho v. Fig 166.,4,10,11. (H.IX.xxx-xxxa) MBII**(x3)**
V Jericho v. Fig 171.6. (H.IX.xxxiva) MBII
V Jericho v. Fig 179.24. (H.XI.xlvi) MBII
V Jericho v. Fig 180.1. (H.XI.xlvi) MBII
V Meg. Pl 7.10. Str XVI EBIV
V Meg. Pl 7.11. Str XVI EBIV
V Meg. Pl 9.19. Str XV MBI
V Meg. Pl 30.5. Str XII MBII
V Shec. Pl 23.a. (Cf A.2). Str XX MBII
V TBM 32. Pl 45.17. Str D MBIII
V TBM 32. Pl 46.16. Str D MBIII

10.44
17802
Jug

P Hazor v. Fig III.8.22. (L) MBII
P Meg. Pl 23.7. Str XII MBII
V Alalakh. Pl 44.98. Str IX MBII
V Ghassil. Pl 20.8. Str X MBII

10.44
17805
Sjar (holemouth)

10.44
17806
Sjar (holemouth)

10.44
17807
Jar (tnn)

V Gezer ii. Pl 13.19. (FI. Str 9B) MBII
V Hazor iii-iv. Pl CCXXXV.17. (BA. Str 15) EBIV
V KEL. Pl 30.CT7. Str 9 MBIII
V Shec. Pl 42.c. (JJ.31). Str XVIII MBII

10.44
17808
Jar (tnn)

V Bosra. Pl 12.40. (Str 26) MBI
V Hazor iii-iv. Pl CCXXXV.31. (BA. Str 14) MBII
V Jericho iv. Fig 137.29. (H.IX.xxxii-iii) MBII
V Lachish. Pl 74.674. L119. (Jugs class B) MBII
V Meg. Pl 10.3. Str XIV MBI

10.44
17810
Sjar (holemouth)

P Hazor i. Pl XCIII.23. (D1 Str 4) MBII
V Aphek 75. Fig 2.8. (Str A VI) MBI
V Gezer ii. Pl 13.28. (FI. Str 9B) MBII

10.44
17811
Jar (tnn)

P Aphek 85. Fig 4.14. (X18) MBI
P Mev. Fig 16.8. Str XV MBI
V Aphek 75. Fig 1.15. (Str A VII) MBI
V Bosra. Pl 12.30. (Str 26) MBI
V Gezer iv. Pl 2.20. (FVI. Str 12) MBI
V Ghassil. Pl 7.13. Str XI MBII
V Hazor v. Fig II.7.14. (A Str 9D) MBI
V Jericho iv. Fig 127.27. (H.XI.xlvia) MBII
V Mev. Fig 14.24. Str XIII MBII

10.44
17812
Jar (tnn)

P Gezer ii. Pl 11.15. (FI. Str 10) MBII
P Gezer iv. Pl 2.20. (FVI. Str 12) MBI
V Aphek 75. Fig 7.10. (A IVb) MBI
V Gezer ii. Pl 13.12. (FI. Str 9B) MBII
V Gezer iv. Pl 5.5. (FVI. Str 11/10) MBII

10.44
17813
Jar (tnn)

P Aphek 85. Fig 4.15. (X18) MBI
P Gezer ii. Pl 11.12. (FI. Str 10) MBII
P Meg. Pl 11.20. Str XIV MBI
V Aphek 75. Fig 7.18. (A IVa) MBI
V Aphek 75. Fig 16.9. (B IV) MBI
V Gezer ii. Pl 13.2. (FI. Str 9B) MBII
V Gezer iv. Pl 2.13. (FVI. Str 12) MBI
V Gezer iv. Pl 5.5. (FVI. Str 11/10) MBII
V Meg. Pl 16.10. Str XIIIB MBI

10.44
17814
Cpot (upright)

P Bosra. Pl 11.14. (Str 28) MBI
V Jericho v. Fig 72.3. (Tr II.XXII.lxxi) MBII
V Jericho v. Fig 164.6. (H.VIII.xxvi) MBII

10.44
17819
Sjar

P Bosra. Pl 12.40. (Str 26) MBI
P Jericho iv. Fig 137.28. (H.XII.l-li) MBII
P Lachish. Pl 74.673. L1542. (Jug class B) MBII
P Meg. Pl 8.4. Str XV MBI
V Bosra. Pl 17.72. (Str 23) MBII
V Hazor iii-iv. Pl CCXXXV.31. (BA. Str 14) MBII
V Jericho iv. Fig 135.19. (H.IX.xxxii-iii) MBII
V Lachish. Pl 87.1013. L.7003. (Sjar class C) MBI
V Meg. Pl 12.21. Str XIV MBI

10.44
17820
Bowl

V Hazor v. Fig II.7.19. (A Str 9D) MBI

10.45
17262
Sjar

V Mev. Fig 16.9. Str XV MBI

10.45
17821
Jar (holemouth)

P	Hazor i. Pl XCIII.23. (D1. Str 4)	MBII

10.45
17822
Sjar (tnn)

P	Gezer ii. Pl 11.14. (FI. Str 10)	MBII
P	Gezer ii. Pl 12.23. (FI. Str 9B)	MBII
P	Mev. Fig 16.22. Str XIV	MBI
P	Shec. Pl 41.g. (JJ.21). Str XX	MBII
V	Aphek 75. Fig 16.8. (B17)	MBI
V	Gezer ii. Pl 13.9. (FI. Str 9B)	MBII
V	Jericho iv. Fig 133.34. (H.IX.xxx-xxia)	MBII
V	Meg. Pl 11.20. Str XIV	MBI
V	Meg. Pl 18.1. Str XIIIA	MBI

10.45
17823
Bowl (rounded)

P	Alalakh. Pl 19.20. Str VIII	MBII
V	Alalakh. Pl 19.19. Str VIII	MBII
V	Sali. Pl 43. TS198 (Str XII)	MBII

10.45
17824
Jar (holemouth)

V	Aphek 75. Fig 2.9. (Str A VI)	MBI

10. 45
17825
Bowl (rounded)

P	Alalakh. Pl 19.20. Str VIII	MBII
V	Alalakh. Pl 19.19. Str VIII	MBII
V	Sali. Pl 43. TS198 (Str XII)	MBII

10.45
17826
Sjar (tnn)

P	Hazor v. Fig II.6.33. (A Str 9D)	MBI
V	Gezer v. Pl 4.9. (FVI. Str 12/11)	MBII
V	Hazor i. Pl C1.7. (D3 Str 2)	MBIII
V	Meg. Pl 40.2. Sr X	MBIII

10.45
17827
Small Jar

P	Meg. Pl 11.11. Str XIV	MBI
V	Gezer iv. Pl 2.14. (FVI. Str 12)	MBI
V	Jericho iv. Fig 126.12. (H.XII.lii-iii)	MBII
V	Meg. Pl 23.6. Str XII	MBII

10.45
17829
Jug

10.45
17830
Bowl (simple)

P	Meg. Pl 16.13. Str XIIIB	MBI
P	Meg. Pl 28.3. Str XII	MBII
V	Gezer iv. Pl 8.6. (FVI. Str 10B)	MBIII
V	Shec. Pl 1.c. (Bp.11). Str XIX	MBII

10.45
17831
Jar (tnn)

P	Jericho iv. Fig 136.8. (H.lii-iii)	MBII
P	Lachish. Pl 87.1007. L119	MBIII
V	Alalakh. Pl 60.48. Str X	MBII
V	Arqa 91. Fig 7.2. Str 16	EBIV
V	Bosra. Pl 12.29. Str 26	MBI
V	Hazor v. Fig III.8.17. (L)	MBII
V	Hazor v. Fig III.13.9. (L)	MBIII
V	TNM Pl 37.10. Ph G	MBIII

10.45
17832
Sjar (holemouth)

10.45
17833
Sjar (tnn)

P	Bosra. Pl 7.1. (Str 26)	MBI
P	Bosra. Pl 14.72. (Str 23)	MBII
P	Jawa. Fig 12.20	MBI
P	Jericho iv. Fig 135.19. (H.IX.xxxii-iii)	MBII
P	Lachish. Pl 74.674. L119. (Jugs class B)	MBIII
V	Aphek 75. Fig 5.15. (A Ivb)	MBI
V	Gezer ii. Pl 13.22. (FI. Str 9B)	MBII
V	Hazor iii-iv. Pl CCXXXV.31. (BA. Str 14)	MBII
V	Jericho iv. Fig 137.29. (H.IX.xxxii-iii)	MBII
V	Jericho iv. Fig 135.30. (H.XI.xlvi-ii)	MBII
V	Meg. Pl 10.3. Str XIV	MBI
V	Meg. Pl 12.22. Str XIV	MBI

10.45
17834
Pithoi

P	Hazor iii-iv. Pl CLVI.26. (A Str XVII)	MBII

10.45
17836
Sjar

V	Alalakh. Pl 21.32. Str 8	MBII
V	Meg. Pl 40.5. Str XIII	MBI

10.45
17838
Sjar (holemouth)

10.45
17839
Jar (tnn)

P	Alalakh. Pl 60.48. Str X	MBII
P	Hazor v. Fig III.8.17. (L)	MBII
P	Jericho iv. Fig 136.8. (H.XII.lii-liii)	MBII
P	Lachish. Pl 87.1007. L.119. (Sjar class A)	MBIII
V	Alalakh. Pl 44.92. Str IX	MBII
V	Bosra. Pl 12.29. (Str 26)	MBI
V	Hazor iii-iv. Pl CCLX.5. (H Str 3)	MBIII
V	Hazor iii-iv. Pl CCXCVI.13. (210/A1 Str 3)	MBIII

10.45
17840
Jar (tnn)

P	Lachish. Pl 87.1006. L.119. (Sjar class A)	MBIII
V	Gezer iv. Pl 4.7. (FVI. Str 12/11)	MBII
V	Ghassil. Pl 20.41. Str X	MBII
V	Sali. Pl 41.8. (Str XII)	MBII
V	Shec. Pl 40.n. (JJ.13). Str XVIII	MBII

10.45
17841
Bowl

P	Alalakh. Pl 80.4. Str XII	MBII
P	Bosra. Pl 12.19. (Str 26)	MBI
P	Gezer ii. Pl 14.7. (F1. Str 9A)	MBII
P	Gezer ii. Pl 15.20. (FI. Tr 4168)	MBII
P	Gezer iv. Pl 2.2. (FVI. Str 12)	MBI
P	Jericho v. Fig 168.3. (H.IX.xxxii)	MBII
P	Meg. Pl 14.26. Str XIV	MBI
P	Meg. Pl 15.2. Str XIV	MBI
V	Aphek 85. Fig 2.3. (X20-19)	MBI
V	Bosra. Pl 18.170. (Str 19)	MBIII
V	Gezer iv. Pl 3.10. (FVI. Str 12)	MBI
V	Hama. Fig 109.3C126. (H5)	MBI
V	Meg. Pl 14.39. Str XIV	MBI
V	Meg. Pl 16.14. Str XIIIB	MBI
V	Meg. Pl 19.3. Str XIIIA	MBI

10.45
17843
Sjar (holemouth)

10.45
17844
Sjar (holemouth)

10.45
17845
Sjar (holemouth)

P	Bosra. Pl 14.7. (Str 23)	MBII
V	Hazor v. Fig V.1.10. (P. Ph c)	MBIII

10.45
17846
Jar (shortnecked)

P	Gezer iv. Pl 7.17. (FVI. Str 10)	MBIII
V	Gezer iv. Pl 4.14. (FVI. Str 12/11)	MBII
V	Hazor i. Pl XCIII.18 (D1. Str 4)	MBII
V	Hazor iii-iv. Pl CCXXXV.3. (BA Str 16)	EBIV
V	Meg. Pl 6.4. Str XVI	EBIV
V	Meg. Pl 13.6. Str XIV	MBI
V	Tyre. Pl LV.1. Str XIX	MBII

10.45
17847
Bowl (globular)

P	Bosra. Pl 12.18. (Str 26)	MBI
P	Gezer ii. Pl 14.20. (FI. Str 9A)	MBII
P	Gezer ii. Pl 19.22. (FI. Tr 4168)	MBII
P	Gezer iv. Pl 5.16. (FVI. Str 11/10)	MBII
P	Mev. Fig 11.2. Str XII	MBIII
P	Meg. Pl 14.26. Str XIV	MBI
P	Meg. Pl 19.3. Str XIIIA	MBI
V	Dan. Fig 4.78. (T23.1). Str XII	MBI
V	Dan. Fig 4.90.2. (T23.1). Str XII	MBI
V	Gezer ii. Pl 16.25. (FI. Tr 4168)	MBII
V	Gezer iv. Pl 2.24. (FVI. Str 12)	MBI
V	Hazor iii-iv. Pl CCXXXV.24. (BA. Str 14)	MBII
V	Hazor v. Fig II.8.3. (A Str 9D)	MBI
V	Meg. Pl 14.39. Str XIV	MBI
V	Meg. Pl 15.2. Str XIV	MBI
V	Shec. Pl 8.a. (Bd C.13). Str XVIII	MBII
V	Shec. Pl 13.a. (Bg B.11). Str XIX	MBII

10.47
17850
Jar (tnn)

P	Gezer ii. Pl 12.17. (FI. Str 9B)	MBII
P	Hazor v. Fig II.8.11. (A Str 9D)	MBI
P	Shec. Pl 41.e. (JJ.21). Str XIX	MBII
V	Aphek 85. Fig 5.12. (XI7)	MBI
V	Aphek 75. Fig 5.7. (Str A IVb)	MBI
V	Gezer ii. Pl 11.16. (FI. Str 10)	MBII
V	Gezer iv. Pl 2.19. (FVI. Str 12)	MBI
V	Gezer iv. Pl 3.5. (FVI. Str 12)	MBI
V	Gezer iv. 4.1. (FVI. Str 12/11)	MBII
V	Gezer iv .Pl 8.3. (FVI. Str 10B)	MBIII
V	Meg. Pl 11.20. Str XIV	MBI
V	Meg. Pl 18.1. Str XIIIA	MBI
V	Sarepta. Pl 20.13 Str L	MBIII

10.47
17856
Jar (tnn)

P	Aphek 85. Fig 4.14. (X18)	MBI
P	Gezer iv. Pl 2.20. (FVI. Str 12)	MBI
P	Hazor v. Fig II.7.14. (A Str 9D)	MBI
V	Bosra. Pl 12.30. (Str 26)	MBI
V	Gezer ii. Pl 11.11. (FI. Str 10)	MBII
V	Gezer ii. Pl 12.17. (FI. Str 9B)	MBII
V	Hazor iii-iv. Pl CCXXXV.19. (BA. Str 15)	EBIV
V	Jericho iv. Fig 128.5. (H.X.xxxvii-iii)	MBII
V	Meg. Pl 18.1. Str XIIIA	MBI

10.47
17857
Bowl (simple)

P	Aphek 85. Fig 4.1. (X18)	MBI
P	Meg. Pl 14.24. Str XIV	MBI
V	Gezer ii. Pl 17.24. (FI. Tr 4168)	MBII
V	Gezer iv. Pl 5.28. (FVI. Str 11/10)	MBII
V	Gezer iv. Pl 8.6. (FVI. Str 10B)	MBIII
V	Meg. Pl 28.3. Str XII	MBII
V	Meg. Pl 16.13. Str XIIIB	MBI
V	Shec. Pl 46.a. (Bp.11). Str XX	MBII

10.53
17869
Bowl (rounded)

P	Jericho v. Fig 106.17. (Tr III.XXI.lxxxvi)	MBI
P	Meg. Pl 9.4. Str XV	MBI
P	Meg. Pl 21.15. Str XIII	MBI
V	Gezer iv. Pl 6.17. (FVI. Str 11/10)	MBII
V	Hazor i. Pl XCIII.1. (D1 Str 5)	MBII
V	Hazor v. Fig III.9.1. (L)	MBII
V	Meg. Pl 37.25. Str XI	MBII
V	Shec. Pl 1.h. (Bp.12). Str XIX	MBII
V	Shec. Pl 2.a. (Bp.21). Str XX	MBII

10.55
17872
Jar (tnn)

P	Aphek 85. Fig 5.12. (X17)	MBI
P	Gezer ii. Pl 11.11. (FI Str 10)	MBII
P	Meg. Pl 11.20. Str XIV	MBI
V	Aphek 85. Fig 4.14. (X18)	MBI
V	Gezer ii. Pl 12.16. (FI Str 9B)	MBII
V	Gezer ii. Pl 14.4. (FI Str 9A)	MBIII
V	Gezer iv. Pl 2.20. (FVI Str 12)	MBI
V	Meg. Pl 18.1. Str XIIIA	MBI

10.55
17874
Bowl (simple)

V Shec. Pl 7.g. (Bd C.11). Str XVIII MBII

10.57
17878
Sjar

V Alalakh. Pl 21.32. Str 8 MBII
V Meg. Pl 20.19. Str XIII MBI

10.58
17880
Sjar (holemouth)

V Bosra. Pl 14.91. (Str 23) MBII

10.58
17882
Pithoi?

P Mev. Fig 15.20. Str XIV MBI
V Gezer iv. Pl 8.20. (FVI. Str 10B) MBIII
V Jericho v. Fig 165.11. (H.VIII.xxxixa) MBII
V Meg. Pl 35.5. Str XII-XI MBII
V Shec. Pl 32.n. (Jl.22). Str XVIII MBII
V Shec. Pl 33.f. (Jl.23). Str XVIII MBII
V Shec. Pl 35.a. (Jl.43). Str XIX MBII
V Shec. Pl 46.n.(Bf A.2). Str XX MBII
V TNM Pl 41.8. Ph F MBIII

10.58
17888
Sjar

V Gezer iv. Pl 1.1. (FVI. Str 12) MBI
V Jericho iv. Fig 127.9. (H.IX.xxxa) MBII
V Meg. Pl 28.1. Str XII MBII
V Shec. Pl 32.k. (Jl.22). Str XIX MBII

10.59
17889
Jar (tnn)

P Alalakh. Pl 60.50. Str X MBII
P Jericho iv. Fig 127.22. (H.VI.xvii) MBI
V Ghassil. Pl 20.31. Str X MBII
V Ghassil. Pl 7.12. Str XI MBII

10.59
17890
Jar (tnn)

P Gezer ii. Pl 12.17. (FI. Str 9B) MBII
P Hazor v. Fig II.7.14. (A Str 9D) MBI
P Mev. Fig 14.24. Str XIII MBII
P Mev. Fig 16.8. Str XV MBI
V Ghassil. Pl 7.13. Str XI MBII

10.59
17891
Jar (tnn)

P Gezer ii. Pl 12.17. (FI. Str 9B) MBII
V Hazor iii-iv. Pl CCXXXV.19. (BA. Str 15) EBIV

10.59
17892
Jug

V KEL. 44.T141. Str 8 MBIII

10.59
17896
Jar (tnn)

P Ghassil. Pl 20.26. Str X MBII
P TBM 33. Pl 7.7. Str E MBII
P TNM. Pl 37.10. Ph G MBIII
V Hazor i. Pl CXVII.2. (D5 Str 5) MBII
V Jericho iv. Fig 138.14. (H.XII.lii-iii) MBII
V Shec. Pl 40.m. (JJ.13). Str XIX MBII

10.59
17897
Jug

V Lachish. Pl 74.675. L.7003. (Jugs class B) MBII
V Meg. Pl 11.11. Str XIV MBII

10.64
17900
Jar (tnn)

P Aphek 85. Fig 4.14. (X18) MBI
P Jericho iv. Fig 127.27. (H.XI.xlvia) MBII
P Mev. Fig 16.8. Str XV MBI
V Aphek 75. Fig 7.8. (A IVb) MBI
V Aphek 85. Fig 2.12. (X20-19) MBI
V Bosra. Pl 12.30. (Str 26) MBI
V Gezer ii. Pl 13.8. (FI. Str 9B) MBII
V Ghassil. Pl 7.13. Str XI MBII
V Hazor v. Fig II.7.14. (A Str 9D) MBI
V Mev. Fig 14.24. Str XIII MBII
V Shec. Pl 33.d. (Jl.23). Str XVII MBII

10.64
17904
Sjar (holemouth)

10.64
17905
Sjar (holemouth)

10.64
17908
Bowl (simple)

V Hazor i. Pl XCIII.18 (D1 Str 4) MBIII
V Jericho iv. Fig 118.4. (H.VII.xxixa) MBII
V Shec. Pl 7.g. (Bd C.11). Str XVIII MBII

10. 65
17912
Bowl (almond)

P Meg. Pl 14.17. Str XIV MBI
P Meg. Pl 22.2. Str XIII MBI
V Jericho iv. Fig 114.9. (H.IX.xxxiiia) MBII

10.65
17913
Bowl (almond)

P Ghassil. Pl 1.7. Str XI MBII
P Ghassil. Pl 18.12. Str IX MBII
V Meg. Pl 14.8. Str XIV MBI
V Shec. Pl 4.I. (Bp 52). Str XX MBII

10.65
17914
Cpot (upright)

P	Hazor v. Fig V.1.24. (P. Ph c)	MBIII
P	Jericho iv. Fig 144.1-15. (H.IX.xxxii)	MBII**(x15)**
P	Jericho v. Fig 106.24. (Tr III.XXI.lxxxvi)	MBI
P	Jericho v. Fig 108.35. (Tr III.XXI.lxxxiiia)	MBI
P	Jericho v. Fig 167.1,10. (H.IX.xxxi)	MBII**(x2)**
P	Jericho v. Fig 168.6. (H.IX.xxxii)	MBII
P	Jericho v. Fig 173.10,12. (H.IX.xxxvc)	MBII**(x2)**
V	Aphek 85. Fig 2.10. (X20-19)	MBI
V	Gezer ii. Pl 11.25. (FI. Str 10)	MBII
V	Gezer iv. Pl 4.15. (FVI. Str 12/11)	MBI
V	Gezer iv. Pl 4.22. (FVI. Str 12/11)	MBI
V	Gezer iv. Pl 6.11. (FVI. Str 11/10)	MBII
V	Hazor iii-iv. Pl CLVI.24. (A Str XVII)	MBII
V	Hazor iii-iv. Pl CLVI.25. (A Str XVII)	MBII
V	Hazor v. Fig III.8.12. (L)	MBII
V	Jericho v. Fig 70.1. (Tr II.XXI-XXII)	MBI
V	Jericho v. Fig 73.3 (Tr II.XXI-XXII)	MBI
V	Jericho v. Fig 106.11. (Tr III.XXI.lxxxi)	MBI
V	Jericho v. Fig 106.22. (Tr III.XXI.lxxxvi)	MBI
V	Jericho v. Fig 160.3-4. (H.VI-VII.xxia)	MBI**(x2)**
V	Jericho v. Fig 163.4-11. (H.VII.xxiv-xxva)	MBII**(x8)**
V	Jericho v. Fig 164.7-10. (H.VIII.xxvi)	MBII**(x4)**
V	Jericho v. Fig 165.3,7,8,10. (H.VIII.xxviia-xxixa)	MBII**(x4)**
V	Jericho v. Fig 166.3,4,10-12,17. (H.IX)	MBII**(x6)**
V	Jericho v. Fig 179.24. (H.XI.xlvi)	MBII
V	Jericho v. Fig 186.13,14. (H.XII.li)	MBII**(x2)**
V	KEL. Pl 24. CT16. Str 8	MBIII
V	Meg. Pl 7.10. Str XVI	EBIV
V	Meg. Pl 7.11. Str XVI	EBIV
V	Meg. Pl 9.19. Str XV	MBI
V	Meg. Pl 30.5. Str XII	MBII
V	Shec. Pl 23.c. (Cf A.3). Str XIX	MBII
V	TBM 32. Pl 41.6. Str E	MBII
V	TBM 32. Pl 43.9. Str D	MBIII
V	TBM 32. Pl 44.10-14. Str D	MBIII**(x5)**
V	TBM 33. Pl 13.3-6. Str D	MBIII**(x4)**

10.69
17919
Jar (tnn)

P	Hazor v. Fig II.6.40. (A Str 9D)	MBI
P	Jawa. Fig 12.22	MBI
P	TNM. Pl 38.4. Ph G-B	MBIII
V	Bosra. Pl 15.97. (Str 20)	MBIII
V	Hazor iii-iv. Pl CCXXXV.18. (BA. Str 15)	EBIV
V	Jericho iv. Fig 132.14. (H.XII.l)	MBII

10.75
17924
Jar (tnn)

P	Aphek 75. Fig 7.9. (A IVb)	MBI
P	Gezer ii. Pl 13.20. (FI. Str 9B)	MBII
V	Gezer iv. Pl 2.20. (FVI. Str 12)	MBI
V	Gezer iv. Pl 6.2. (FVI. Str 11/10)	MBII
V	Sarepta. Pl 20.14. Str L	MBIII

10.75
17925
Bowl (simple)

P	Gezer ii. Pl 11.34. (Fl. Str 10)	MBII
V	Alalakh. Pl 39.59. Str IX	MBII
V	Gezer i. Pl 30.6. (Fl. Str 7)	MBIII
V	Meg. Pl 15.9. Str XIV	MBI
V	Shec. Pl 13.a. (Bg 13.11). Str XIX	MBII
V	TBM 33. Pl 4.9. Str G-F	MBI

10.75
17927
Krater

P	Hazor i. Pl XCIII.23. (D1 Str 4)	MBII

10.75
17928
Jar (tnn)

P	Gezer ii. Pl 13.3. (FI. Str 9B)	MBII
V	Shec. Pl 40.1. (JJ.11). Str XVIII	MBII

10.75
17932
Jar (shortnecked)

P	Gezer ii. Pl 18.11. (FI. Tr 4168)	MBII
P	Jericho iv. Fig 139.10. (H.XIII.liii)	MBIII
P	Jericho v. Fig 108.36. (Tr III.XXI. lxxxiiia)	MBI
V	Bosra. Pl 15.104. (Str 20)	MBIII
V	Lachish. Pl 87.1007. L119. (Sjar class A)	MBII
V	Meg. Pl 20.19. Str XIII	MBI

10.75
17935
Sjar

P	Shec. Pl 33.o. (Jl.33). Str XVIII	MBII
V	Meg. Pl 27.7. Str XII	MBII

Appendix C: Rukeis, Area 05, Comparative Pottery Analysis

0705 206
DCR 1616
Sjar 094

P	Hazor i. Pl XCIII.20. (D)	MBII
P	Sali. Pl 41.10. Str XII	MBII
P	Sali. Pl 42.101. Str XII	MBII
V	Sali. Pl 40.66. Str XI	MBIII
V	Jericho iv. Fig 138.12. (H.XIII.liii)	MBIII

0705 206
DCR 1618
Sjar 195

P	Mev. Fig 14.26. Str XIII	MBII
P	Mev. Fig 15.21-23. Str XIV	MBI
V	Bosra. Pl 17.141. Str 17	MBIII
V	Sali. Pl 42.101. Str XII	MBII

0705 206
DCR 1620
Sbowl 001

P	Lachish. Pl 66.430. (L.2278)	EBIV
P	TBM 33. Pl 3.7. Str H	EBIV
V	Aphek 85. Fig 5.8. (Str X17)	MBI
V	Gezer ii. Pl 14.8. (FI Str 9A)	MBII
V	Mev. Fig 16.3. Str XV	MBI

0705 207
DCR 1429
Sjar 096

V	Aphek 75. Fig 7.15. (Str A Iva)	MBI
V	Bosra. Pl 12.31. Str 26	MBI
V	Gezer ii. Pl 14.17. (FI Str 9A)	MBII
V	Gezer ii. Pl 13.21. (FI Str 9B)	MBII
V	Jawa. Fig 156.646	MBI
V	Jericho iv. Fig 127.5. (H.HX.xxxvii-iii)	MBIII
V	Jericho v. Fig 168.16. (H.IX.xxxiia)	MBII
V	Meg. Pl 28.1. Str XII	MBII
V	Mev. Fig 14.24. Str XIII	MBII
V	Sarepta. Pl 20.13. Str L	MBIII

0705 207
DCR 1635
Xjar 059

P	Jawa. Fig 156.648,649	MBI**(x2)**
P	Meg. Pl 12.19. Str XIV	MBII
P	Mev. Fig 15.22. Str XIV	MBI
P	Sali. Pl 42.102. Str XII	MBII
V	Gezer ii. Pl 15.4. (FI Tr4168)	MBIII
V	Jericho iv. Fig 136.24. (H.VI.xv-xvia)	MBI
V	Mev. Fig 15.25. Str XIV	MBI

0705 207
DCR 1638
Cpot 016

P	Jawa. Fig 154.622	MBI

0705 207
DCR 1750
Sjar 094

P	Hazor i. Pl XCIII.20. (D)	MBII
P	Sali. Pl 41.10. Str XII	MBII
P	Sali. Pl 42.101. Str XII	MBII

0705 207
DCR 1751
Xjar 117

P	Jawa. Fig 156.648	MBI
P	Meg. Pl 12.19. Str XIV	MBII
P	Mev. Fig 15.22. Str XIV	MBI
P	Sali. Pl 42.102. Str XII	MBII
V	Gezer ii. Pl 15.4. (FI Tr4168)	MBIII
V	Jericho iv. Fig 136.24. (H.VI.xv-xvia)	MBI
V	Mev. Fig 15.25. Str XIV	MBI
V	TBM 32. Pl 45.8. Str D	MBIII
V	TBM 33. Pl 14.10. Str D	MBIII

0705 207
DCR 1753
Sbowl 118

V	Aphek 85. Fig 5.8. (Str X17)	MBI
V	Hazor iii-iv. Pl CCLXXXVII.11. (K Str 3)	MBIII
V	Jericho v. Fig 21.12. (Tr I.XLIII.lvii)	MBII

0705 209
DCR 1536
Xjar 054

P	Hazor iii-iv. Pl CXCII.9. (A Tr500)	MBII
V	Aphek 75. Fig 15.13. (Str B Vb)	MBI
V	Gezer ii. Pl 12.20. (FI Str 9B)	MBII
V	Sali. Pl 41.8. Str XII	MBII
V	TBM 32. Pl 46.10,13. Str D	MBIII**(x2)**
V	TBM 33. Pl 11.6. Str E	MBII

0705 210
DCR 995
Fjar 018

P	Bosra. Pl 12.17. Str 26	MBI
P	Meg. Pl 6.7. Str XVI	EBIV
P	Sali. Pl 42.76. Str XII	MBII
P	Sali. Pl 40.9. Str XI	MBIII
V	Gezer ii. Pl 10.18. (FI. Str 11)	MBI
V	Jawa. Fig 155.636	MBI
V	Tyre. Pl LIV.1. Str XIX	EBIV

0705 210
DCR 1236
Sjar 082

V	Gezer ii. Pl 11.25. (FI Str 10)	MBII
V	Jawa. Fig 156.653	MBI

0705 210
DCR 1270
Sjar 084

P	Aphek 75. Fig 1.1. (Str AVII)	MBI
P	Hazor i. Pl CXII.13. (D)	MBII
P	Sali. Pl 42.83. Str XII	MBII
P	Tyre. Pl LVI.1. Str XX	EBIV
V	Hazor iii-iv. Pl CCLIX.20. (H Str 3)	MBIII
V	Jawa. Fig 156.652	MBI
V	Jericho iv. Fig 116.32. (H.IX.xxxiv)	MBII
V	Jericho iv. Fig 138.9. (H.VI.xx-xxi)	MBI
V	Jericho v. Fig 198.12,14. (H.XIII.liii)	MBIII**(x2)**
V	Meg. Pl 36.2. Str XI	MBIII
V	Shec. Pl 34.b. (Jl.41). Str XIX	MBII

0705 210
DCR 1485
Sjar 103

P	Bosra. Pl 11.6. Str 28	MBI
P	Gezer ii. Pl 12.21. (FI Str 9B)	MBII
P	Gezer ii. Pl 11.10. (FI Str 10)	MBII
P	Hazor iii-iv. Pl CXCII:18. (A Tr500)	MBII
P	Jericho iv. Fig 137.32. (H.IX.xxxiv)	MBII
V	Aphek 85. Fig 4.17. (Str X18)	MBI
V	Arqa 78. Fig 49.1. Str 11	MBIII
V	Bosra. Pl 12.39. Str 28	MBI
V	Bosra. Pl 13.71. Str 25	MBII
V	Jawa. Fig 156.641,643	MBI**(x2)**
V	Jericho iv. Fig 135.18. (H.IX.xxxiia)	MBII
V	Jericho v. Fig 180.5. (H.XI.xlvii)	MBII
V	Lachish. Pl 87.1017. (L.4402)	MBIII
V	Sali. Pl 41.8. Str XII	MBII
V	Shec. Pl 32.b. (JI.11). Str XVII	MBII
V	Shec. Pl 40.j. (JJ.13). Str XVIII	MBII
V	TBM 33. Pl 14.4. Str D	MBIII

0705 210
DCR 1689
Sjar 112

P	Aphek 75. Fig 1.1. (Str AVII)	MBI
P	Hazor i. Pl CXII.13. (D)	MBII
P	Jericho v. Fig 173.6. (H.IX.xxxvc)	MBII
P	Sali. Pl 42.83. Str XII	MBII
P	Tyre. Pl LVI.1. Str XX	EBIV
V	Hazor iii-iv. Pl CCLIX.20. (H Str 3)	MBIII
V	Jawa. Fig 156.652	MBI
V	Jericho iv. Fig 116.32. (H.IX.xxxiv)	MBII
V	Jericho iv. Fig 138.9. (H.VI.xx-xxi)	MBI
V	Meg. Pl 36.2. Str XI	MBIII
V	Shec. Pl 34.b. (JI.41). Str XIX	MBII

0705 210
DCR 1709
Sjar 084

P	Aphek 75. Fig 1.1. (Str AVII)	MBI
P	Hazor i. Pl CXII.13. (D)	MBII
P	Meg. Pl 13.5. Str XIV	MBI
P	Sali. Pl 42.83. Str XII	MBII
P	Tyre. Pl LVI.1. Str XX	EBIV
V	Hazor iii-iv. Pl CCLIX.20. (H Str 3)	MBIII
V	Jawa. Fig 156.652	MBI
V	Jericho iv. Fig 116.32. (H.IX.xxxiv)	MBII
V	Jericho iv. Fig 138.9. (H.VI.xx-xxi)	MBI
V	Jericho v. Fig 198.12,14. (H.XIII.liii)	MBIII**(x2)**
V	Meg. Pl 36.2. Str XI	MBIII
V	Shec. Pl 34.b. (JI.41). Str XIX	MBII

0705 210
DCR 1738
Sjar 056

P	Aphek 75. Fig 5.15,16. (Str A IV)	MBI**(x2)**
P	Gezer ii. Pl 13.3. (FI Str 9B)	MBII
P	Hazor i. Pl CXIII.1. (D)	MBII
P	Hazor ii. Pl CXIV.15. (C Str 3)	MBIII
P	Hazor iii-iv. Pl CCXXXV.31. (BA Str 14)	MBII
P	Jawa. Fig 156.642	MBI
P	Jericho iv. Fig 136.26. (H.XIII.liii)	MBIII
P	Jericho v. Fig 166.14. (H.IX.xxxa)	MBII
P	Lachish. Pl 87.1017. (L.4402)	MBII
P	Meg. Pl 8.4. Str XV	MBI
P	Sali. Pl 41.8. Str XII	MBII
P	TBM 33. Pl 14.4. Str D	MBIII
V	Bosra. Pl 11.6. Str 28	MBI
V	Shec. Pl 40.l. (JJ.13). Str XVIII	MBII

0705 210
DCR 1745
Sbowl 001

P	Jericho iv. Fig 112.14. (H.XI.xlixa)	MBIII
P	Lachish. Pl 66.430. (L.2278)	EBIV
P	TBM 33. Pl 3.7. Str H	EBIV
V	Aphek 85. Fig 5.8. (Str X17)	MBI
V	Gezer ii. Pl 21.13. (FI GL8012)	MBIII
V	Gezer ii. Pl 14.8. (FI Str 9A)	MBII
V	Mev. Fig 16.3. Str XV	MBI

0705 212
DCR 1286
Sjar 085

P	Hazor i. Pl XCIII.17. (D Str 4)	MBII
V	Hazor i. Pl CXII.11. (D)	MBII
V	Hazor iii-iv. Pl CCLXXXVI.9. (K Str 4)	MBII
V	Jericho iv. Fig 134.15. (H.XI.xlvi-ii)	MBIII
V	Mev. Fig 13.1. Str XIII	MBII
V	Sali. Pl 42.101. Str XII	MBII

0705 212
DCR 1759
Sbowl 037

V	Sali. Pl 40.66. Str XI	MBIII

0705 213
DCR 312
Sjar 034

V	Aphek 75. Fig 5.14. (Str A IVb)	MBI
V	Gezer ii. Pl 14.26,27. (FI Str 9A)	MBII**(x2)**
V	Gezer ii. Pl 20.9. (FI GL8012)	MBIII
V	Jericho iv. Fig 131.20. (H.XIII.liii)	MBIII
V	Jericho v. Fig 106.4. (Tr III.XXI.lxxxi)	MBI
V	Jericho v. Fig 170.3. (H.IX.xxxiii)	MBII
V	Jericho v. Fig 171.8. (H.IX.xxxiva)	MBII
V	Mev. Fig 16.10. Str XV	MBI
V	Sali. Pl 40.66. Str XI	MBIII

0705 215
DCR 1033
Cpot 016

P	Jawa. Fig 154.622	MBI

0705 215
DCR 1786
Sjar 126

P	Hazor iii-iv. Pl CXCII.17. (A Tr500)	MBII
V	Bosra. Pl 13.70. Str 25	MBI
V	Jericho v. Fig 22.30. (Tr I.XLIV.lxvi)	MBIII

0705 215
DCR 1787
Sjar (tnn) 050

P	TBM 33. Pl 7.9. Str E	MBII
V	Bosra. Pl 12.29. Str 26	MBI
V	Bosra. Pl 22.272. Str 19	MBIII
V	Jericho iv. Fig 137.12. (H.XII.l-li)	MBIII
V	Lachish. Pl 87.1006. (L.619)	MBII
V	Meg. Pl 32.1. Str XI	MBIII**0705**

215DCR 1788
Sjar (tnn) 023

P	Sali. Pl 42.41. Str XII	MBII
P	TBM 33. Pl 7.1. Str E	MBII

- P Tyre. Pl LVI.3. Str XX — EBIV
- V Hazor i. Pl CXIII.1. (D) — MBII
- V Jawa. Fig 156.644 — MBI
- V Jericho iv. Fig 137.33. (Tr I.XLIII.lvii) — MBIII
- V Shec. Pl 40.l. (JJ.13). Str XVIII — MBII

0705 215
DCR 1789
Sbowl 030

- P Sali. Pl 42.98. Str XII — MBII
- V Hazor i. Pl CXII.11. (D) — MBII

0705 215
DCR 1791
Xjar 062

- P Hazor iii-iv. Pl CCXXXV.32. (BA Str 14) — MBII
- P Jericho iv. Fig 129.23. (H.XIII.liii) — MBIII
- P TBM 33. Pl 7.7. Str E — MBII
- V Bosra. Pl 12.29. Str 26 — MBI
- V Bosra. Pl 22.272. Str 19 — MBIII
- V Gezer ii. Pl 18.4. (FI Tr 4168) — MBII
- V Lachish. Pl 87.1006. (L.619) — MBI
- V Sali. Pl 40.43. Str XII — MBIII
- V Sarepta. Pl 20.15. Str L — MBIII
- V Shec. Pl 32.a. (Jl.11). Str XIX — MBII
- V TBM 32. Pl 41.14. Str D — MBIII
- V Tyre. Pl LVI.1. Str XX — EBIV

0705 215
DCR 2173
Xjar 077

- V Gezer ii. Pl 18.2. (FI Tr4168) — MBII
- V Hazor i. Pl XCIII.20. (D Str 4) — MBII
- V Jericho iv. Fig 151.11. (H.XIII.li-liii) — MBIII
- V Mev. Fig 14.27. Str XIII — MBII
- V Sali. Pl 41.1. Str XII — MBII

0705 216
DCR 2
Sbowl 001

- P Lachish. Pl 66.430. (L.2278) — EBIV
- P TBM 33. Pl 3.7. Str H — EBIV
- V Aphek 85. Fig 5.8. (Str X17) — MBI
- V Gezer ii. Pl 21.13. (FI GL8012) — MBIII
- V Gezer ii. Pl 14.8. (FI Str 9A) — MBII
- V Mev. Fig 16.3. Str XV — MBI

0705 216
DCR 1641
Sjar 108

- P Bosra. Pl 12.40. Str 26 — MBI
- P Meg. Pl 12.22. Str XIV — MBII
- P Meg. Pl 13.5. Str XIV — MBII
- V Bosra. Pl 13.71. Str 25 — MBII
- V Jawa. Fig 156.653 — MBI
- V Jericho iv. Fig 136.27. (H.XIII.liii) — MBIII
- V Lachish. Pl 87.1020. (L.3621) — MBIII
- V Mev. Fig 16.16. Str XIV — MBI

0705 216
DCR 1897
Xjar 068

- P Jericho iv. Fig 125.2. (H.VIII.xxviia) — MBII
- P Jericho v. Fig 173.5. (H.X.xliia) — MBII
- P Shec. Pl 8.a. (Bd C.13). Str XVIII — MBII
- V Hazor i. Pl CXII.15. (D) — MBII
- V Jawa. Fig 155.636,637 — MBI**(x2)**
- V Jericho v. Fig 167.7,8. (H.IX.xxxia) — MBII**(x2)**

0705 216
DCR 1899
Sjar 136

- P Hazor iii-iv. Pl CCXLVI.14. (G) — MBII
- P Shec. Pl 32.I. (Jl.21). Str XVIII — MBII
- P Shec. Pl 35.b. (Jl.43). Str XVIII — MBII
- P Shec. Pl 35.c,d. (Jl.43). Str XX — MBII**(x2)**
- V Aphek 75. Fig 7.14. (Str A IVa) — MBI
- V Arqa 78. Fig 4 — MBII
- V Gezer i. Pl 31.16. (FI Str 8) — MBIII
- V Gezer ii. Pl 20.13. (FI L8012) — MBIII
- V Hazor i. Pl XCIV.14. (D Str 3) — MBIII
- V Jericho iv. Fig 127.28. (H.XI.xlv) — MBII
- V Jericho iv. Fig 132.3. (H.VI.xx-xxi) — MBI
- V Lachish. Pl 87.1024,1025. (L.3970,L.3694) — MBIII**(x2)**
- V Meg. Pl 12.18. Str XIV — MBII
- V Sarepta. Pl 20.13. Str L — MBIII

0705 217
DCR 1485
Sjar 103

- P Bosra. Pl 11.6. Str 28 — MBI
- P Gezer ii. Pl 12.21. (FI Str 9B) — MBII
- P Gezer ii. Pl 11.10. (FI Str 10) — MBII
- P Hazor iii-iv. Pl CXCII:18. (A Tr500) — MBII
- P Jericho iv. Fig 137.32. (H.IX.xxxiv) — MBII
- V Aphek 85. Fig 4.17. (Str X18) — MBI
- V Arqa 78. Fig 49.1. Str 11 — MBIII
- V Bosra. Pl 12.39. Str 28 — MBI
- V Bosra. Pl 13.71. Str 25 — MBII
- V Jawa. Fig 156.641,643 — MBI**(x2)**
- V Jericho iv. Fig 135.18. (H.IX.xxxiia) — MBII
- V Jericho v. Fig 180.5. (H.XI.xlvii) — MBII
- V Lachish. Pl 87.1017. (L.4402) — MBIII
- V Sali. Pl 41.8. Str XII — MBII
- V Shec. Pl 32.b. (Jl.11). Str XVII — MBII
- V Shec. Pl 40.j. (JJ.13). Str XVIII — MBII
- V TBM 33. Pl 14.4. Str D — MBIII

0705 217
DCR 1850
Xjar 065

- P Shec. Pl 40.g,h. (JJ.12). Str XX-XIX — MBII**(x2)**
- V Bosra. Pl 12.20. Str 26 — MBI
- V Lachish. Pl 87.1015. (L.5193) — MBIII

0705 219
DCR 1866
Xjar 066

- P Aphek 75. Fig 4.20,21. (Str A IVb) — MBI**(x2)**
- P Aphek 75. Fig 6.17. (Str A IVb) — MBI
- P Aphek 85. Fig 4.8. (Str X17) — MBI
- P Aphek 85. Fig 5.10. (Str X17) — MBI
- P Gezer ii. Pl 17.10. (FI Tr 4168) — MBII
- P Hazor iii-iv. Pl CCXLVI:9. (G) — MBII
- P TBM 33. Pl 4.214. Str G-F — MBI
- V Aphek 75. Fig 2.11,13. (Str A VI) — MBI**(x2)**
- V Hazor i. Pl CXVI.2. (D Str 4) — MBII
- V Jawa. Fig 155,636,637 — MBI**(x2)**
- V Shec. Pl 7.e. (Bd C.II). Str XVII — MBII

Appendix D: Rukeis, Area 08, Comparative Pottery Analysis

0708 058
DCR 2043
Sbowl 001

V Jericho iv. Fig 112.14. (H.XI.xlixa) MBIII
V Aphek 85. Fig 5.8. (Str X17) MBI
V Gezer ii. Pl 21.13. (F1 GL8012) MBIII
V Gezer ii. Pl 14.8. (F1 Str 9A) MBIII
V Lachish. Pl 66.430. (T.2001) EBIV
V Mev. Fig 16.3. Str XV MBI
V TBM 33. Pl 3.7. Str H EBIV

0708 058
DCR 2044
Sjar 077

V Jericho iv. Fig 135.30. (H.VIII.xxviiia) MBII

0708 058
DCR 2045
Bowl 044

V Jericho iv. Fig 103.28. (H.XII.lii-liii) MBIII
V KEL. T12. IC14. Str 10 MBIII

0708 058
DCR 2061
Sjar 142

P Ghassil. Pl 22.12. Str X MBII
V Ghassil. Pl 37.66. Str IX MBIII
V Jericho iv. Fig 126.19. (H.XIII.liii) MBIII
V Mev. Fig 14.27. Str XIII MBIII
V Ghassil. Pl 36.22. Str IX MBIII
V Hama. Fig 14.27. (J3) EBIV
V Mev. Fig 11.13. Str XII MBIII

0708 058
DCR 2062
Sjar (tnn) 018

P Aphek 75. Fig 5.16. (Str A IVb) MBI
P Aphek 75. Fig 15.13. (Str B Vb) MBI
P Bosra. Pl 11.6. Str 28 MBI
P Gezer ii. Pl 13.3. (FI Str 9B) MBII
P Hazor i. Pl CXVII.1. (D. Str 5) MBII
P Hazor i. Pl CXVII.2. (D. Str 5) MBII
P Hazor i. Pl CXVII.4. (D. Str 5) MBII
P Hazor iii-iv. Pl CCXXXV.31. Str 14 MBII
V Bosra. Pl 16.115. Str 19 MBIII
V Jericho iv. Fig 137.12. (H.XII.l-li) MBIII
V Lachish. Pl 87.1013. (Sjar Class C) MBII

0708 058
DCR 2063
Sjar 103

P Bosra. Pl 11.6. Str 28 MBI
P Gezer ii. Pl 12.21. (FI Str 9B) MBII
P Hama. Fig 64.3G120. (J6) EBIV
P Jawa. Fig 156.641,643 MBI**(x2)**
P Lachish. Pl 87.1017. (L.4402) MBII
P Sali. Pl 41.8. Str XII MBII
P Shec. Pl 32. (J1.11b). Str XVII MBII
V Jericho iv. Fig 135.19. (H.IX.xxxii-xxxiii) MBII
V TBM 33. Pl 14.4. Str D MBIII

0708 058
DCR 2064
Xjar 174

V Hazor iii-iv. Pl CLVI.16. Str XVIII EBIV
V Hazor iii-iv. Pl CLVI.17. Str XVIII EBIV

0708 058
DCR 2065
Sbowl 001

V Jericho iv. Fig 112.14. (H.XI.xlixa) MBIII
V Aphek 85. Fig 5.8. (Str X17) MBI
V Gezer ii. Pl 21.13. (F1 GL8012) MBIII
V Gezer ii. Pl 14.8. (F1 Str 9A) MBIII
V Lachish. Pl 66.430. (T.2001) EBIV
V Mev. Fig 16.3. Str XV MBI
V TBM 33. Pl 3.7. Str H EBIV

0708 058
DCR 2067
Sjar 155

V Bosra. Pl 12.31. Str 26 MBII
V Gezer ii. Pl 13.21. (FI Str 9B) MBII

0708 058
DCR 2068
Sjar 123

P Mev. Fig 13.4. Str XIII MBII
V KEL. T 18. IC14. Str 10 MBIII
V Sali. Pl 40.65. Str XI MBIII
V Alalakh. Pl 26.55. Str VIII MBII
V Hazor iii-iv. Pl CLVI.22. Str XVII MBII
V Mev. Fig 14.3. Str XIII MBII

0708 058
DCR 2075
Bowl 029

P Bosra. Pl 13.53. Str 26 MBI
P Dan. Fig 4.92.5. (T.4663). Str X MBII
P Hazor v. Fig III.9.1. (A Str 9C) MBII
P Hazor v. Fig III.12.8. (A Str 9C) MBII
P Jericho v. Fig 106.7. (Tr III.XXI.lxxxvi) MBI
P Jericho v. Fig 177.15. (H.X.xlia) MBII
P Jericho v. Fig 181.16. (H.XI.xlixa) MBII
P Meg. Pl 14.14. Str XIV MBI
P Meg. Pl 29.24. Str XII MBII
P Sali. Pl 43.TS195. Str XII MBII
P Shec. Pl 3.h. (Bp.41). Str XVIII MBII
V Gezer ii. Pl 16.10. (FI. Tr 4168) MBIII
V Alalakh. Pl 55.6. Str X MBI
V Dan. Fig 4.92.3. (T.4663). Str X MBII
V Gezer i. Pl 31.19. (FI. Str 8) MBIII
V Gezer ii. Pl 18.17. (FI. Tr 4168) MBIII
V Ghassil. Pl 13.18. Str X MBII
V Ghassil. Pl 28.7. Str IX MBIII
V Jericho iv. Fig 103.1. (H.XI.xlvi-xlvii) MBIII
V Jericho v. Fig 175.3. (H.X.xxxviia) MBII
V Jericho v. Fig 176.1. (H.X.xxxvii) MBII
V Jericho v. Fig 177.1. (H.X.xxxxviii) MBII
V Jericho v. Fig 178.11. (H.XI.xliva) MBII
V Jericho v. Fig 186.5. (H.XII.li) MBII
V Jericho v. Fig 189.3. (H.XII.lii) MBIII
V Jericho v. Fig 189.8. (H.XII.lii) MBIII
V Jericho v. Fig 190.1. (H.XII.lii) MBIII
V Meg. Pl 9.4. Str XV MBI
V Meg. Pl 37.21. Str XI MBII
V Meg. Pl 38.2. Str XII MBII
V Mev. Fig 15.12. Str XIV MBI

0708 058
DCR 2176
Xjar 004

P Bosra. Pl 11.5. Str 28 MBI
V TNM. Fig 37.4. Ph G MBIII
V TNM. Fig 37.5. Ph G MBIII

0708 060
DCR 2095
Sjar 157

P Hazor v. Fig II.6.26. (A Str 9D) MBI
V Gezer ii. Pl 15.11. (FI Tr 4168) MBIII
V Hazor. iii-iv. Pl CXCVII.18 MBII
V Sali. Pl 42.102. Str XII MBII

0708 060
DCR 2098
Fjar 028

P Aphek 75. Fig 11.4. (Str A.II) MBI
P Hazor i. Pl CXXI.12 (D Str 4) MBII
V Sali. Pl 40.12. Str XI MBIII
V Sali. Pl 40.13. Str XI MBIII
V Sali. Pl 40.14. Str XI MBIII
V Hazor i. Pl CXXI.1 (D Str 4) MBII
V Hazor i. Pl CXXI.11 (D Str 4) MBII
V Hazor v. Fig III.6.8. (L) MBII
V Megiddo. Pl 33.12. Str XI MBII
V Megiddo. Pl 41.7. Str XIII-X MBII

0708 060
DCR 2099
Sjar 158

P Bosra. Pl 11.13. Str 28 MBI
V TNM. Fig 47.7. Ph G MBIII
V Bosra. Pl 12.37. Str 26 MBI
V Gezer ii. Pl 13.28. (FI Str 9B) MBII
V Hazor iii-iv. Pl CCLXXXVI.6. (K Str 4) MBII

0708 062
DCR 2093
Sjar 157

P Hazor v. Fig II.6.26. (A Str 9D) MBI
V Gezer ii. Pl 15.11. (FI Tr 4168) MBIII
V Hazor. iii-iv. Pl CXCVII.18 MBII
V Sali. Pl 42.102. Str XII MBII

0708 062
DCR 2094
Sjar (tnn) 009

V Ghassil. Pl 36.28 MBIII
V Jericho v. Fig 200.1,2. (H.XII.lii) MBIII
V Lachish. Pl 87.1007. (Sjar Class A) MBIII
V Jericho iv. Fig 135.1. (H.VI.xx-xxi) MBI
V Sali. Pl 40.47. Str XI MBIII

0708 063
DCR 2080
Sjar 068

V Jericho v. Fig 72.3. (Tr II.XXII.lxxii) MBIII
V Jericho iv. Fig 147.9. (H.XIII.lii-liii) MBIII
V Mev. Fig 16.19. Str XIV MBI

0708 063
DCR 2081
Sjar 156

V Jericho iv. Fig 148.1. (H.XII.l-li) MBIII
V Ghassil. Pl 22.15. Str X MBII

0708 063
DCR 2082
Sjar 023

0708 063
DCR 2088
Xjar 2088

V Arqa 91. Fig 6.4. Str 16-15 EBIV
V Hazor i. Pl C.24. (D. Str 3) MBIII

0708 063
DCR 2089
Sbowl 001

V Jericho iv. Fig 112.14. (H.XI.xlixa) MBIII
V Aphek 85. Fig 5.8. (Str X17) MBI
V Gezer ii. Pl 21.13. (F1 GL8012) MBIII
V Gezer ii. Pl 14.8. (F1 Str 9A) MBIII
V Lachish. Pl 66.430. (T.2001) EBIV
V Mev. Fig 16.3. Str XV MBI
V TBM 33. Pl 3.7. Str H EBIV

0708 063
DCR 2090
Sbowl 008

P Hazor v. Fig II.6.7. (A Str 9D) MBII
P Jericho iv. Fig 104.2. (H.IX.xxxa) MBII
P Shec. Pl 2.c. (Bp.21). Str XX MBII
V Bosra. Pl 18.178. Str 19 MBIII
V Dan. Fig 4.95.11. (T 187b). Str IX MBIII
V Dan. Fig 4.94.9. (T.187a). Str IX MBIII
V Gezer ii. Pl 16.11. (Fl. Tr 4168) MBIII
V Hazor i. Pl XCIV.2. (D Str 3) MBIII
V Hazor v. Fig II.7.9. (A Str 9D) MBII
V Jericho v. Fig 166.5. (H.IX.xxxa) MBII
V Meg. Pl 21.21 Str XIV-X MBII

0708 063
DCR 2091
Bowl 041

P Aphek. 75. Fig 6.11. (Str AIVb) MBI
P Aphek. 75. Fig 8.10. (Str AIVa) MBI
P Hazor v. Fig II.6.6. (A Str 9D) MBI
P Hazor v. Fig II.6.9. (A Str 9D) MBI
P Jericho iv. Fig 105.6. (H.X.xliia) MBIII
P Jericho v. Fig 183.3. (H.XI.l) MBII
P Jericho v. Fig 184.4. (H.XI.l) MBII
P Meg. Pl 14.9. Str XIV MBI
P Meg. Pl 19.11. Str XIV-XIVA MBI
P Meg. Pl 29.21. Str XIII-X MBII
P Shec. Pl 2.l. (Bp.22). Str XX MBII
P Shec. Pl 2.m. (Bp. 22). Str XIX MBII
V Dan. Fig 4.92.4. (T.4663). Str X MBII
V Dan. Fig 4.104.1. (T.1025). Str XII MBI
V Gezer i. Pl 31.19. (Fl. Str 8) MBIII
V Gezer ii. Pl 18.14. (Fl. Tr 4168) MBIII
V Hazor i. Pl CXIX.2. (D5 Str 4) MBII
V Jericho v. Fig 72.1. (Tr II.XXII.lxxi) MBIII
V Jericho v. Fig 106.27. (Tr III.XXI.lxxxvii) MBI
V Jericho v. Fig 172.2. (H.IX.xxxvb) MBII
V Jericho v. Fig 177.16. (H.X.xlia) MBII
V Jericho v. Fig 181.23. (H.XI.xlixa) MBII

V Jericho v. Fig 189.14. (H.XIII.liii) MBIII
V Jericho v. Fig 190.4,5. (H.XII.lii) MBII**(x2)**
V KEL. T179. IE15. Str 7 MBII
V Meg. Pl 19.10. Str XV-XIIIA MBI

0708 063
DCR 2092
Xjar 037

P KEL. T288. IIIA16. Str 4 MBI
P Tyre. Pl LV.16. (Str XIX) MBII
V Hazor iii-iv. Pl CXCVIII.9 MBIII
V Mev. Fig 14.14. Str XIII MBII
V Sali. Pl 40.42. Str XI MBIII
V Shec. Pl 30.b. (Jg.B). Str XIX MBII
V Shec. Pl 30.c. (Jg.B). Str XIX MBII
V Shec. Pl 40.f. (JJ.12). Str XIX MBII
V Shec. Pl 40.g. (JJ.12). Str XIX MBII
V Shec. Pl 40.h. (JJ.12). Str XX MBII
V Shec. Pl 47.x. (JJ.12). Str XIX MBII

0708 063
DCR 2101
Sjar 196

P Jericho iv. Fig 122.12. (H.XI.xliv) MBII
V Jericho v. Fig 176.11,13. (H.X.xxxvii) MBII**(x2)**

0708 063
DCR 2102
Xjar 025

P Sali. Pl 42.83. Str XII MBII
V Jericho iv. Fig 134.14. (H.X.xxxviia) MBII
V TNM. Fig 39.2. Ph G MBIII

0708 064
DCR 2135
Sjar 163

0708 064
DCR 2138
Xjar 041

V Bosra. Pl 15.111. Str 20 MBIII

0708 064
DCR 2147
Sjar 107

P Bosra. Pl 14.75,6. Str 23 MBII
P Jericho iv.Fig 94.7,9-12,15.(Tr I.XLII.iv-lviii; Tr II.XI.lxviii; TrIII.XXI.lxxxi) EBIV**(x6)**
P Jericho v. Fig 108.38. (Tr III.XXI.lxxxv) MBI
P Sali. Pl 42.70. Str XII MBII
V Sali. Pl 40.46. Str XI MBIII
V Alalakh. Pl 61.55. Str X MBII
V Alalakh. Pl 64.68-71. Str X MBII
V Aphek 75. Fig 5.13. (A IVb) MBI
V Aphek 75. Fig 7.20. (A Iva) MBI
V Bosra. Pl 15.109. Str 20 MBIII
V Dan. Fig 4.103. (T.902d). Str XII MBI
V Dan. Fig 4.96.9. (T.367). Str IX MBIII
V Hazor i. Pl CXII.14. (D) MBIII
V Hazor i. Pl CXIII.4. (D) MBIII
V Hazor ii. Pl CX.2. (C. Str 3) MBIII
V Hazor ii. Pl CX.6-8. (C. Str 3) MBIII**(x3)**
V Hazor iii-iv. Pl CCXXXIX.17. (F) MBII
V Jawa. Fig 13.34 MBI
V KEL. CT14. ID15. Str 8 MBII

V KEL. T307. IJ15. Str 4 MBI
V Lachish. Pl 66.397,404. (L.1529) EBIV**(x2)**
V Shec. Pl 38.f. (Jl). Str XVIII MBII
V TBM 33. Pl 3.3,6. Str H EBIV**(x2)**

0708 064
DCR 2148
Sjar (tnn) 026

P Meg. Pl 18.3. Str XIIIA MBI
P Shec. Pl 41.n. (JJ.22) Str XVIII MBII
V Gezer ii. Pl 20.1. (FI. GL8012) MBIII
V Hazor v. Fig V.5.26. (P. Ph C) MBIII
V Jericho iv. Fig 127.3. (H.IX.xxxiva) MBII
V KEL. T36. Str 10 MBIII
V KEL. T279. IJ15. Str 5a MBI
V KEL. T292. IE155. Str 7 MBII
V Meg. Pl 12.18. Str XIV MBI
V Mev. Fig 15.20. Str XIV MBI
V Shec. Pl 33.a. (Jl.23). Str XVII MBII
V Shec. Pl 33.b. (Jl.23). Str XVIII MBII
V Shec. Pl 33.d. (Jl.23). Str XVII MBII
V TBM 33. Pl 14.10. Str D MBIII

0708 064
DCR 2149
Fjar 018

P Bosra. Pl 12.17. Str 26 MBI
P Jericho v. Fig 178.17. (H.X.xliia) MBII
P Jericho v. Fif 183.16. (H.XI.la) MBII
P Meg. Pl 6.7. Str XVI EBIV
P Sali. Pl 42.76. Str XII MBII
P Tyre. Pl LVIIA.16. Str XXI-XXV EBIV
V Bosra. Pl 16. 121. Str 18 MBIII
V Bosra. Pl 18. 170. Str 19 MBIII
V Jawa. Fig 155.636 MBI
V Jericho v. Fig 161.1,2. (H.VII.xxii) MBII**(x2)**
V Sali. Pl 40.9. Str XI MBIII

0708 065
DCR 2137
Sjar 093

P Aphek 75. Fig 1.15. (Str A VII) MBI
P Aphek 75. Fig 3.1. (Str A VI) MBI
P Gezer ii. Pl 20.1. (FI GL8012) MBIII
P Ghassil. Pl 7.13. Str XI MBI
P Jericho iv. Fig 127.25. (H.IX.xxixa) MBII
P KEL. T42.IC14. Str 10 MBIII
P Meg. Pl 18.3. Str XIIIA MBI
P Mev. Fig 16.8. Str XV MBI
P Shec. Pl 33.G. (Jl.23). Str XVIII MBII
V Aphek 75 Fig 5.4. (Str A IVb) MBI
V Bosra. Pl 12.30. Str 26 MBI
V Hazor iii-iv. Pl CCXLVI.15 MBIII
V Shec. Pl 33.D. (Jl.23). Str XVII MBII

0708 066
DCR 2108
Sjar 074

P Aphek 85. Fig 4.13. (Str XI) MBI
P Bosra. Pl 15.96. Str 20 MBIII
P Hazor v. Fig III.6.2. (L) MBII
P Hazor v. Fig V.1.11. (P. Ph C) MBIII
P Jericho iv. Fig 128.29. (H.VII.xxiva) MBII
P Jericho iv. Fig 131.4. (H.XII.l-li) MBIII
P Mev. Fig 14.24. Str XIII MBII
P Sarepta. Pl 20.13. Str L MBIII
P Shec. Pl 41.g. (JJ.21). Str XX MBII
V Aphek 75. Fig 1.14. (Str A VII) MBI

V Gezer ii. Pl 20.1. (FI GL8012) MBIII
V Jericho v. Fig 171.10. (H.IX.xxxiva) MBII
V Jericho v. Fig 176.18. (H.X.xxxviii) MBII
V Jericho v. Fig 185.2. (H.XI.l) MBII
V Lachish. Pl 87.1025. (L.216). Class E MBIII
V Shec. Pl 41.e. (JJ.21). Str XIX MBII

0708 066
DCR 2110
Sjar 133

P Aphek 75. Fig 2.7. (Str A VI) MBI
P Aphek 75. Fig 6.15. (Str A IVb) MBI
P Hazor i. Pl XCIII.23. (D1 Str 4) MBII
P Shec. Pl 10.6 (Bd C.32). Str XVII MBII
V KEL. T13. I.C14. Str 10 MBIII
V Mev. Fig 11.13. Str XII MBIII
V Shec. Pl 7.6. (Bd B.23). Str XIX MBII

0708 066
DCR 2111
Sjar 160

P Mev. Fig 11.13. Str XII MBIII
P Sali. Pl 42.99. Str XII MBII
V Aphek 75. Fig 16.2. (Str B IV) MBI
V Shec. Pl 10.d. (Bd C.32). Str XVIII MBII

0708 066
DCR 2112
Sjar 115

P Sali. Pl 40.39. Str XI MBIII
V Ghassil. Pl 9.9. Str XI MBI
V Jericho iv. Fig 137.4. (H.VI.xx-xxi) MBI
V Jericho v. Fig 170.3. (H.IX.xxxiiii) MBII
V Jericho v. Fig 171.8. (H.IX.xxxiva) MBII
V TNM. Fig 48.7. Ph G MBIII

0708 066
DCR 2114
Bowl 019

P Jawa. Fig 11.17 MBI
P TBM 32. Pl 41.8. Str D MBIII

0708 066
DCR 2116
Sjar 161

0708 066
DCR 2117
Sbowl 039

0708 066
DCR 2155
Xjar 037

P KEL. T288. IIIA16. Str 4 MBI
P Tyre. Pl LV.16. (Str XIX) MBII
V Hazor iii-iv. Pl CXCVIII.9 MBIII
V Mev. Fig 14.14. Str XIII MBII
V Sali. Pl 40.42. Str XI MBIII
V Shec. Pl 30.b. (Jg.B). Str XIX MBII
V Shec. Pl 30.c. (Jg.B). Str XIX MBII
V Shec. Pl 40.f. (JJ.12). Str XIX MBII
V Shec. Pl 40.g. (JJ.12). Str XIX MBII
V Shec. Pl 40.h. (JJ.12). Str XX MBII
V Shec. Pl 47.x. (JJ.12). Str XIX MBII

0708 067
DCR 2126
Sjar 048

P Hazor iii-iv. Pl CCLXXXVI.9. (K. Str 4) MBII
P Mev. Fig 13.3. Str XIII MBII
V Sali. Pl 40.67. Str XI MBIII
V Bosra. Pl 11.11. Str 28 MBI
V Hazor iii-iv. Pl CCXXXV.15. (BA. Str 15) EBIV
V Sali. Pl 41.9. Str XII MBII
V Shec. Pl 9.c. (Bd C.24). Str XIX MBII

0708 067
DCR 2127
Sjar (tnn) 026

P Meg. Pl 18.3. Str XIIIA MBI
P Shec. Pl 41.n. (JJ.22) Str XVIII MBII
V Gezer ii. Pl 20.1. (FI. GL8012) MBIII
V Hazor v. Fig V.5.26. (P. Ph C) MBIII
V Jericho iv. Fig 127.3. (H.IX.xxxiva) MBII
V KEL. T36. Str 10 MBIII
V KEL. T279. IJ15. Str 5a MBI
V KEL. T292. IE155. Str 7 MBII
V Meg. Pl 12.18. Str XIV MBI
V Mev. Fig 15.20. Str XIV MBI
V Shec. Pl 33.a. (JI.23). Str XVII MBII
V Shec. Pl 33.b. (JI.23). Str XVIII MBII
V Shec. Pl 33.d. (JI.23). Str XVII MBII
V TBM 33. Pl 14.10. Str D MBIII

0708 067
DCR 2128
Bowl 001

V Meg. Pl 15.18. Str XIV MBI

0708 067
DCR 2131
Sjar 044

V Jericho iv. Fig 147.1. (H.VII.xxiva) MBI
V Jericho iv. Fig 94.13,14. (Tr I.XLII.lvi-ii) EBIV**(x2)**
V Jericho v. Fig 176.13. (H.X.xxxvii) MBII

0708 067
DCR 2132
Fjar 018

P Bosra. Pl 12.17. Str 26 MBI
P Jericho v. Fig 178.17. (H.X.xliia) MBII
P Jericho v. Fif 183.16. (H.XI.la) MBII
P Meg. Pl 6.7. Str XVI EBIV
P Tyre. Pl LVIIA.16. Str XXI-XXV EBIV
P Sali. Pl 42.76. Str XII MBII
V Sali. Pl 40.9. Str XI MBIII
V Bosra. Pl 16. 121. Str 18 MBIII
V Bosra. Pl 18. 170. Str 19 MBIII
V Jawa. Fig 155.636 MBI
V Jericho v. Fig 161.1,2. (H.VII.xxii) MBII**(x2)**

0708 067
DCR 2154
Sjar 024

P Dan. Fig 4.79. (T.902d). Str XII MBI
P Hama. Fig 120. Z0884. (H3) MBII
P Hazor v. Fig III.7.6. (L) MBII
P Sali. Pl 41.11. Str XII MBII
P Sali. Pl 42.85. Str XII MBII
V Bosra. Pl 18.176. Str 19 MBIII
V Gezer ii. Pl 16.4. (FI Tr 4168) MBIII

V	Ghassil. Pl 22.15. Str X	MBII
V	Ghassil. Pl 42.16. Str IX	MBIII
V	Hazor i. Pl XCVIII.20. (D2)	MBIII
V	Sali. Pl 40.67. Str XI	MBIII
V	TNM. Fig 45.6. Ph G	MBIII

0708 067
DCR 2155
Sjar 164

P	Hazor v. Fig II.6.24. (A Str 9D)	MBI
V	Sali. Pl 40.26. Str XI	MBIII
V	Sali. Pl 42.80. Str XI	MBIII
V	Ghassil. Pl 20.16. Str X	MBII
V	Ghassil. Pl 36.23. Str IX	MBIII

0708 067
DCR 2156
Sjar 123

P	Mev. Fig 13.4. Str XIII	MBII
V	KEL. T 18. IC14. Str 10	MBIII
V	Sali. Pl 40.65. Str XI	MBIII
V	Alalakh. Pl 26.55. Str VIII	MBII
V	Hazor iii-iv. Pl CLVI.22. Str XVII	MBII
V	Mev. Fig 14.3. Str XIII	MBII

0708 067
DCR 2157
Sjar 164

P	Hazor v. Fig II.6.24. (A Str 9D)	MBI
V	Sali. Pl 40.26. Str XI	MBIII
V	Sali. Pl 42.80. Str XI	MBIII
V	Ghassil. Pl 20.16. Str X	MBII
V	Ghassil. Pl 36.23. Str IX	MBIII

0708 067
DCR 2158
Xjar 074

P	Shec. Pl 33.a. (Jl.23). Str XVII	MBII
P	Shec. Pl 33.b. (Jl.23). Str XVIII	MBII
V	Alalakh. Pl 60.50. Str X	MBII
V	Gezer ii. Pl 18.4. (Fl Tr 4168)	MBIII
V	Hazor i. Pl XCIII.20. (D1 Str 4)	MBII
V	Hazor i. Pl CXI.1. (D4)	MBIII
V	Sali. Pl 42.71. Str XII	MBII
V	Shec. Pl 32.1. (Jl.22). Str XIX	MBII

0708 067
DCR 2159
Bowl 004

P	Hama. Fig 75.3F69. (J5)	EBIV
V	Bosra. Pl 13.48. Str 26	MBI
V	Meg. Pl 6.16. Str XVI	EBIV

0708 067
DCR 2160
Fjar 018

P	Bosra. Pl 12.17. Str 26	MBI
P	Jericho v. Fig 178.17. (H.X.xliia)	MBII
P	Jericho v. Fif 183.16. (H.XI.la)	MBII
P	Meg. Pl 6.7. Str XVI	EBIV
P	Sali. Pl 42.76. Str XII	MBII
P	Tyre. Pl LVIIA.16. Str XXI-XXV	EBIV
V	Bosra. Pl 16. 121. Str 18	MBIII
V	Bosra. Pl 18. 170. Str 19	MBIII
V	Jawa. Fig 155.636	MBI
V	Jericho v. Fig 161.1,2. (H.VII.xxii)	MBII**(x2)**
V	Sali. Pl 40.9. Str XI	MBIII

0708 068
DCR 2164
Xjar 067

P	Shec. Pl 42.b. (JJ.31). Str XIX	MBII
V	Bosra. Pl 12.33. Str 26	MBI
V	Bosra. Pl 12.31. Str 26	MBI
V	Hazor i. Pl XCIV.14. (D1 Str 3)	MBIII
V	Meg. Pl 35.1. Str XI	MBII
V	Shec. Pl 42.c. (JJ.31). Str XVIII	MBII

0708 068
DCR 2166
Sjar 024

P	Bosra. Pl 18.176. Str 19	MBIII
P	Dan. Fig 4.79. (T.902d). Str XII	MBI
P	Hama. Fig 120. Z0884. (H3)	MBII
P	Hazor i. Pl XCVIII.20. (D2)	MBIII
P	Hazor v. Fig III.7.6. (L)	MBII
P	Sali. Pl 40.67. Str XI	MBIII
P	Sali. Pl 41.11. Str XII	MBII
P	Sali. Pl 42.85. Str XII	MBII
V	Gezer ii. Pl 16.4. (Fl Tr 4168)	MBIII
V	Ghassil. Pl 22.15. Str X	MBII
V	Ghassil. Pl 42.16. Str IX	MBIII
V	TNM. Fig 45.6. Ph G	MBIII

Appendix E: Rukeis, Area 14, Comparative Pottery Analysis

0714 029
DCR 2190
Bowl 022

V	TNM. Fig 45.6. Ph G	MBIII
P	Bosra. Pl 11.10. Str 28	MBI
P	Bosra. Pl 13.57. Str 26	MBI
P	Jericho iv. Fig 105.22. (H.IX.xxxiva)	MBII
P	Jericho v. Fig 183.14. (H.XI.la)	MBII
P	KEL. T102. ID15. Str 8	MBII
V	Jericho v. Fig 204.4. (H.XIII.liii)	MBIII
V	Gezer ii. Pl 13.14. (FI Tr4168)	MBIII
V	Alalakh. Pl 37.37. Str IX	MBII
V	Aphek 75. Fig 8.10. (Str A IVa)	MBI
V	Dan. Fig 4.77. (T.1025). Str XIX	MBI
V	Hama. Fig 124.2C926. (H2)	MBIII
V	Hazor v. Fig III.10.1. (L)	MBII
V	Hazor v. Fig III.10.4. (L)	MBII
V	Jericho v. Fig 168.9. (H.IX.xxxiia)	MBII
V	Jericho v. Fig 177.16. (H.X.xlia)	MBII
V	Jericho v. Fig 186.2. (H.XII.li)	MBII
V	Meg. Pl 14.11. Str XIV	MBI
V	Meg. Pl 37.26. Str XIII-XI	MBII
V	Mev. Fig 12.26. Str XIII	MBII
V	TNM. Fig 28.8 Ph G-C	MBIII

0714 035
DCR 2181
Sjar 166

P	Jericho iv. Fig 116.27. (H.IX.xxxiv)	MBII
P	Meg. Pl 13.10. Str XIV	MBI
P	Sali. Pl 42.102. Str XII	MBII
P	TBM 33. Pl 2.8. Str H	EBIV
V	Bosra. Pl 12.34. Str 26	MBI
V	Bosra. Pl 12.35. Str 26	MBI
V	Hama. Fig 110.3D 580. (H)	MBI
V	Hama. Fig 117.3D 583. (H4)	MBII
V	Hazor iii-iv. Pl CLVI.23. (A. Str XVII)	MBII
V	KEL. T87.ID15. Str 10	MBIII
V	Sali. Pl 41.9. Str XII	MBII
V	Shec. Pl 38.d. (Jl). Str XIX	MBII
V	TBM 32. Pl 45.8. Str D	MBIII

0714 035
DCR 2181
Xjar 047

P	Hazor i. Pl CXVII.5. (D5 Str V)	MBII
V	Gezer ii. Pl 12.21. (FI Str 9B)	MBII
V	Hazor v. Fig III.8.23. (L)	MBII
V	Shec. Pl 32.b. (J1.11). Str XVII	MBII
V	TNM. Fig 37.10. Ph G	MBIII

0714 035
DCR 2188
Xjar 020

P	Aphek 75. Fig 5.16. (Str A IVb)	MBI
P	Aphek. 85. Fig 2.16. (Str X20-X19)	MBI
P	Hazor v. Fig II.6.38. (A Str 9D)	MBI
P	Lachish. Pl 78.804	EBIV
P	Meg. Pl 10.3. Str XIV	MBI
P	Meg. Pl 12.16. Str XIV	MBI
V	Alalakh. Pl 55.8. Str X	MBII
V	Aphek 75. Fig. 5.15. (Str A IVb)	MBI
V	Aphek 85. Fig 4.17. (Str X18)	MBI
V	Gezer i. Pl 31.11. (FI. Str 8)	MBII
V	Hama. Fig 64.3. (J6)	EBIV
V	Hama. Fig 75.5. (J5)	EBIV
V	Hazor i. Pl CXVII.1. (D Str 5)	MBII
V	Hazor i. Pl CXVII.4. (D Str 5)	MBII
V	Meg. Pl 12.22. Str XIV	MBI
V	Meg. P 13.1. Str XIV	MBI
V	Meg. Pl 13.5. Str XIV	MBI

0714 035
DCR 2191
XTNN 028

P	Bosra. Pl 12.29. Str 26	MBI
P	Shec. Pl 32.b. (Jl.11). Str XVII	MBII
P	Shec. Pl 49.x. (Jl.11). Str XVII	MBII
V	Hazor v. Fig III.8.15. (L)	MBII
V	Sali. Pl 40.45. Str XI	MBIII

0714 035
DCR 2211
Sjar 191

P	KEL. T307.IJ15. Str 4	MBI
P	Mev. Fig 15.22. Str XIV	MBI
P	Mev. Fig 16.10. Str XV	MBI
P	Sali. Pl 40.5. Str XII	MBII
P	Shec. Pl 32.n. (Jl.22). Str XVIII	MBII
P	Shec. Pl 34.m. (Jl.42). Str XVIII	MBII
P	Shec. Pl 34.g. (Jl.42). Str XIX	MBII
V	Bosra. Pl 15.109. Str 20	MBIII
V	Aphek 75. Fig 7.20. (Str A IVa)	MBI
V	Gezer i. Pl 31.10. (FI Str 8)	MBIII
V	Ghassil. Pl 20.45. Str X	MBII
V	Hama. Fig 120.2. (H3)	MBII
V	Jericho iv. Fig 127.9. (H.IX.xxxa)	MBII
V	Sali. Pl 40.46. Str XI	MBIII
V	Sali. Pl 42.70. Str XII	MBII
V	Shec. Pl 32.k. (Jl.22). Str XIX	MBII
V	Shec. Pl 34.f. (Jl.42). Str XX	MBII
V	Shec. Pl 34.l. (Jl.42). Str XVIII	MBII

0714 035
DCR 2214
Sjar 190

P	Sali. Pl 42.98. Str XII	MBII
V	Mev. Fig 11.22. Str XIII	MBII

0714 037
DCR 2197
JugT 202

P	Jericho iv. Fig. 178.2. (T.B48.131)	MBII
P	Meg. Pl 24.13. Str XII-IX	MBII
P	Shec. Pl 28.l. (Jp.O.62). Str XVIII	MBII
P	Shec. Pl 28.m. (Jp.O.62). Str XVIII	MBII
P	Shec. Pl 28.p. (Jp.O.O.12). Str XIX	MBII
V	Hazor i. Pl CXXI.1. (D5 Str 4)	MBII
V	Meg. Pl 17.4. Str XIIIA - XII	MBI
V	Meg. Pl 19.30. Str XIII - XI	MBI
V	Meg. Pl 32.23. Str XI	MBII
V	Meg. Pl 41.31. Str X	MBIII
V	Bosra. Pl 18.161. Str 19	MBIII
V	Mev. Fig 14.7. Str XIII	MBII
V	TBM 33. Pl 15.4. Str D	MBIII

0714 043
DCR 2199
Cpot 018

P	Jawa. Fig 10.1-7	MBI**(x7)**
V	Bosra. Pl 14.89. Str 22	MBIII

0714 043
DCR 2209
Cpot 018

P	Jawa. Fig 10.1-7	MBI**(x7)**

0714 044
DCR 2223
JugT 204

P	Meg. Pl 24.2. Str XIII-XII	MBI
P	Meg. Pl 24.10. Str XII-XI	MBII
P	Mev. Fig 14.7. Str XIII	MBII
V	Aphek 75. Fig 9.2. (Str A IVa)	MBI
V	Hazor i. Pl CXXI.3. (D5 Str 4)	MBII
V	Hazor i. Pl CXXI.4. (D5 Str 4)	MBII
V	Hazor i. Pl CXXI.7. (D5 Str 4)	MBII
V	Hazor i. Pl CXXI.10. (D5 Str 4)	MBII
V	Dan. Fig 4.93.9. (T.371c). Str IX	MBIII
V	Meg. Pl 17.4. Str XIIIA-XII	MBI
V	Meg. Pl 19.29. Str XIII-X	MBI
V	Meg. Pl 33.5. Str XII-X	MBII
V	Meg. Pl 33.8. Str XIIIA-XI	MBI
V	Meg. Pl 41.4. Str X	MBIII
V	Meg. Pl 41.7. Str XIII-X	MBII

0714 045
DCR 2198
Sjar 180

0714 046
DCR 2224
Xjar 081

V	Gezer ii. Pl 15.6. (FI. Tr 4168)	MBIII
V	Gezer ii. Pl 15.29. (FI. Tr 4168)	MBIII
V	Shec. Pl 35.b. (Jl.43). Str XVIII	MBII

0714 046
DCR 2225
JugT 203

P	Hazor i. Pl CXII.18. (D5)	MBII
P	Shec. Pl 28.l. (Jp.O.62). Str XIX	MBII
P	Shec. Pl 28.m. (Jp.O.62). Str XVIII	MBII
V	Meg. Pl 24.23. Str XII	MBII
V	Meg. Pl 41.4. Str X	MBIII
V	Bosra. Pl 18.161. Str 19	MBIII

0717 017
DCR 9911
Fjar 036

P	Alalakh. Pl 40.68. Str IX	MBII
P	Jericho v. Fig 165.9. (H.VIII.xxixa)	MBII
P	Jericho v. Fig 169.4. (H.IX.xxxii.)	MBII
P	Sali. Pl 43.TS157. Str XII	MBII
P	TNM. Pl 51.6. Ph H-F	MBIII
V	Alalakh. Pl 5.25. Str VII	MBIII**(x2)**
V	Alalakh. Pl 5.19-21. Str VII	MBIII**(x4)**
V	Alalakh. Pl 6.27-30. Str VII	MBIII
V	Aphek 75. Fig 11.2. (Str A II)	MBI
V	Aphek 75. Fig 12.2. (Str A II)	MBI
V	Jericho iv. Fig 118.3. (H.IX.xxxii)	MBII
V	Jericho v. Fig 194.3. (H.XII.lii)	MBII
V	Meg. Pl 36.12. Str XI	MBII
V	Mev. Fig 16.12. Str XIV	MBI
V	Shec. Pl 12.c. (Bd F.11). Str XVIII	MBII
V	Shec. Pl 14.f,g,h,l. (Bn B.2). Str XVIII	MBII**(x4)**
V	TNM. Pl 51.7. Ph G-E	MBIII

0717 019
DCR 9914
Sjar 097

V	Alalakh. Pl 41.76. Str IX	MBII
V	Jericho iv. Fig 151.11. (H.XII.lii)	MBIII
V	Shec. Pl 9.d. (Bd C.24). Str XVIII	MBII

0717 021
DCR 9913
Sbowl 032

V	Hazor v. Fig V.5.6. (P Ph C)	MBIII
V	Alalakh. Pl 55.13. Str XI	MBII
V	Jericho iv. Fig 103.15. (H.XI.li)	MBII
V	Jericho iv. Fig 122.6. (H.XIII.liii)	MBIII
V	Jericho v. Fig 73.15. (Tr II.XXII.lxxiii)	MBII
V	Sali. Pl 40.28. Str XI	MBIII

0717 022
DCR 9903
Xjar 046

V	Alalakh. Pl 41.70. Str IX	MBII
V	Hazor iii-iv. Pl CCLXXXVI.5. (K. Str 4)	MBII

0717 022
DCR 9907
Fjar 039

P	Alalakh. Pl 60.44. Str X	MBII
P	Bosra. Pl 12.18. (Str 26)	MBI
P	Dan. Fig 4.104.13. (T.1025). Str XII	MBI
P	Ghassil. Pl 17.3,4. Str X	MBII
P	Jericho iv. Fig 115.5. (H.VIII.xxvi)	MBII
P	Jericho v. Fig 174.2. (H.IX.xxxvb)	MBII
P	Jericho v. Fig 178.17. (H.X.xliva)	MBII
P	Meg. Pl 15.2. Str XIV	MBI
P	Sali. Pl 43.XII1. Str XII	MBII
P	Shec. Pl 16.d. (Bc B.11). Str XIX	MBII
P	TBM 33. Pl 8.4. Str E	MBII
V	Sali. Pl 40.9. Str XI	MBIII
V	Gezer ii. Pl 15.20. (F1. Tr 4168)	MBIII
V	Alalakh. Pl 40.66. Str IX	MBII
V	Alalakh. Pl 4.17. Str VII	MBIII
V	Dan. Fig 104.7,8,14. (T.1025). Str XII	MBI**(x3)**
V	Gezer ii. Pl 21.14. (F1. Gl 8012)	MBIII
V	Hazor i. Pl XCIII.18. (D1. Str 5)	MBII
V	Jericho iv. Fig 119.2. (Tr II.XXII.lxxii)	MBII
V	Jericho iv. Fig 119.19. (H.XI.xlvi)	MBII
V	Jericho v. Fig 161.1. (H.VII.xxii)	MBII
V	Mev. Fig 11.2. Str XII	MBIII
V	TNM. Pl 51.7. Ph G-E	MBIII

0717 023
DCR 9904
Xjar 031

P	Jericho iv. Fig 135.1,4. (H.VI.xx-xxi)	MBI**(x2)**
V	Alalakh. Pl 84.11. Str XIII	MBI
V	Ghassil. Pl 44.4. Str IX	MBIII
V	Hazor v. Fig III.8.17. (L)	MBII
V	Jericho iv. Fig 139.10. (H.XIII.liii)	MBIII
V	Shec. Pl 40.a. (JJ.11). Str XVIII	MBII
V	TNM. Pl 51.5. Ph G-F	MBIII

0717 023
DCR 9905
Xtnn 017

P	Sali. Pl 42.79. Str XII	MBII
V	TNM Pl 36.5. Ph G-F	MBIII
V	Ghassil. Pl 7.7. Str XI	MBII

V Hazor v. Fig V.1.13. (P. Ph C) MBIII
V KEL. Pl 47.T101. Str 8 MBII
V Sali. Pl 40.36. Str XI MBIII

0717 023
DCR 9906
Sjar 144

V Gezer i. Pl 30.5. (F1 Str 7) MBIII
V Jericho iv. Fig 135.16. (H.XI.l) MBII
V Jericho iv. Fig 137.20. (H.X.xliia) MBII
V Mev. Fig 14.15. Str XIII MBII
V Sali. Pl 42.32. Str XII MBII

0717 024
DCR 9917
Xjar 002

V Alalakh. Pl 41.70. Str IX MBII
V Hazor iii-iv. Pl CCLXXXVI.5. (K. Str 4) MBIII

0717 024
DCR 9922
Sjar 213

V Alalakh. Pl 74.16. Str XII MBII
V Alalakh. Pl 21.33. Str VIII MBII
V Hazor v. Fig V.2.9. (P. Ph d) MBII

0717 024
9926
Sjar 097

P Sali. Pl 42.102. Str XII MBII
P Mev. Fig 15.22. Str XIV MBI
V Bosra. Pl 15.109. (Str 20) MBIII
V Gezer ii. Pl 15.25,26. (F1. Tr 4168) MBIII**(x2)**
V Ghassil. Pl 20.45. Str X MBII
V Hama. Fig 117.3D583. (H4) MBII
V Hazor v. Fig II.8.10. (A. Str 9D) MBI
V Jericho v. Fig 198.12,14. (H.XIII.liii) MBIII**(x2)**
V Mev. Fig 15.23,25. Str XIV MBI**(x2)**
V Mev. Fig 16.10. Str XV MBI
V Shec. Pl 34.a. (Jl.41). Str XIX MBII

Appendix F: Rukeis,Area 16, Comparative Pottery Analysis

0717006
DCR 9941
Sjar 151

V	Gezer iv. Pl 4.17. (FVI. Str 12/11)	MBI
V	Meg. Pl 28.1. Str XII	MBII
V	Sali. Pl 41.5. Str XII	MBII

0717006
DCR 9943
Xjar 027

P	Ghassil. Pl 5.3. Str XI	MBII
P	Jawa. Fig 12.20,22	MBI**(x2)**
P	Jericho v. Fig 202.2. (H.XII.lii)	MBII
v	TBM 33. Pl 14.4. Str D	MBIII
V	Bosra. Pl 14.72,93. (Str 23-21)	MBIII**(x2)**
V	Hazor v. Fig II.7.7. (A. Str 9D)	MBI
V	Jericho iv. Fig 135.19. (H.IX.xxxii)	MBII
V	Jericho iv. Fig 136.13,17. (H.XI.xlvi)	MBII**(x2)**
V	Jericho iv. Fig 137.29. (H.IX.xxxii)	MBII
V	KEL. Pl 48.T86. Str 8	MBIII
V	Gezer i. Pl 30.9. (F1 Str 7)	MBIII
V	Meg. Pl 27.5. Str XII	MBII

0717006
DCR 9948
Sjar 072

V	Alalakh. Pl 60.46. Str X	MBII

0717006
DCR 9949
Sjar 157

0717007
DCR 9952
Bowl 007

P	Hazor v. Fig V.5.7. (P. Ph C)	MBII
P	Jericho iv. Fig 103.18. (H.IX.xxxii)	MBII
P	Jericho iv. Fig 106.20. (H.XI.l)	MBII
P	Jericho v. Fig 179.10. (H.XI.xlvi)	MBII
P	Jericho v. Fig 210.25. (H.XII.liib)	MBII
P	Mev. Fig 12.22. Str XII	MBII
P	Sali. Pl 42.28. Str XII	MBII
V	Alalakh. Pl 2.4. Str VII	MBIII
V	Dan. Fig 4.99.8. (T.8096). Str IX	MBIII
V	Hazor i. Pl CIII.3. (D3. Str 1)	MBIII
V	Hazor v. Fig III.12.9. (L. Str XVI)	MBIII
V	Dan. Fig 104.5. (T.1025). Str XII	MBI
V	Gezer iv. Pl 3.24. (FVI. Str 12)	MBI
V	Hama. Fig 127.4B178. (H1)	MBIII
V	Jericho iv. Fig 103.1-3,8,11,16. (H.VIII-XII.xxix-lii)	MBII**(x6)**
V	Jericho v. Fig 167.3. (H.IX.xxxia)	MBII
V	Jericho v. Fig 179.9. (H.XI.xlvi)	MBII
V	Jericho v. Fig 186.1. (H.XII.li)	MBII
V	Meg. Pl 29.23,24. Str XII	MBII**(x2)**
V	Shec. Pl 4.k. (Bp 52). Str XX	MBII
V	Shec. Pl 5.f. (Bp 61). Str XX	MBII
V	Shec. Pl 5.h. (Bp 61). Str XIX	MBII
V	TNM. Pl 30.1. Ph F	MBIII

0717007
DCR 9953
Sjar 134

0717009
DCR 9940
Sbowl 034

Figures

Figure 1: Location of Important MBA sites in Jordan

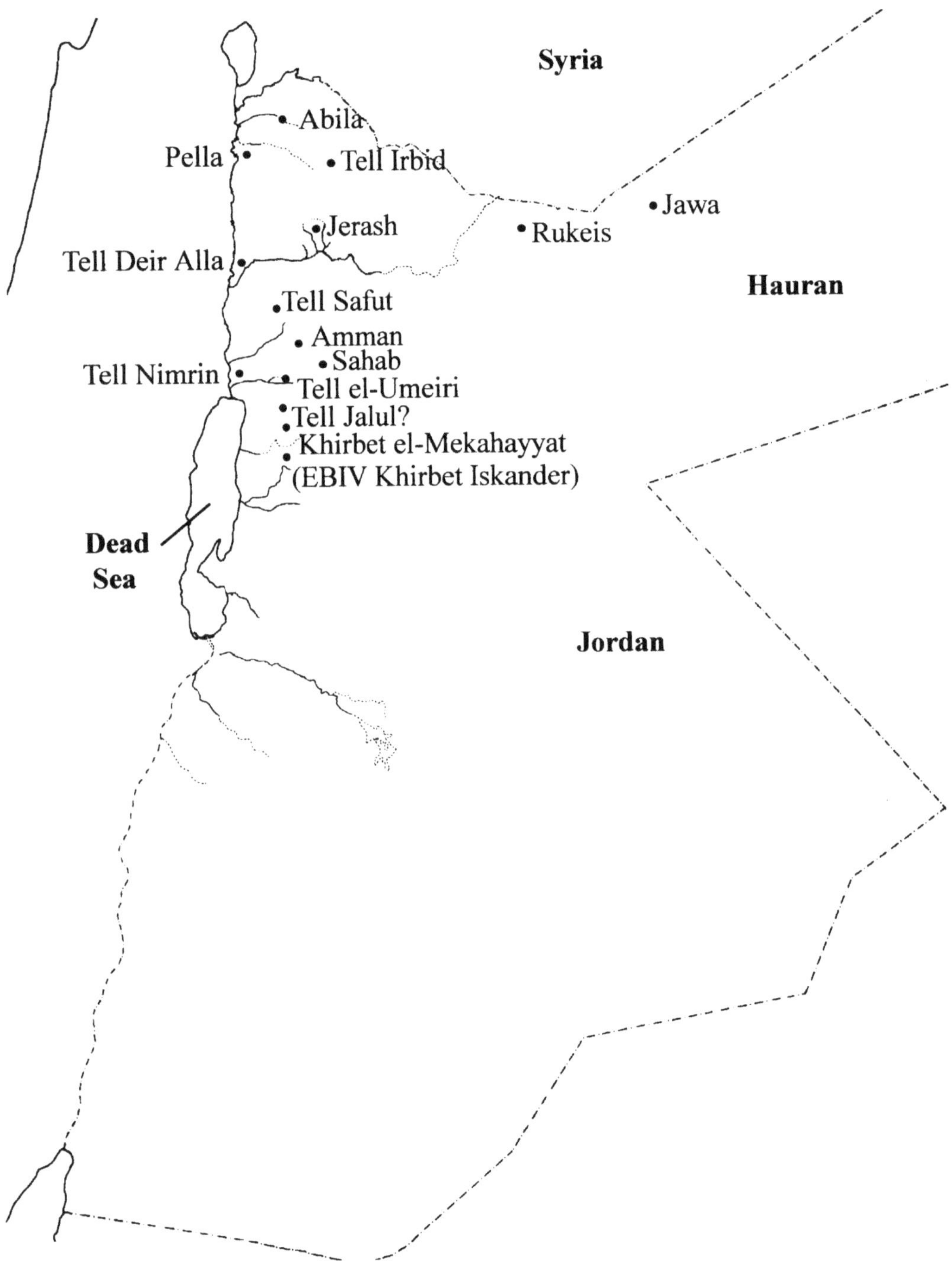

Figure 2: Location of Important MBA Fortified Sites of the Southern Levant

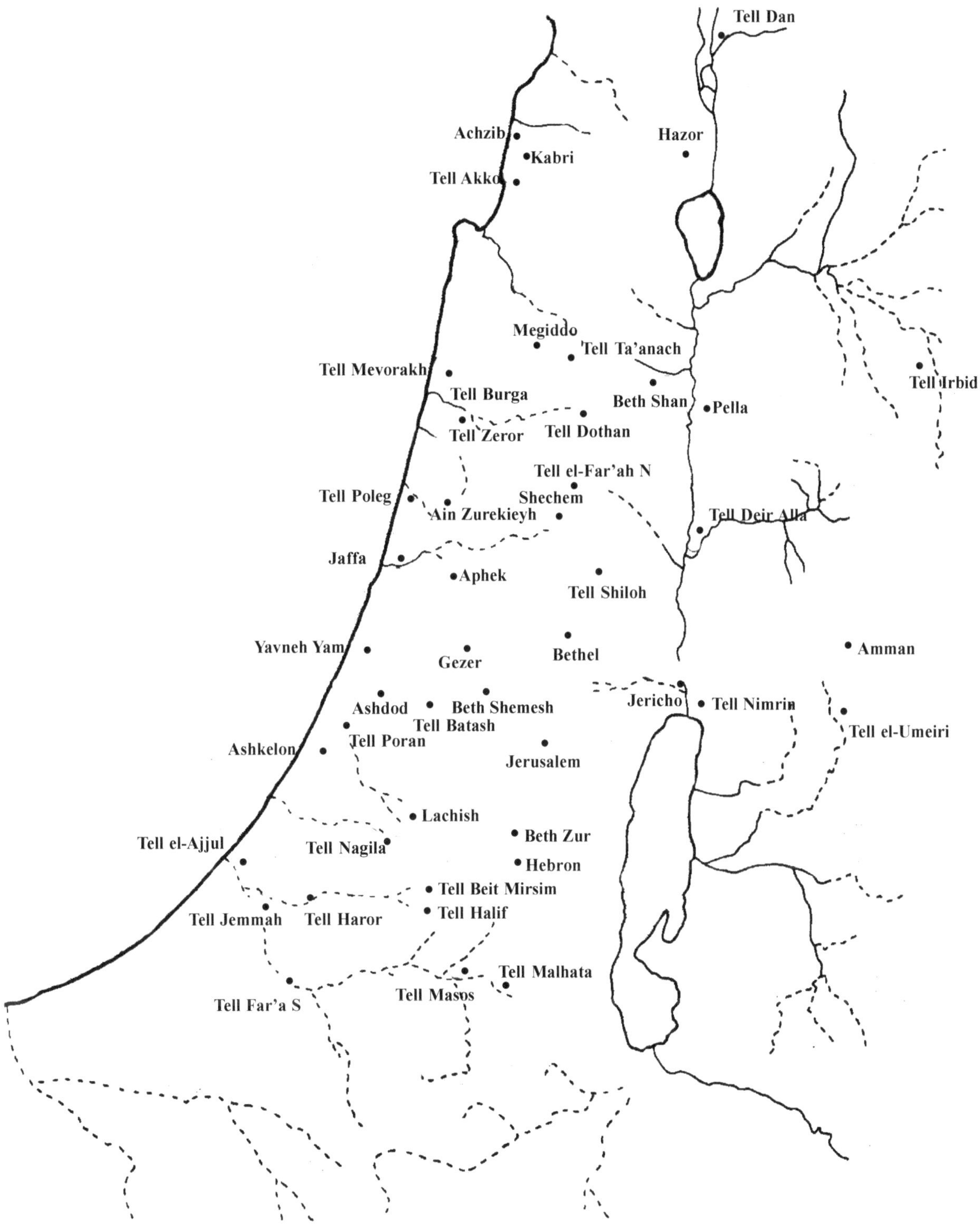

Figure 3a-c: EBA Examples of Standarisation in Military Architecture: Straddle Towers, Semi-Circular Towers and Bent-Axis Gates

a.Straddle Towers

Tell Dothan

Tell Ta'anach

Ai

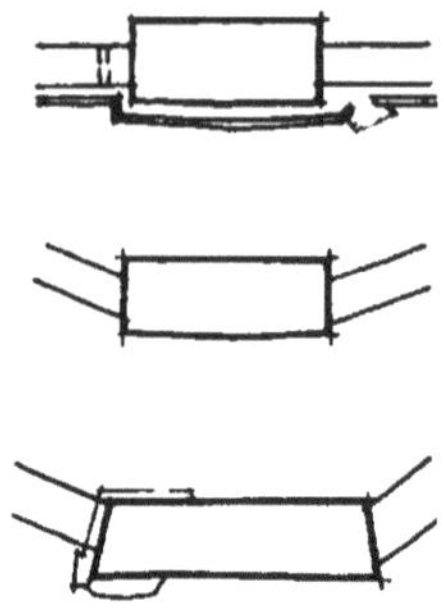

After Helms 1977 fig 4

b.Semi-circular Towers

Arad
After Amiran 1978 figure 1

Ai
After Callaway 1980 figure 6

c.Bent-axis Gates

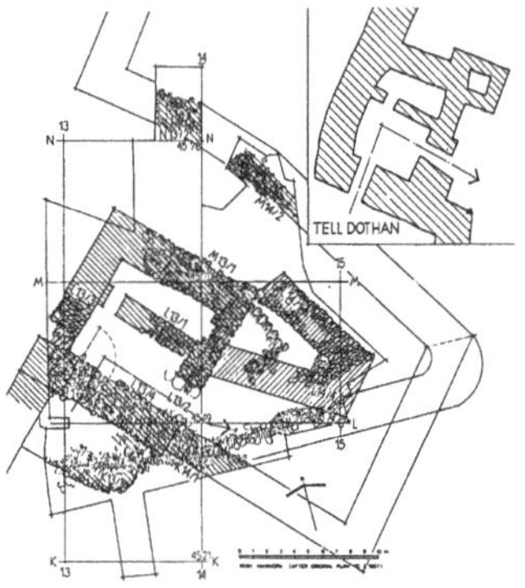

Rosh Hanniqra and Tell Dothan
After Helms 1976 figure 2

Figure 4a-b: MBA Examples of Standardisation in Military Architecture: Triple-Entry Gates and Sandwich Ramparts

4a. Triple-entry gates

Megiddo

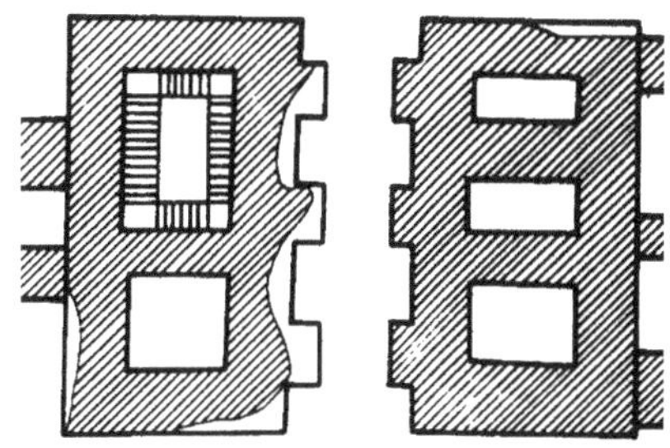

Yavneh-Yam

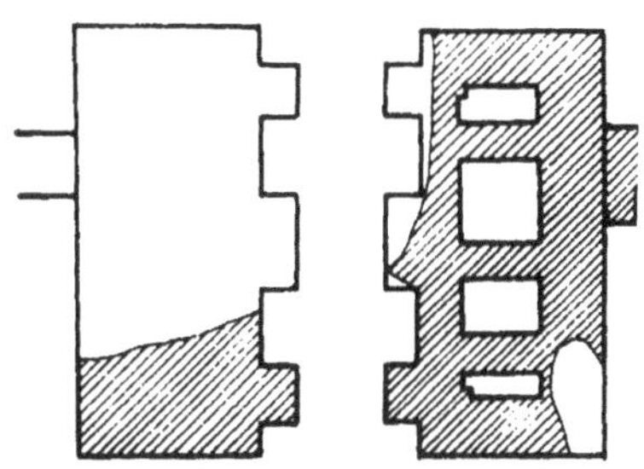

Hazor

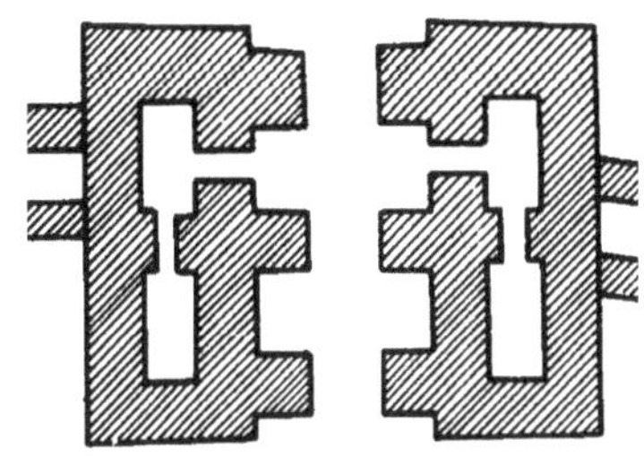

Tell el-Far'ah south

After Gregori 1986

4b.Example of 'sandwich' rampart at Gezer Field I

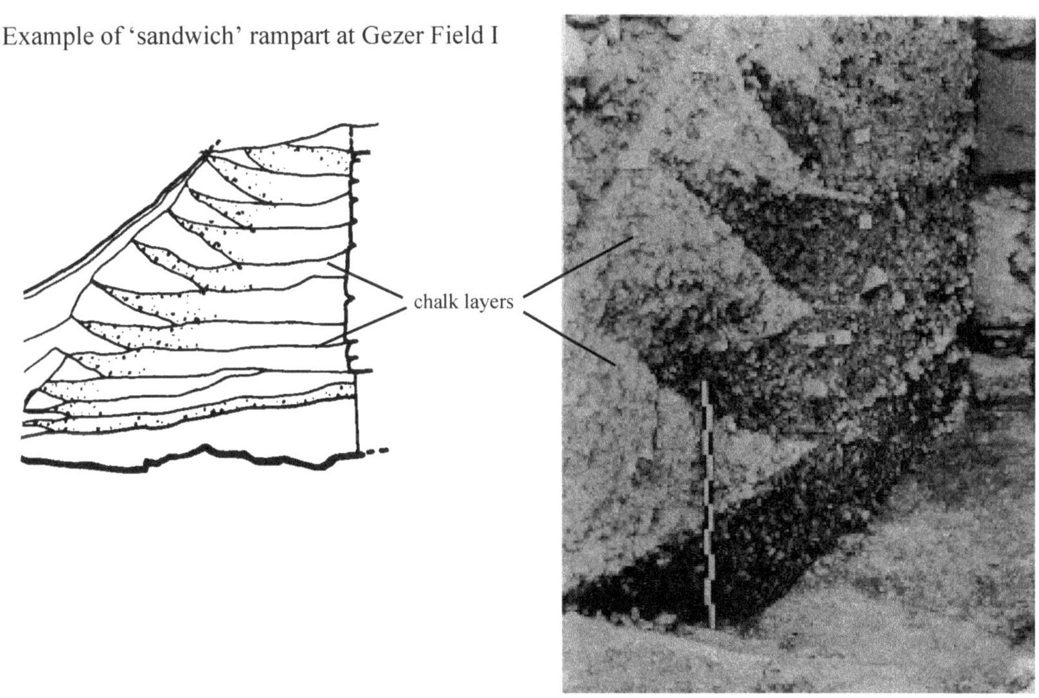

After Dever, Lance and Wright,1970 Plan 2 and Plate 5.B

Figure 5a: Geographic Entities of Jordan and the Southern Levant

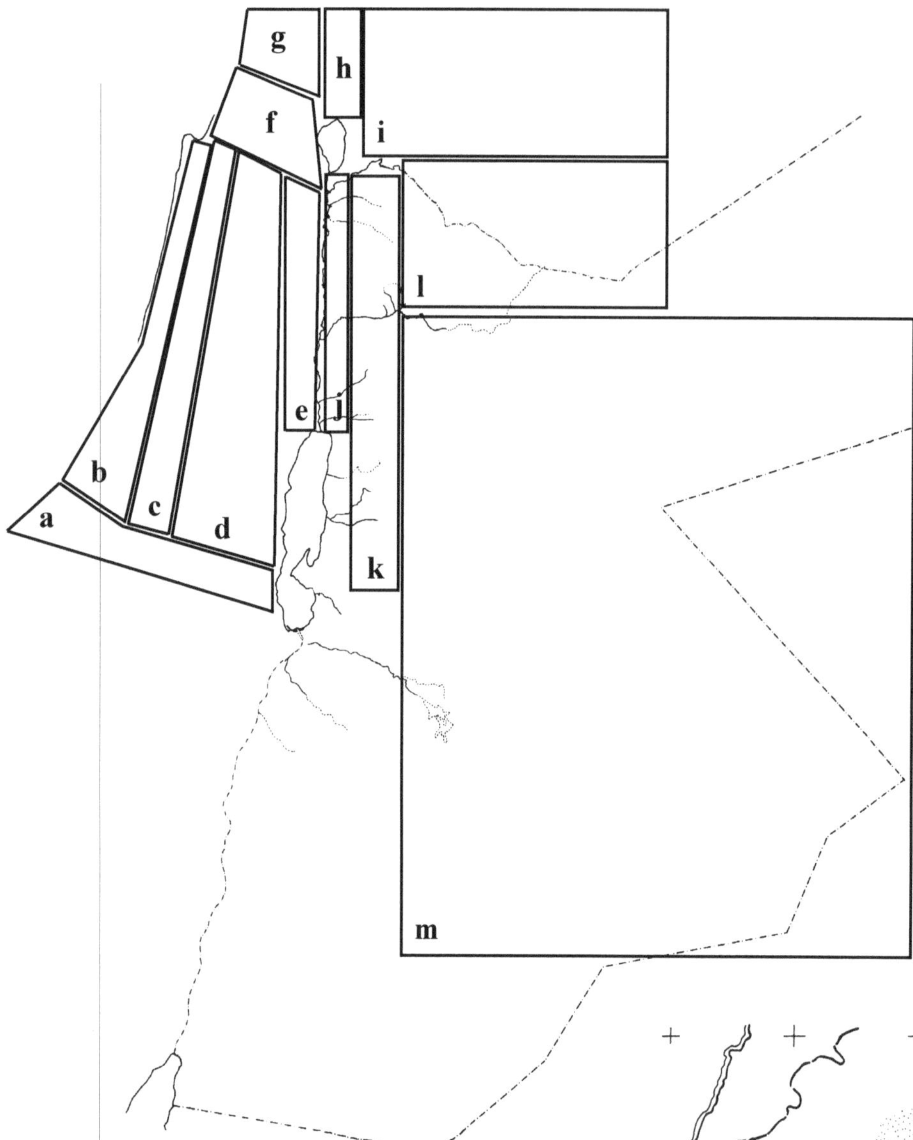

a. Northern Negev
b. Coastal Plain
c. Shephelah
d. Central Highlands
e. West Jordan Valley
f. Esdraelon
g. Southern Lebanon
h. Huleh Valley
i. Damascene
j. East Jordan Valley
k. Jordanian Plateau
l. Plains of Hauran
m. Eastern Desert

Figure 5b: Jordan. Zones of Precipitation

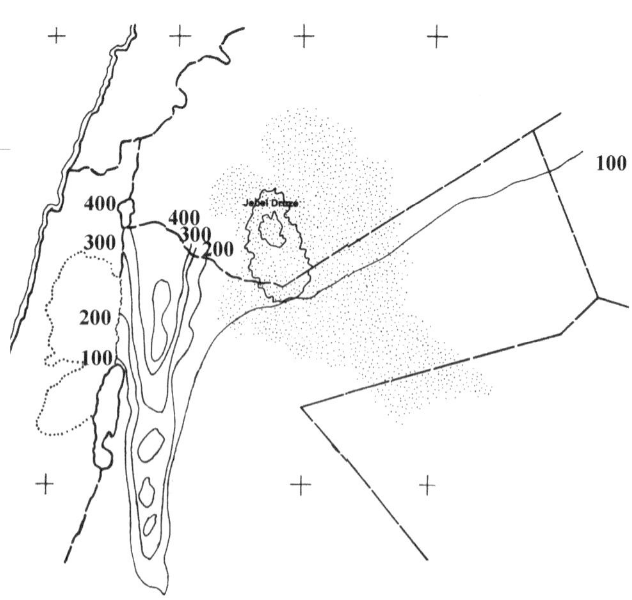

Figure 6a-d: Iconographic Representations of Bronze Age Military Architecture

a. Beni Hasan. 20th century B.C.
After Yadin 1963 page 21

b. 'Narmer Palette'. Predynastic Egyptian
After Yadin 1963 page 124

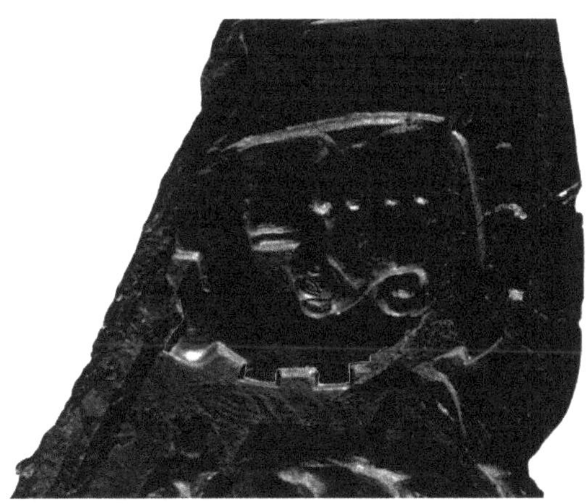

c. 'Bull Palette'. c2900 B.C.
After Yadin 1963 page 123

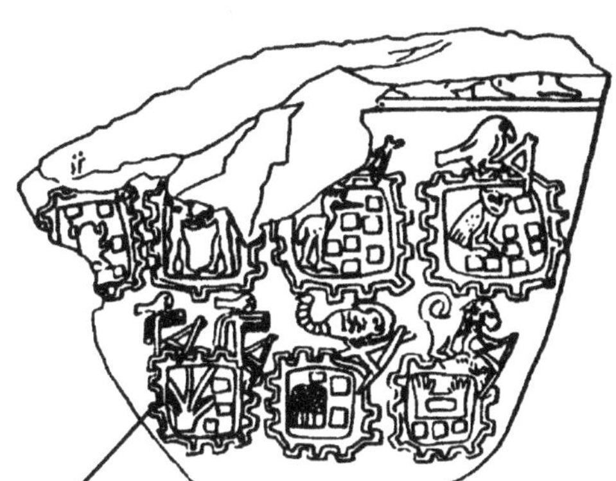

d. 'Cities Palette'. Predynastic Egyptian
After Yadin 1963 page 51

Figure 7: Pella, Site Plan with Areas of Excavation

Figure 8a: Pella, Area III, Plan

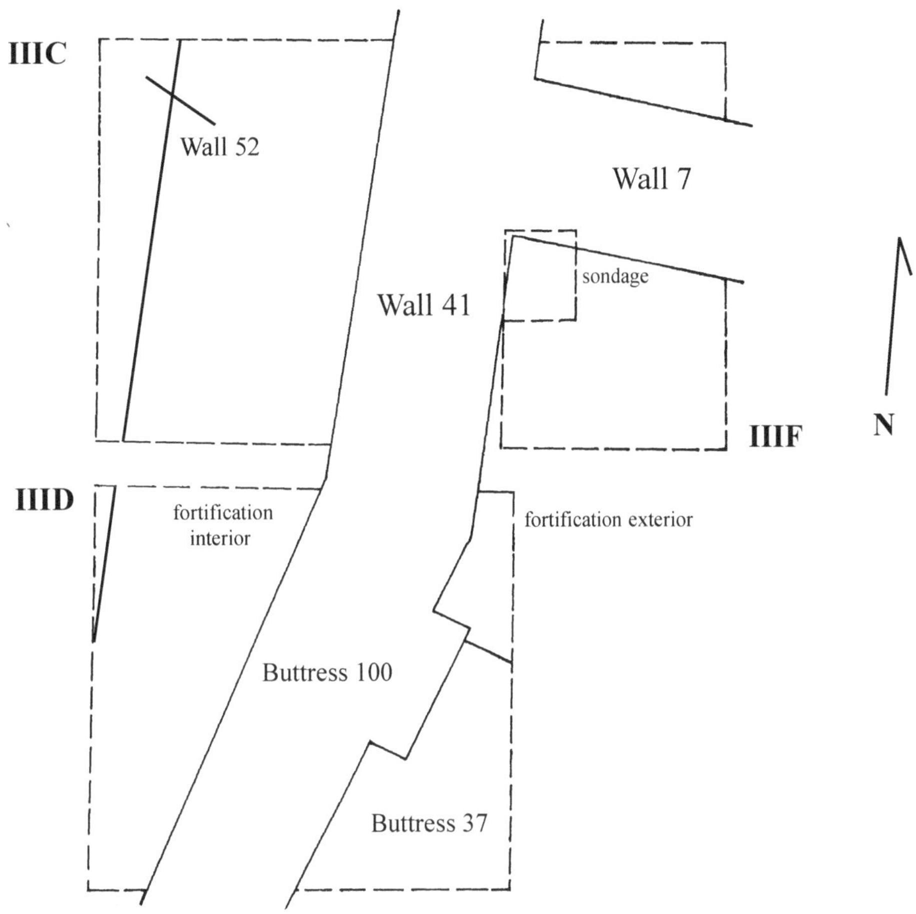

Figure 8b: Pella, North Section of Area IIIC, Wall 41 Foundations

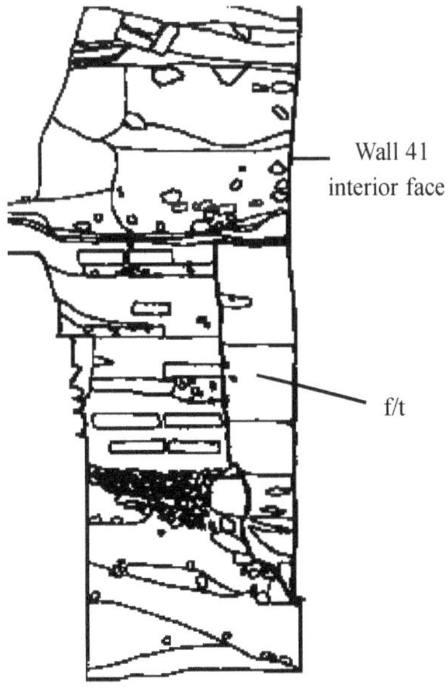

Figure 9: Pella, Area IIIF, West and North Sections, Stratigraphy

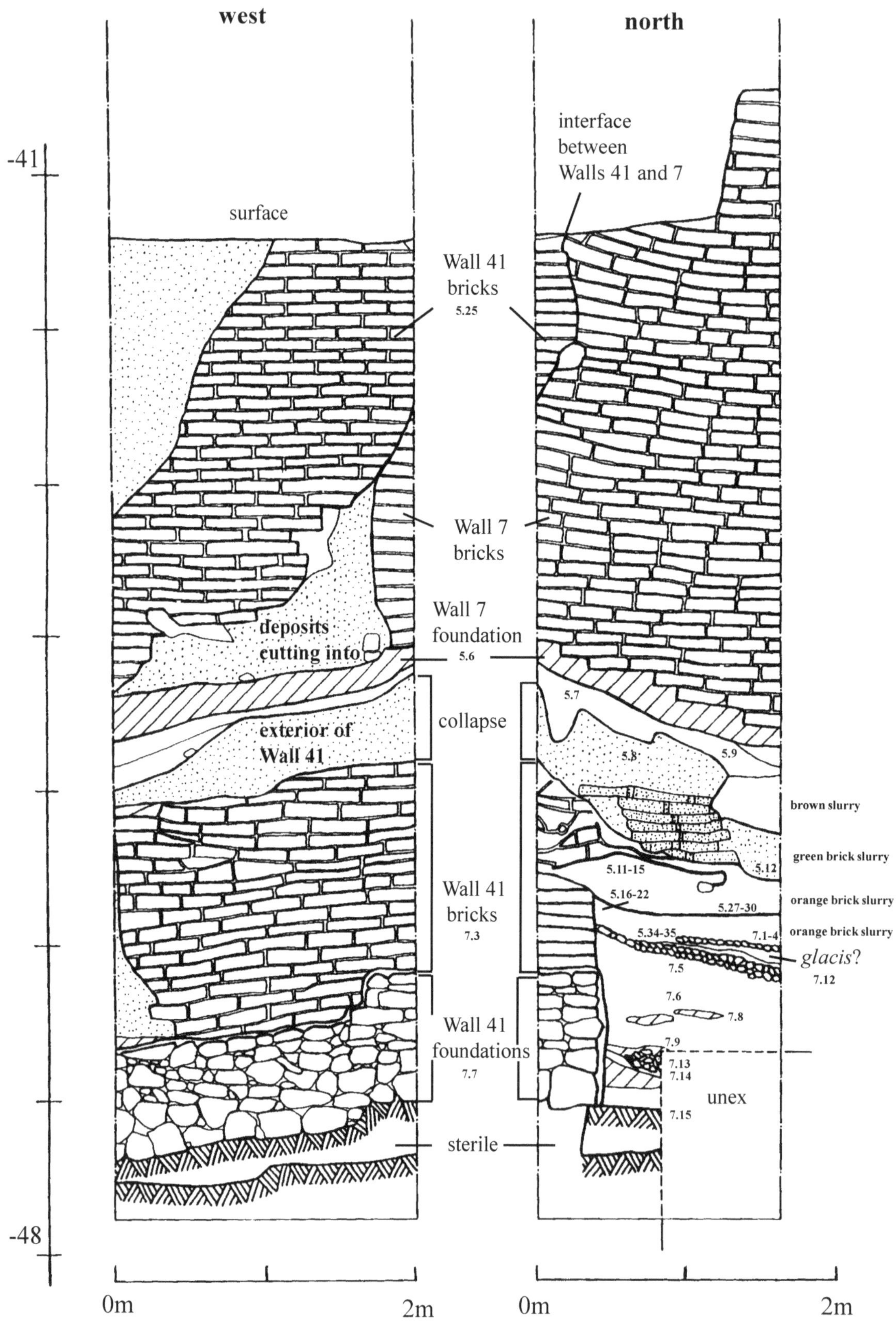

Figure 10: Pella, Area IIIF, West and North Sections, Phasing

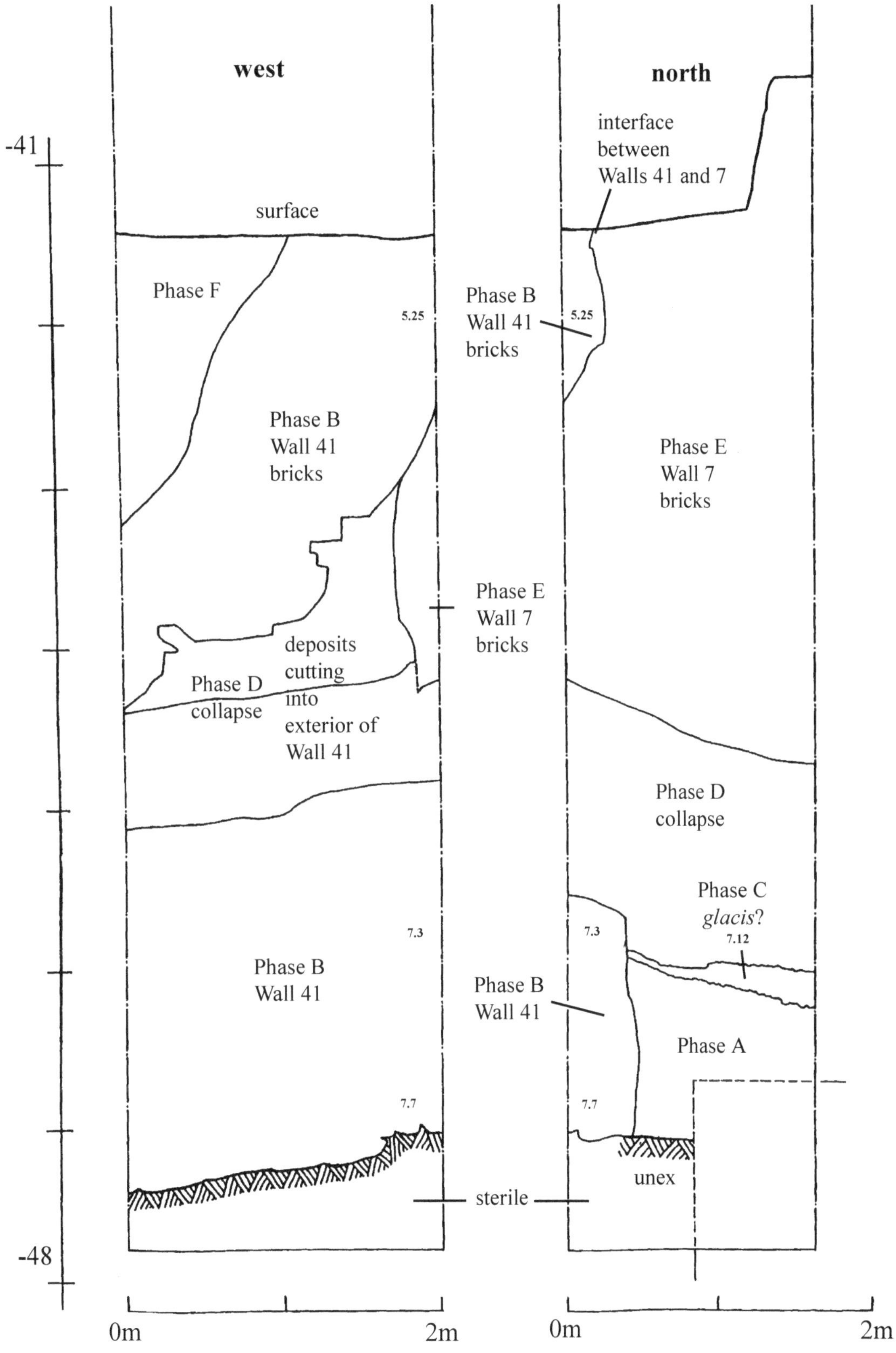

Figure 11: Pella, Area IIIF, South and East Sections, Stratigraphy

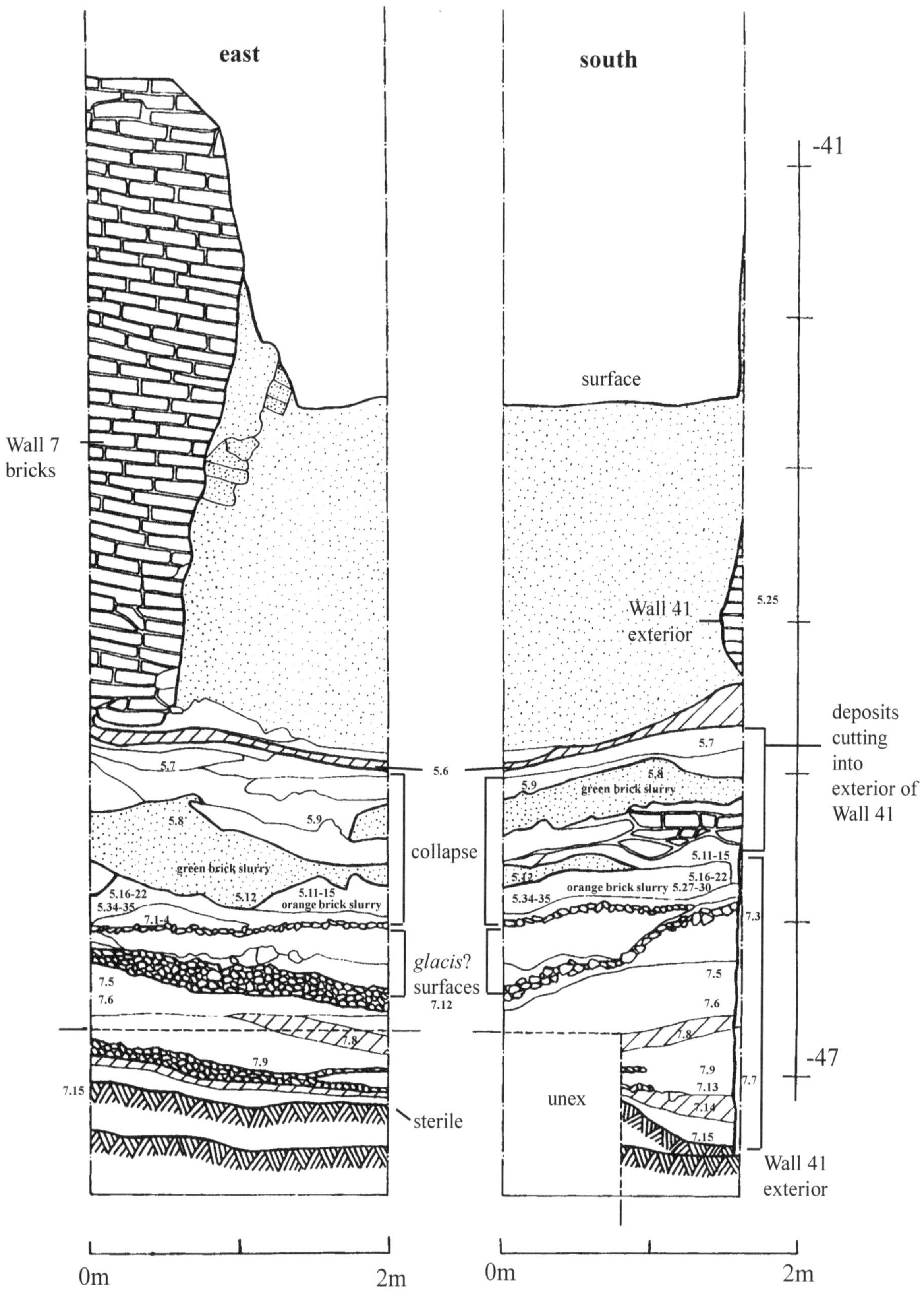

Figure 12: Pella, Area IIIF, South and East Sections, Phasing

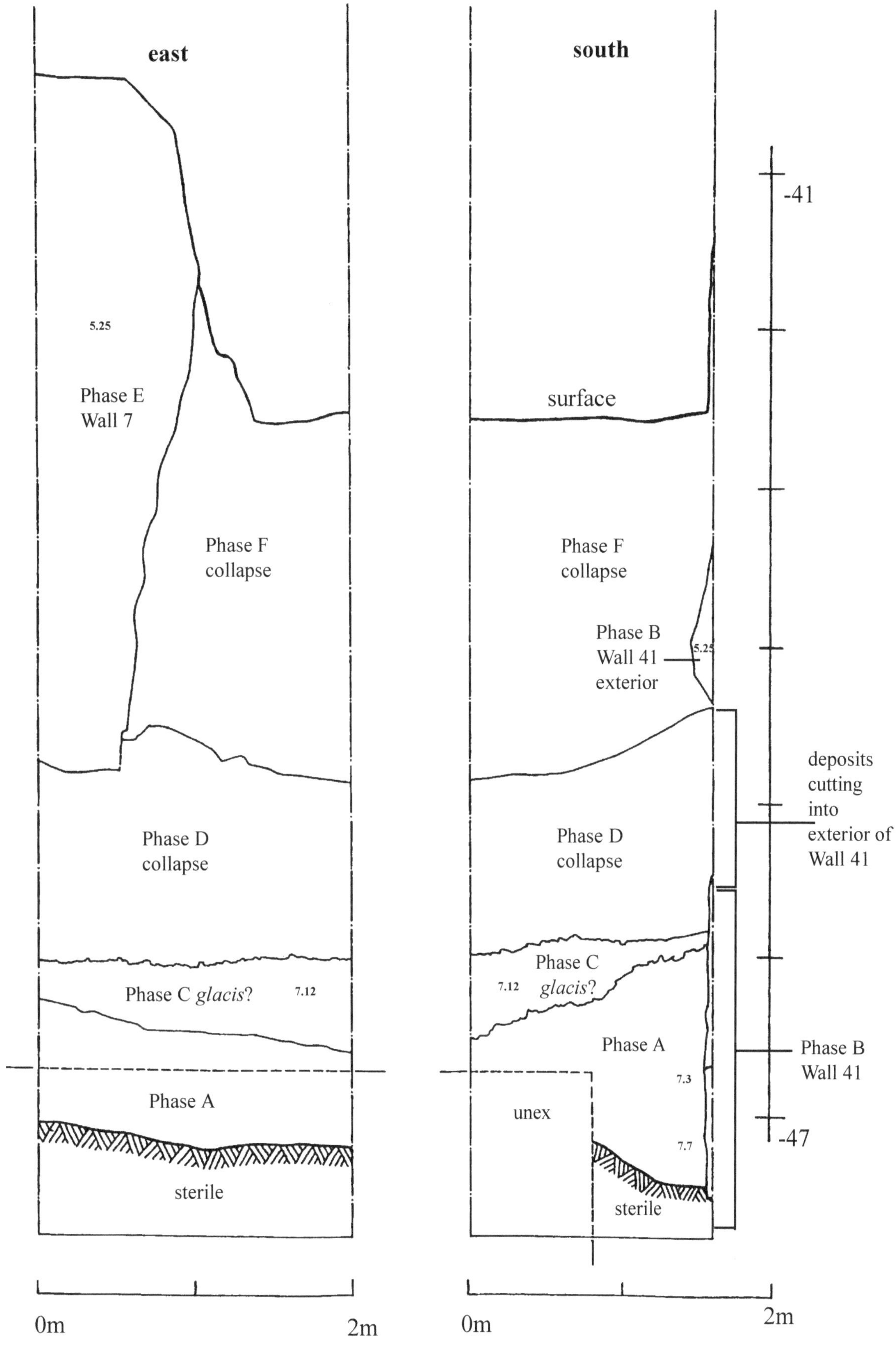

Figure 13: Pella, Area IIIF, Phase A, Pottery and Comparative Ceramic Analysis

Period	Parallel	Variation	Total	%
EBIV		3	3	2.2
MBI	9	23	32	23.4
MBII	36	59	95	69.3
MBIII	1	6	7	5.1
Total	46	91	137	

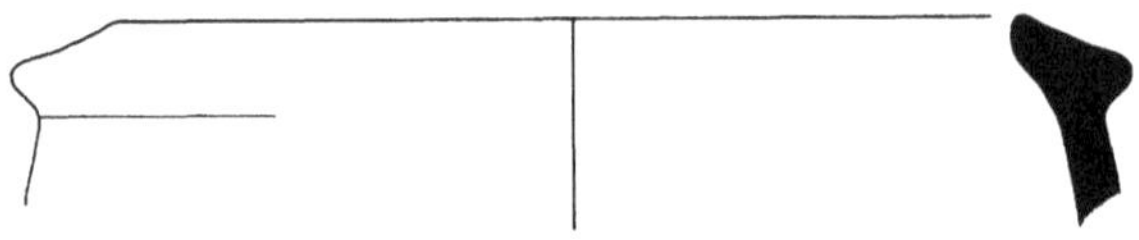

7.5 15474

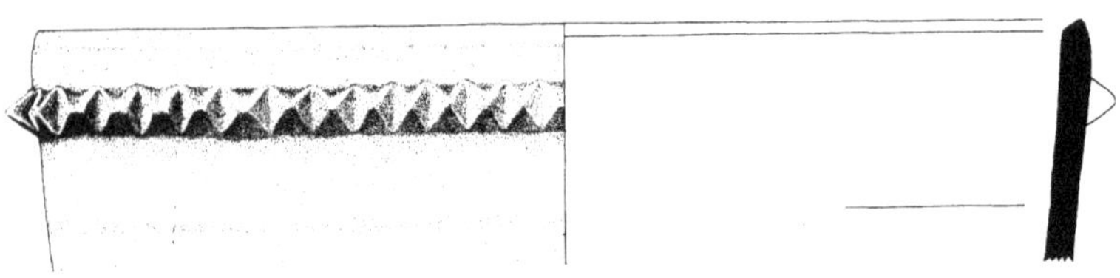

7.8 15466

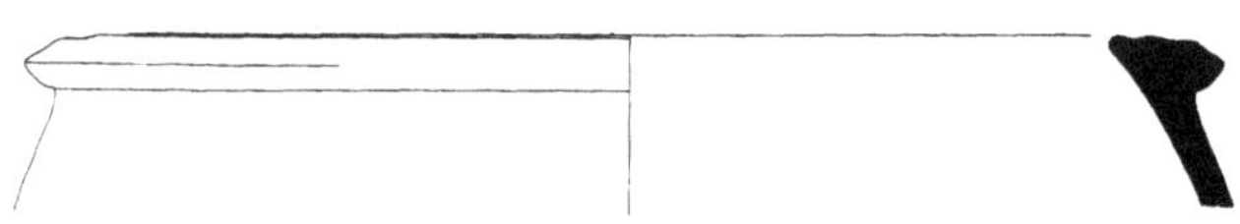

7.8 15467

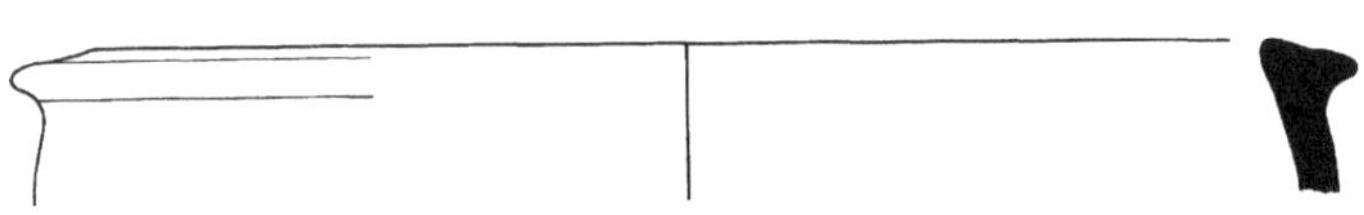

7.8 15468

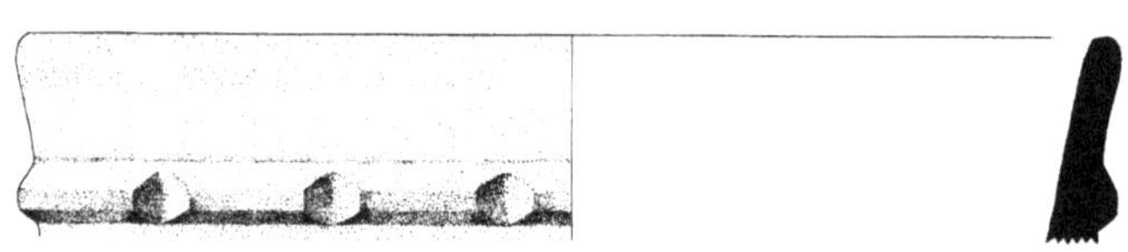

7.8 15470

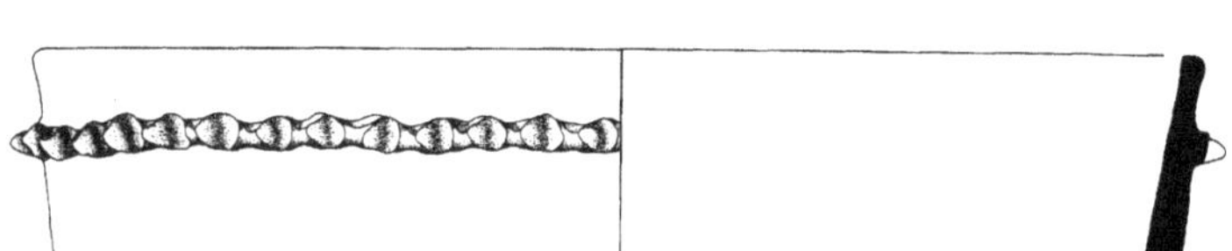

7.13 15482

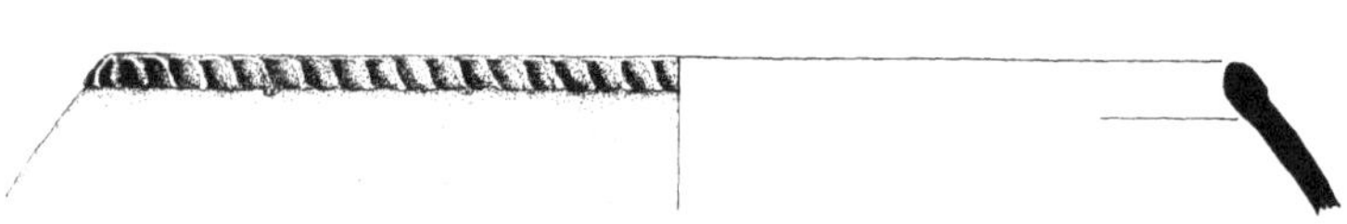

7.15 15481

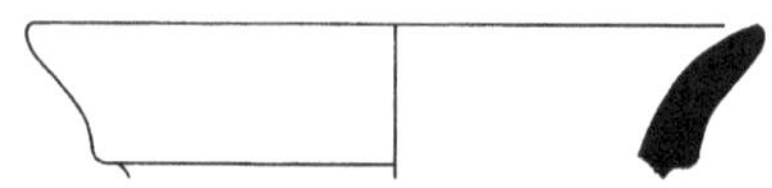

7.8 15471

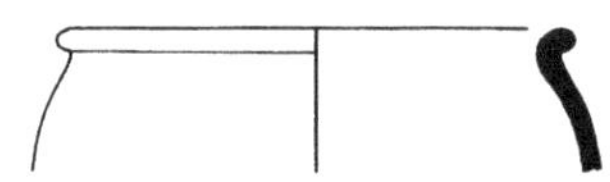

7.6 15473

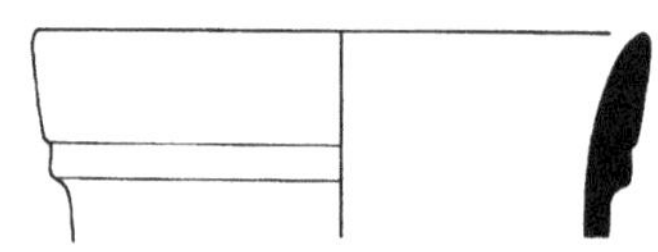

7.8 17074

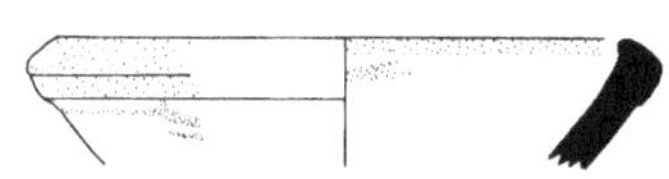

7.13 15463

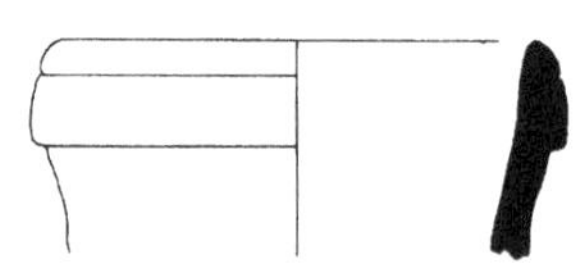

7.13 15464

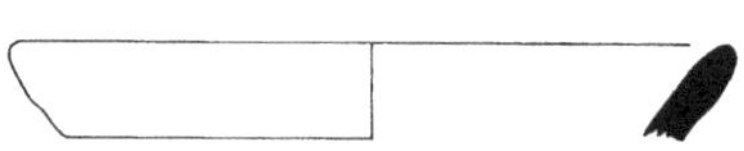

7.15 15479

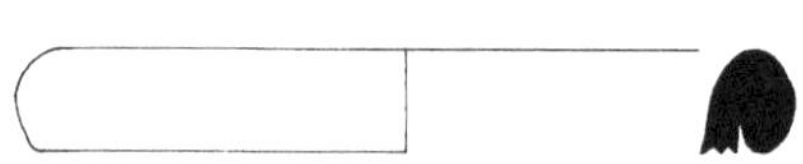

7.15 15480

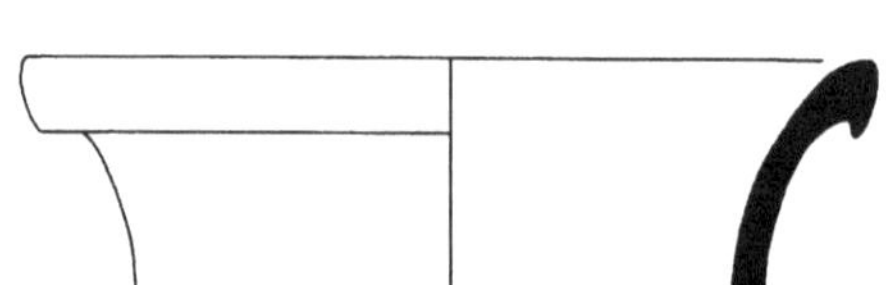

7.13 15483

Figure 14: Pella, Area IIIF, Phase D, Pottery and Comparative Ceramic Analysis

Period	Parallel	Variation	Total	%
EBIV	1	4	5	1.2
MBI	44	99	143	35.1
MBII	91	136	227	55.8
MBIII	3	29	32	7.8
Total	139	268	407	

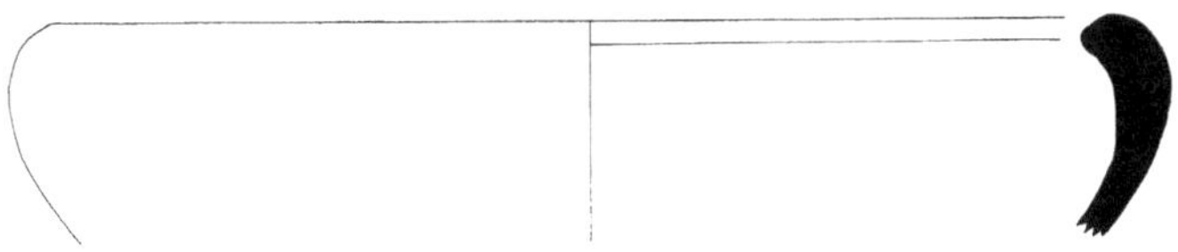

7.4 15475

7.4 15476

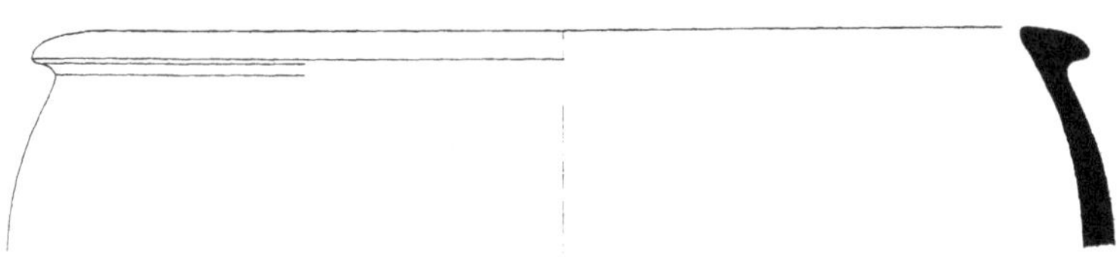

5.7 17084

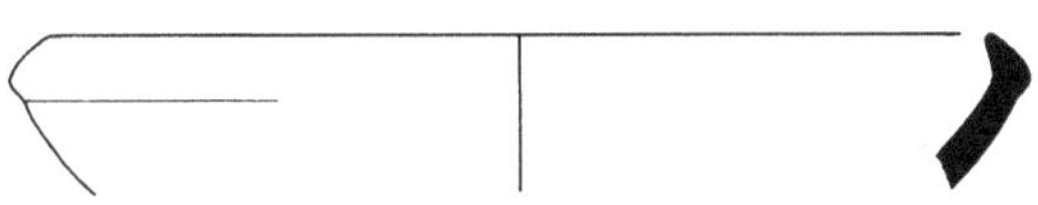

5.7 17085

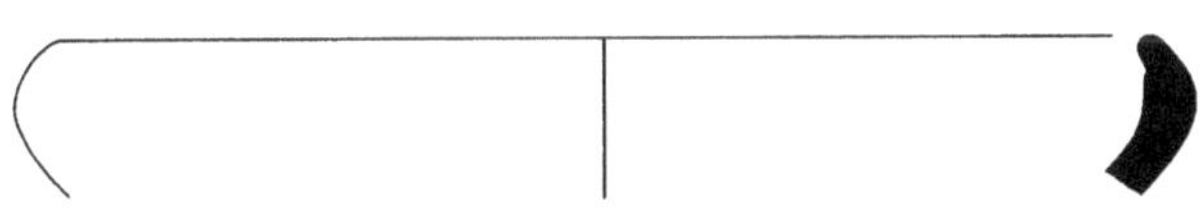

5.7 17086

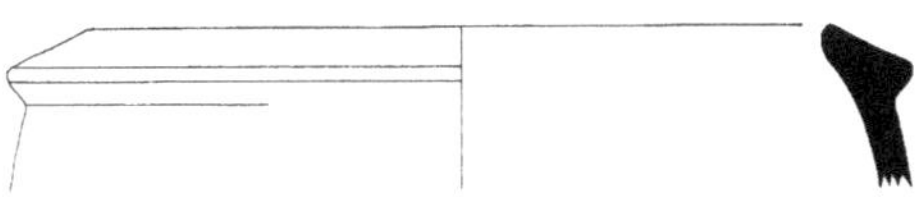

7.2 15474

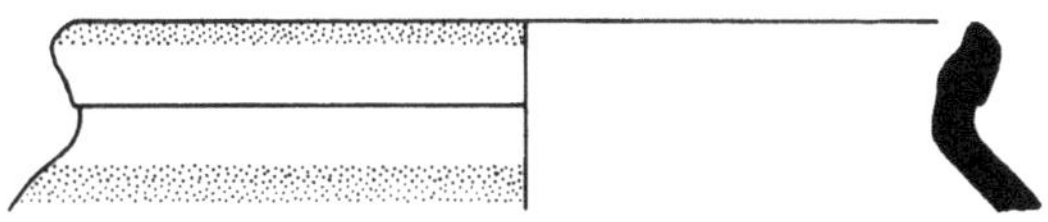

7.10 15460

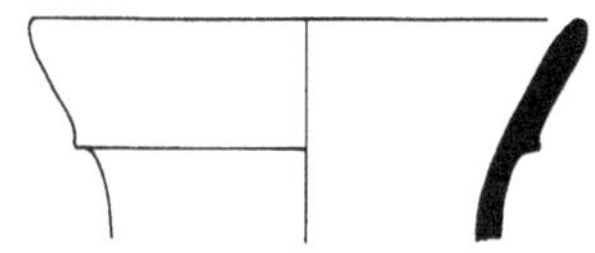

5.9 17107

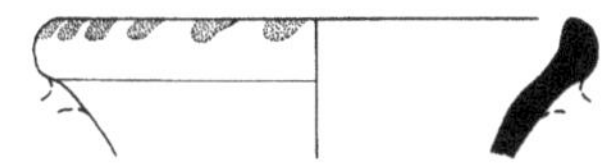

5.11 17989

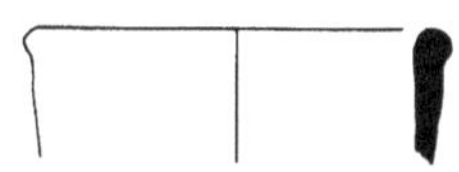

5.30 17945

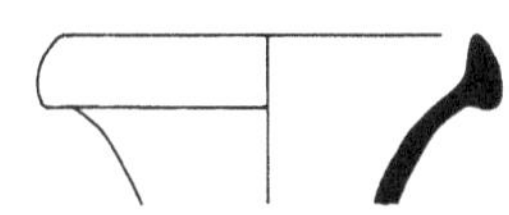

5.7 17097

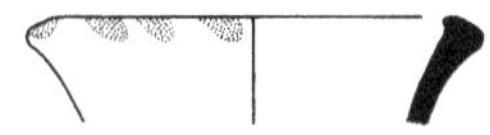

7.10 15461

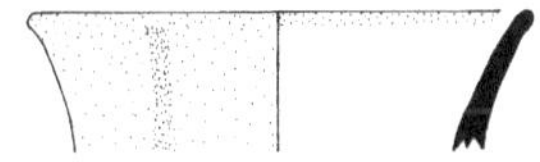

7.10 15462

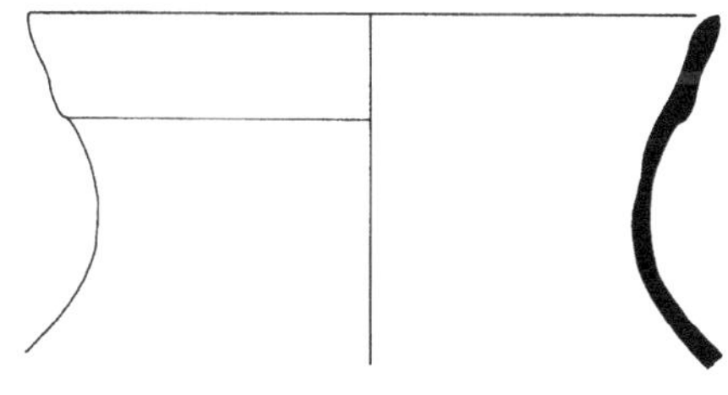

5.16 17986

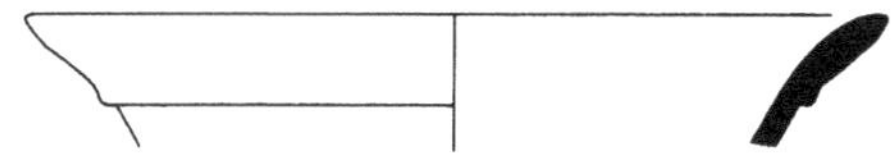

5.15 17972

Figure 15a-c: Pella, Area IIIF, Phase E, Comparative Functions for Wall 7

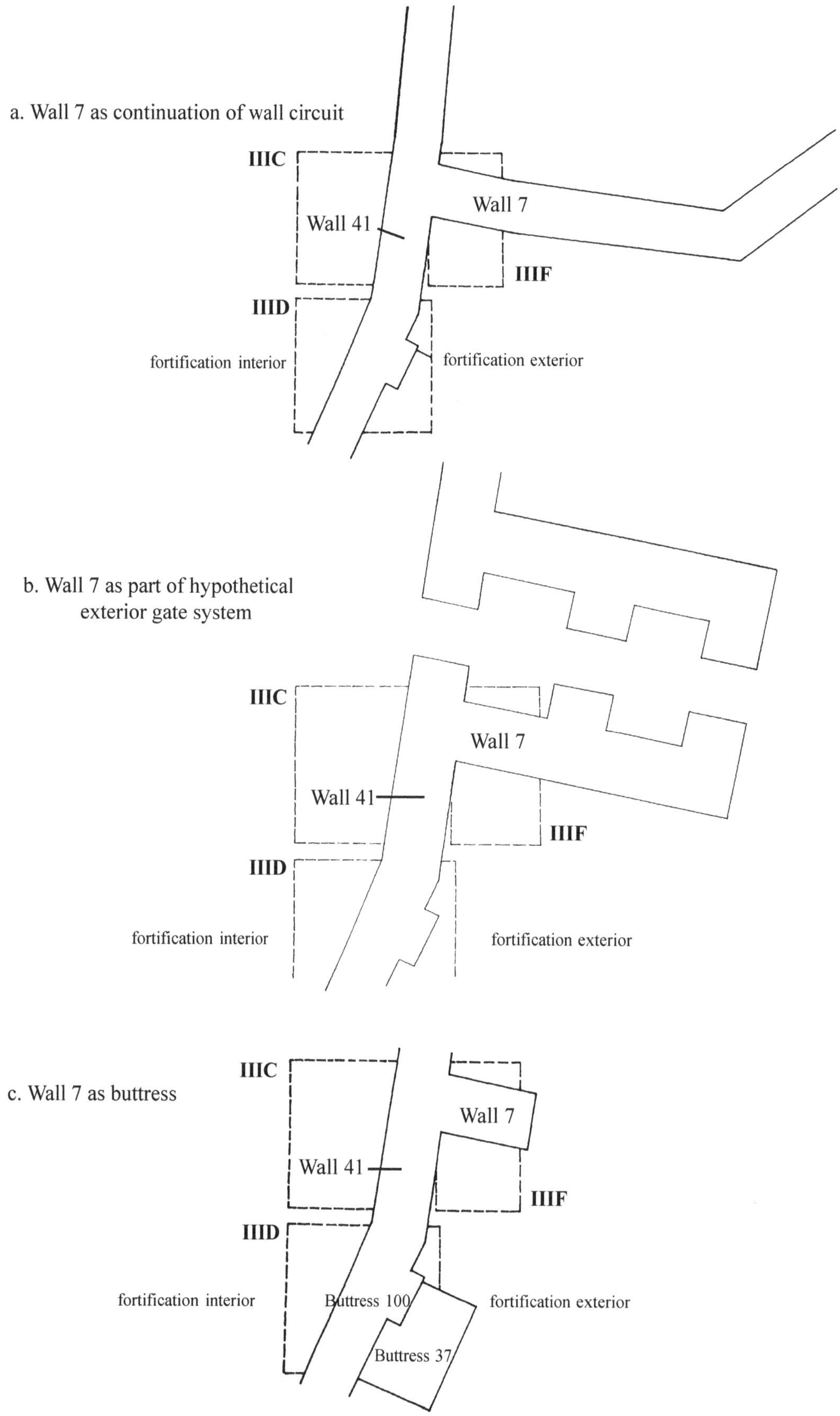

Figure 16: Pella, Area IIIF, Phase F, Pottery and Comparative Ceramic Analysis

Period	Parallel	Variation	Total	%
EBIV	1	3	4	3.7
MBI	5	22	27	24.7
MBII	28	46	74	67.9
MBIII	2	2	4	3.7
Total	36	73	109	

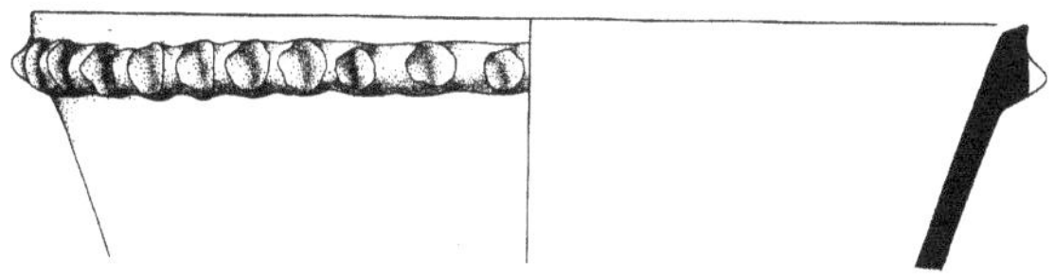

5.5 17077

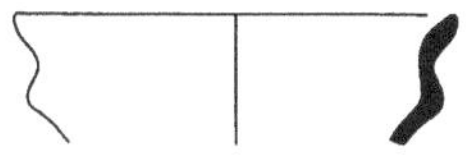

5.4 18002

5.5 17080

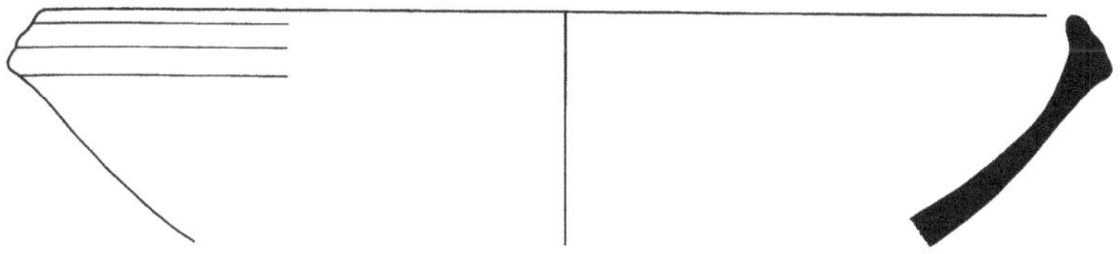

5.5 17079

Figure 17: Pella, Area XXVIIIC, Plan

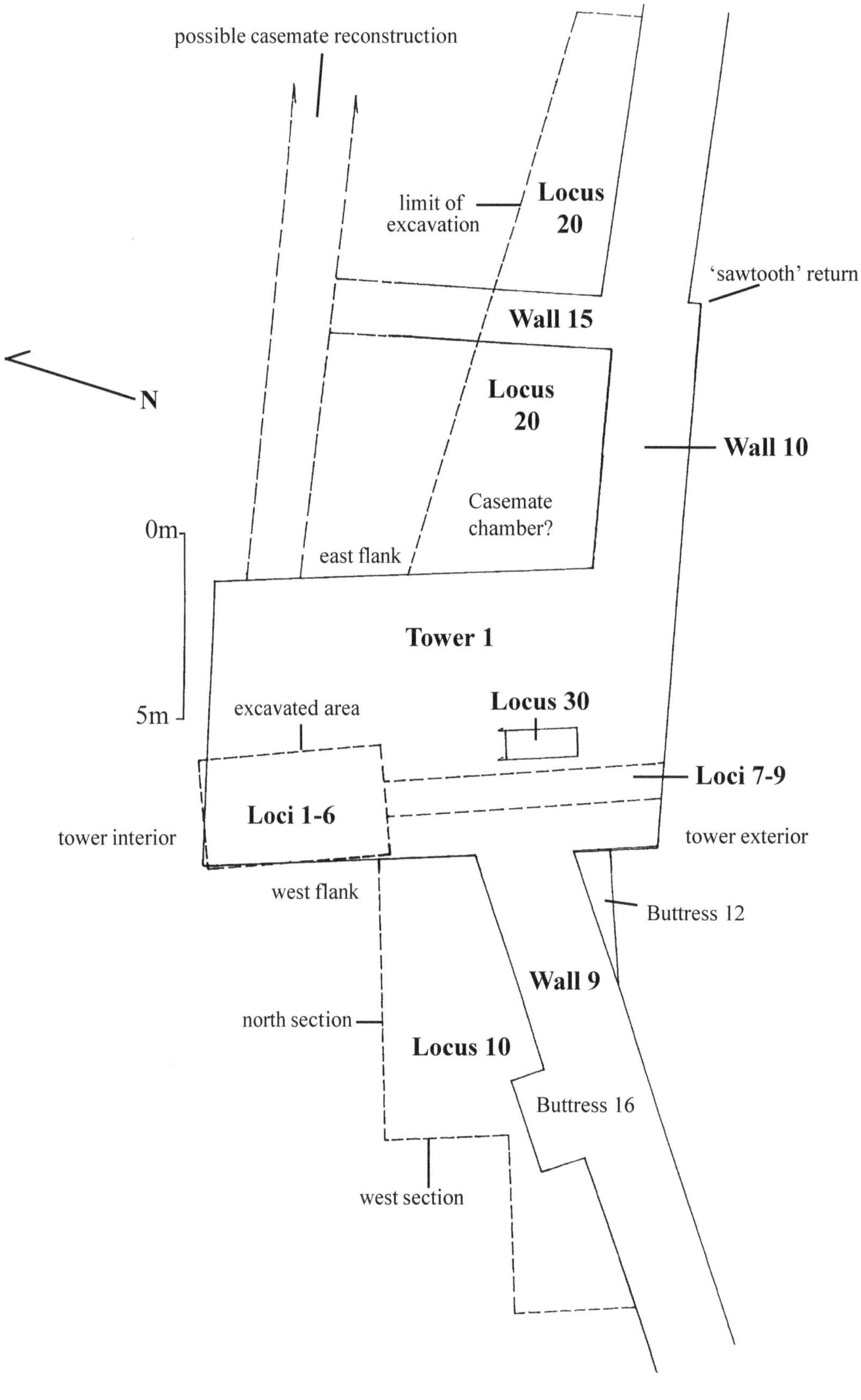

Figure 18: Pella, Area XXVIIIC, Locus 10, North Section, Stratigraphy and Phasing

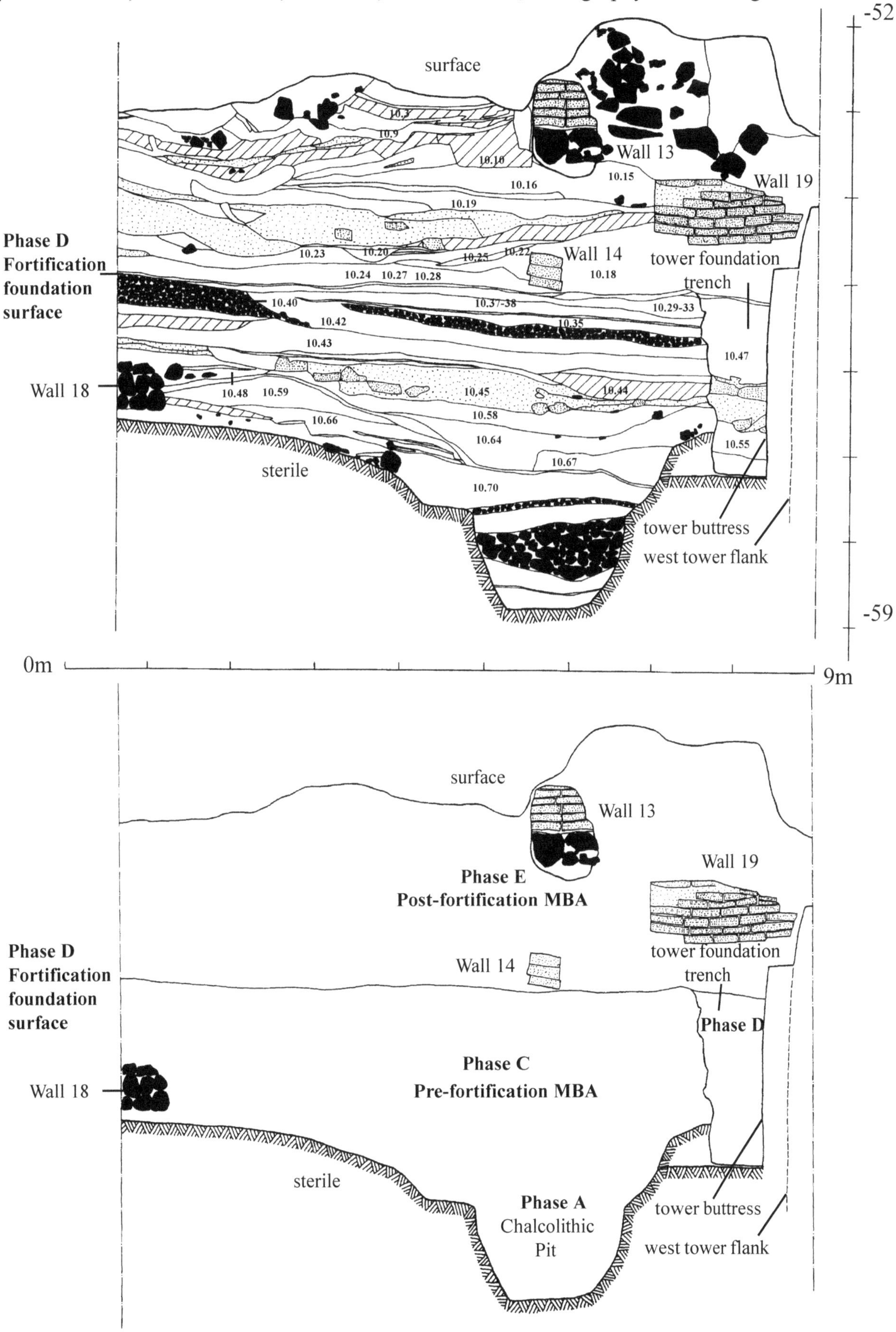

Figure 19: Pella, Area XXVIIIC, Locus 10, West Section, Stratigraphy and Phasing

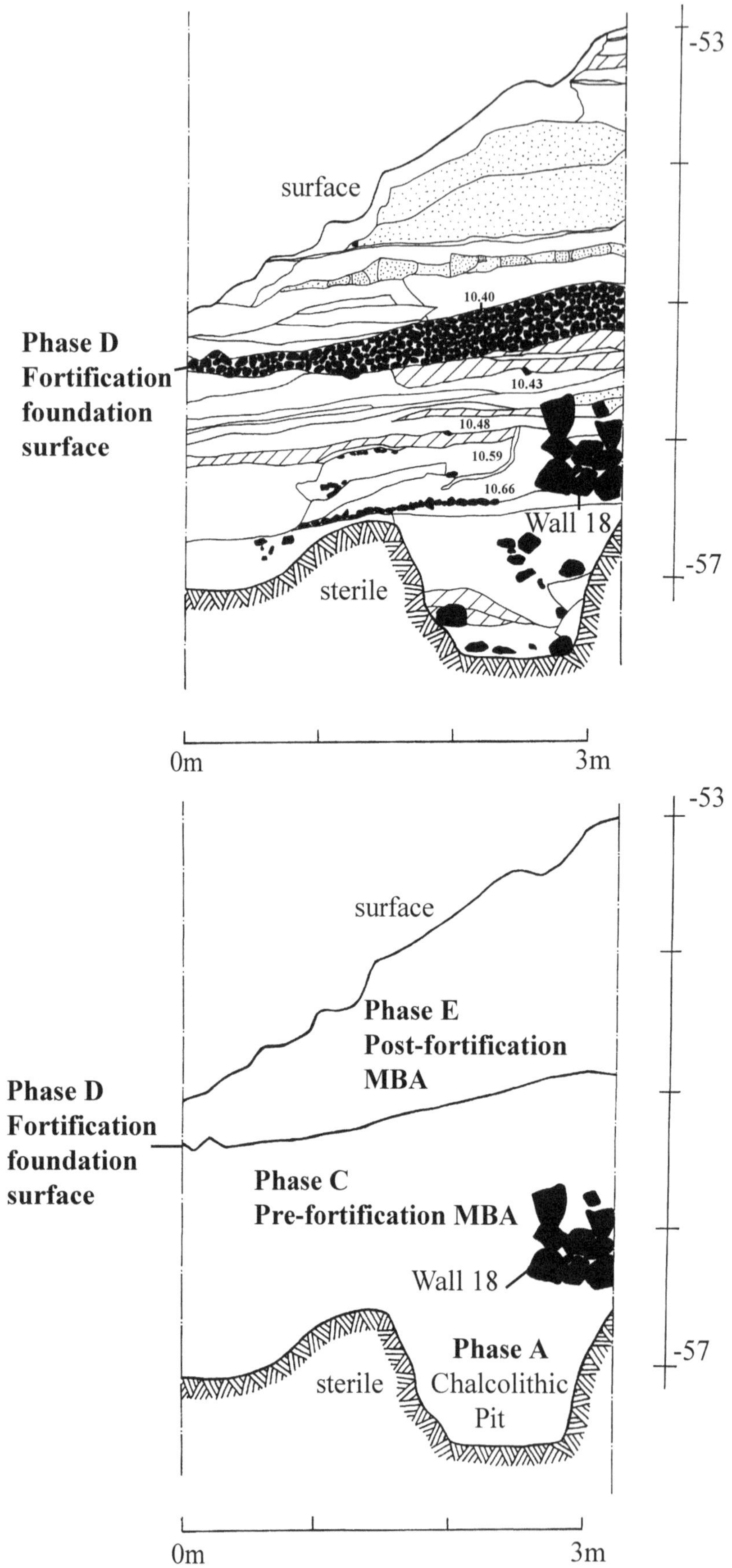

Figure 20: Pella, Area XXVIIIC, Locus 10, Phase C, Pottery and Comparative Ceramic Analysis

Period	Parallel	Variation	Total	%
EBIV	1	16	17	2
MBI	78	165	243	28.7
MBII	172	319	491	58
MBIII	12	83	95	11.2
Total	263	583	846	

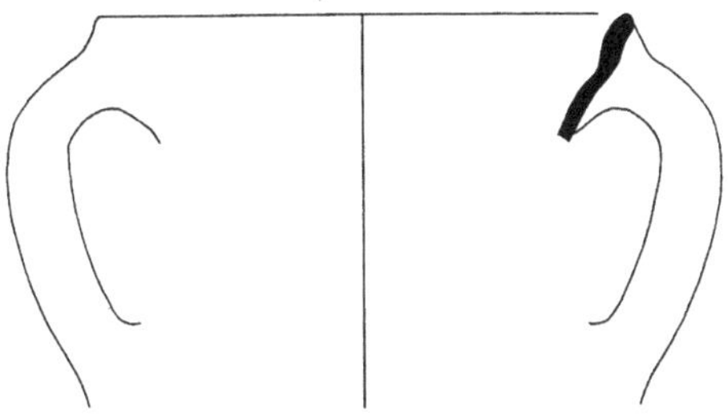

10.30 17704

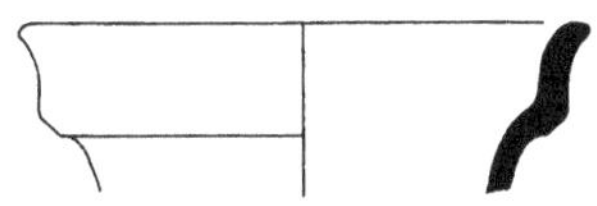

10.30 17705

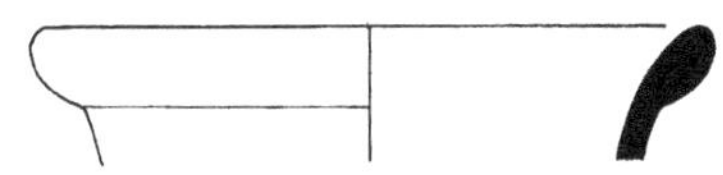

10.35 17732

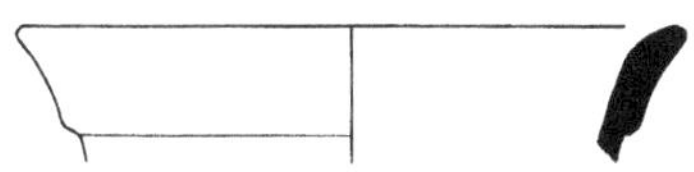

10.35 17731

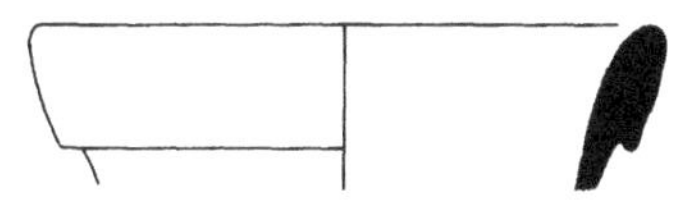

10.40 17761

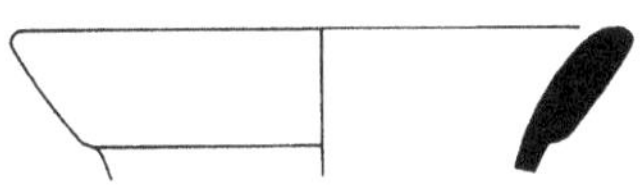

10.44 17795

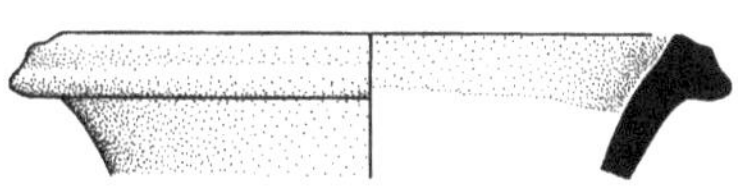

10.35 17739

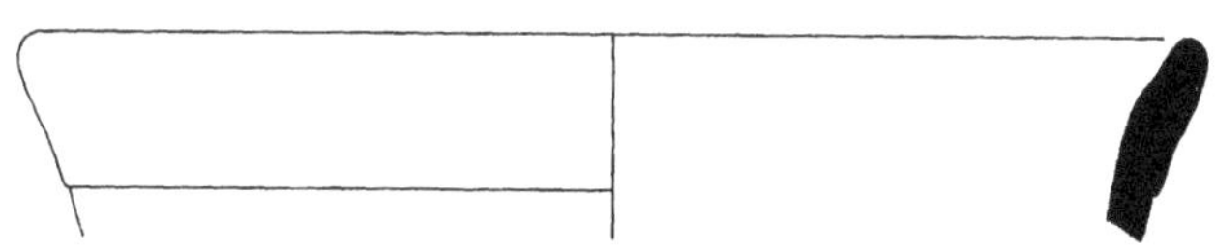

10.38 17754

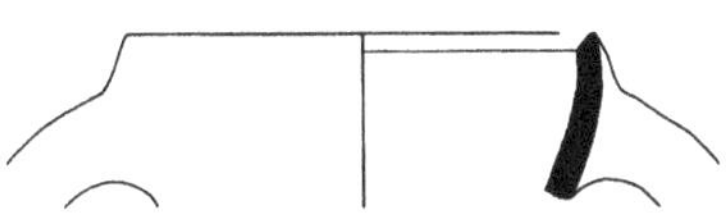

10.33 17730

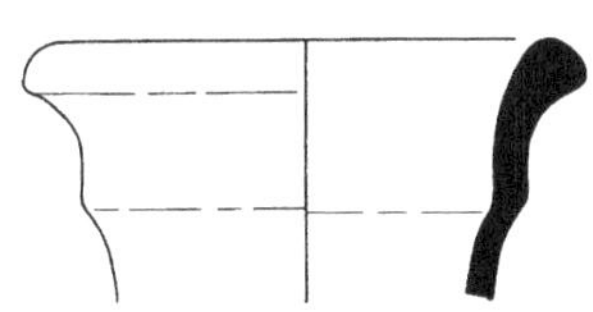

10.42 17768

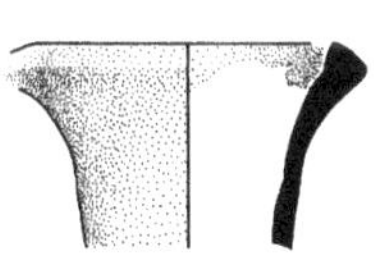

10.44 17797

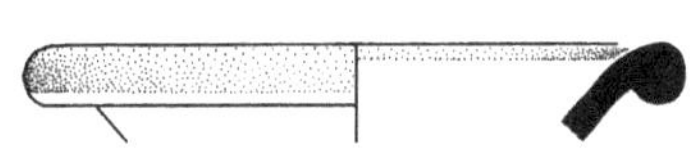

10.42 17769

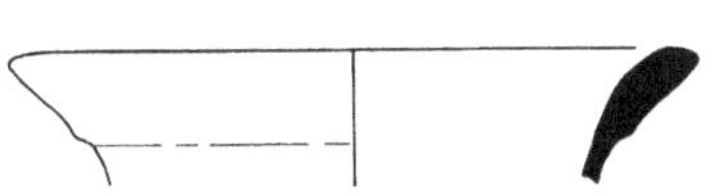

10.64 17900

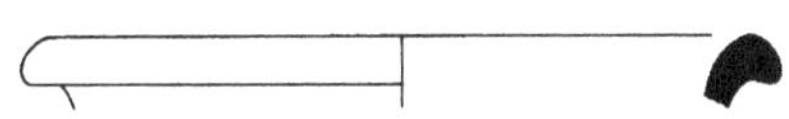

10.42 17787

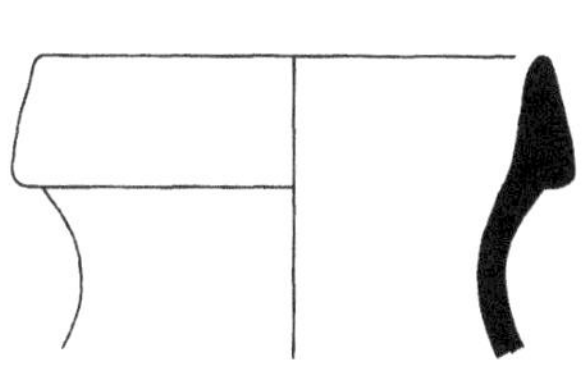

10.44 17808

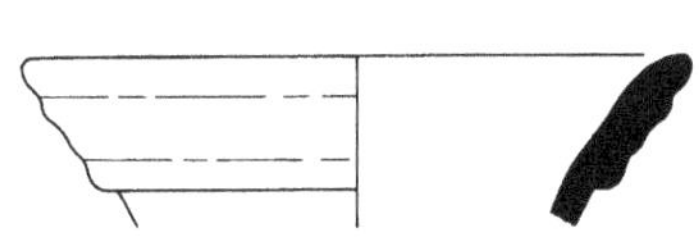

10.44 17807

Figure 21: Pella, Area XXVIIIC, Locus 10, Phase C, Pottery and Comparative Ceramic Analysis (cont.)

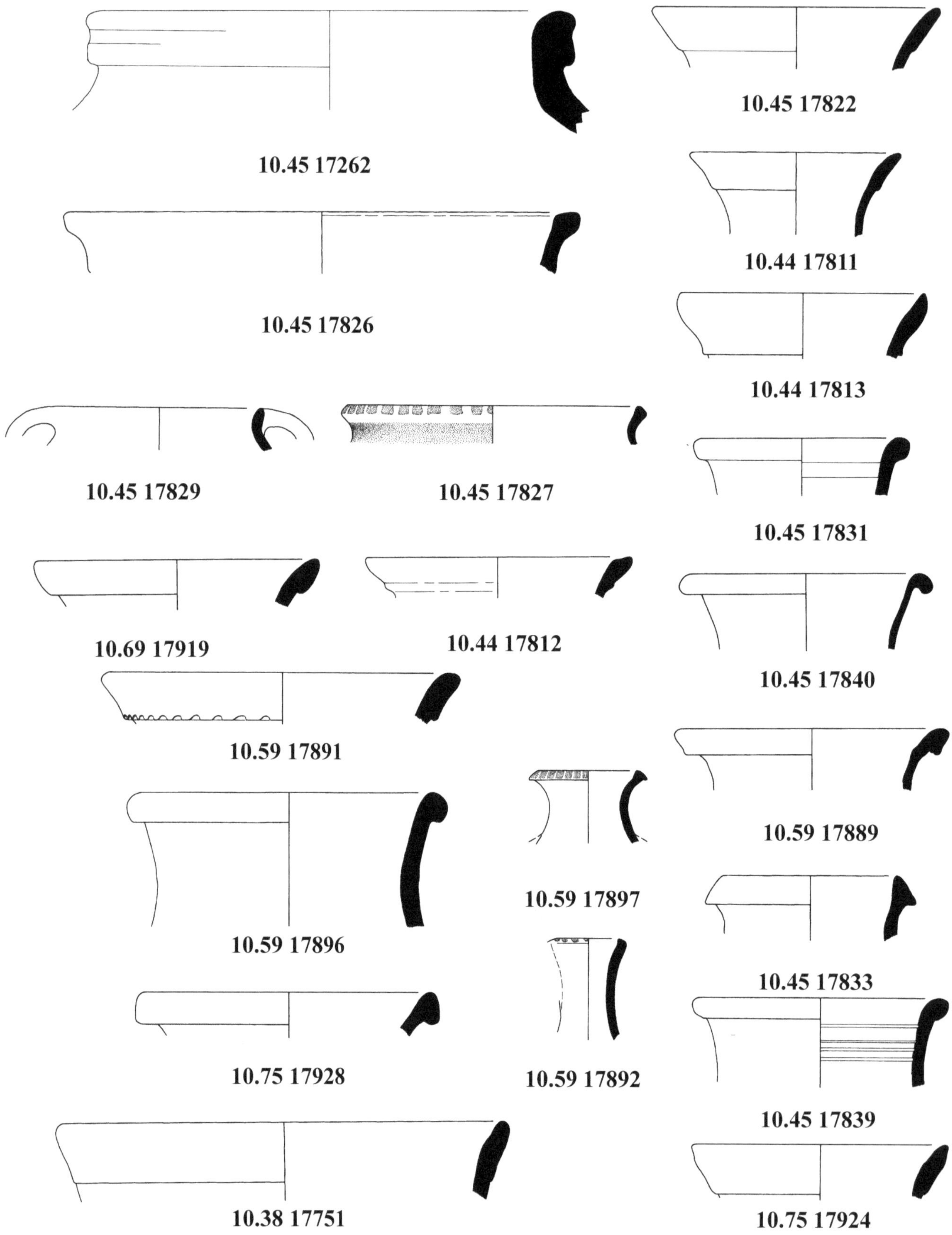

Figure 22a: Pella, Area XXVIIIC, Tower 1

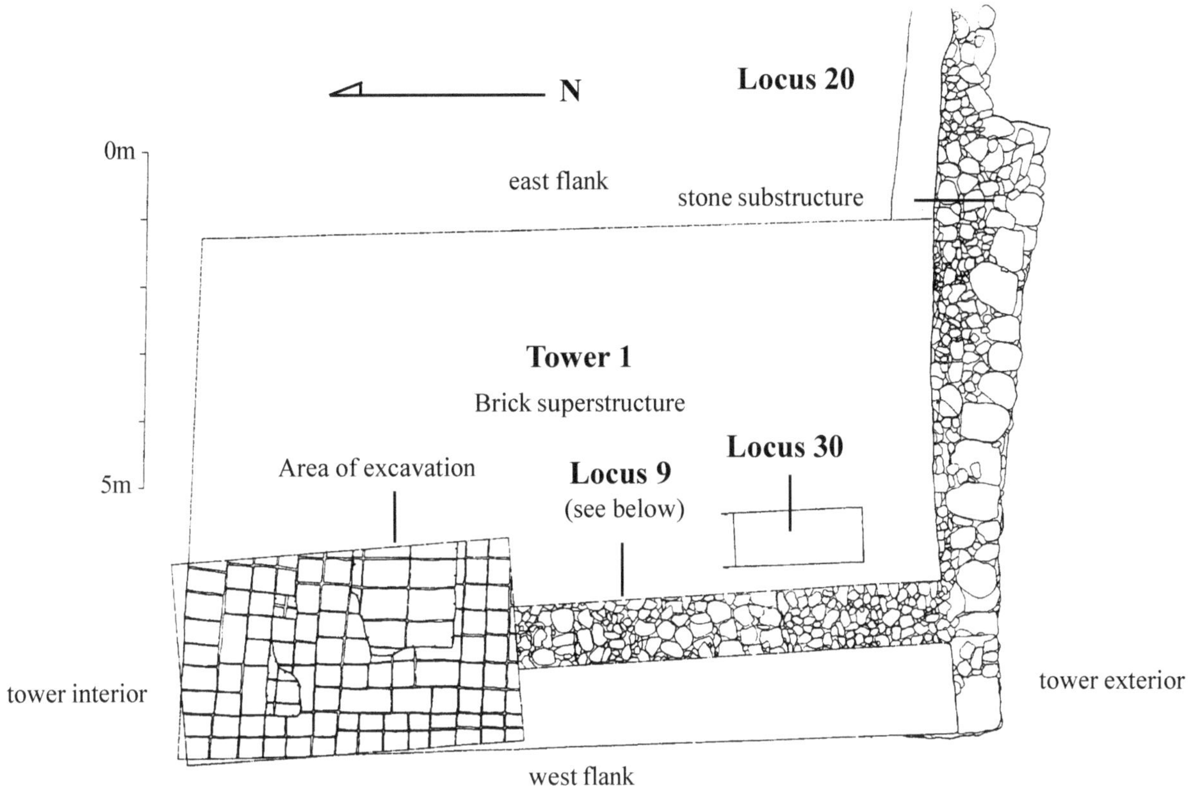

Figure 22b: Pella, Area XXVIIIC, Tower Composition, Locus 9

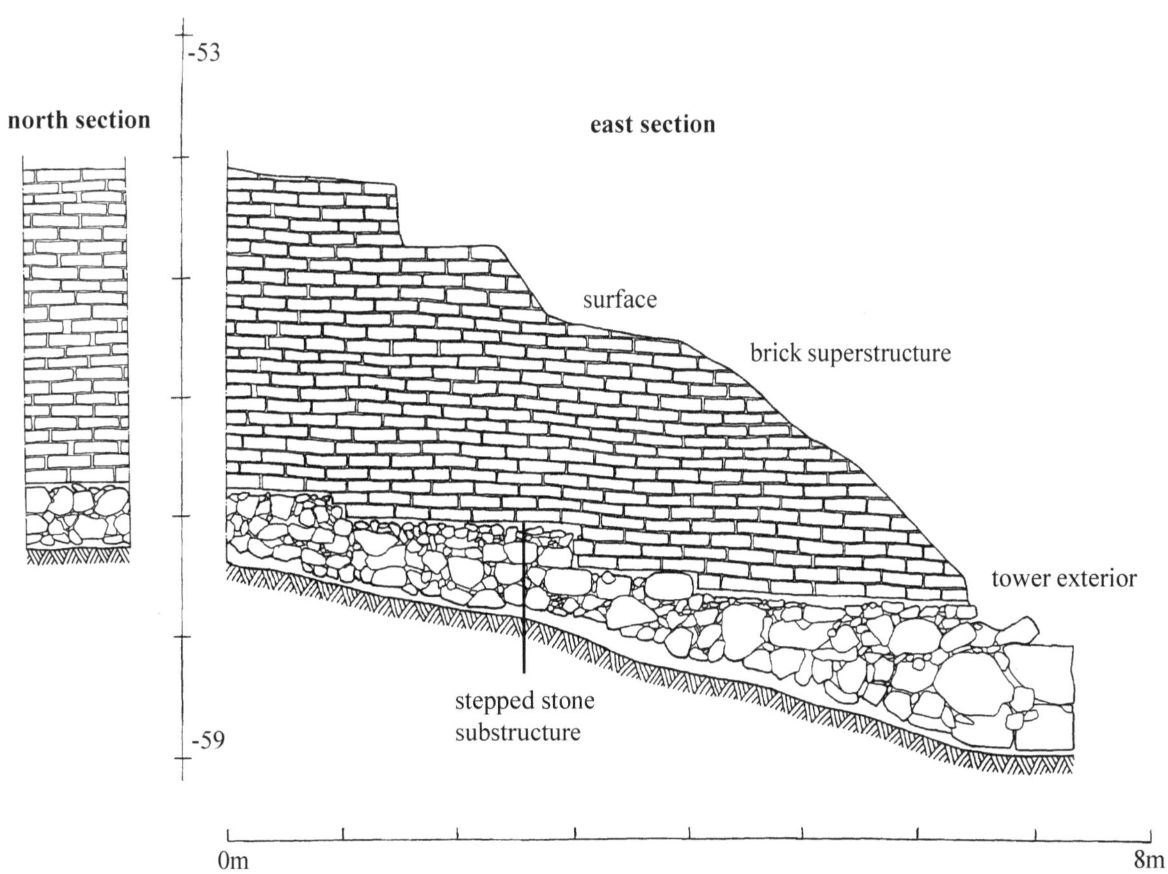

Figure 23a: Pella, Area XXVIIIC, Tower 1, West flank and Buttress 10.26

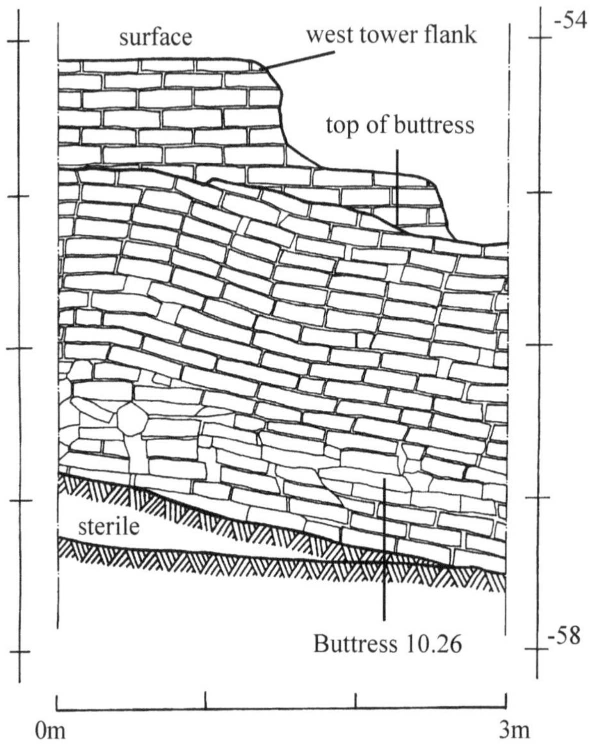

Figure 23b: Pella, Area XXVIIIC, Tower 1 Chamber, Locus 30

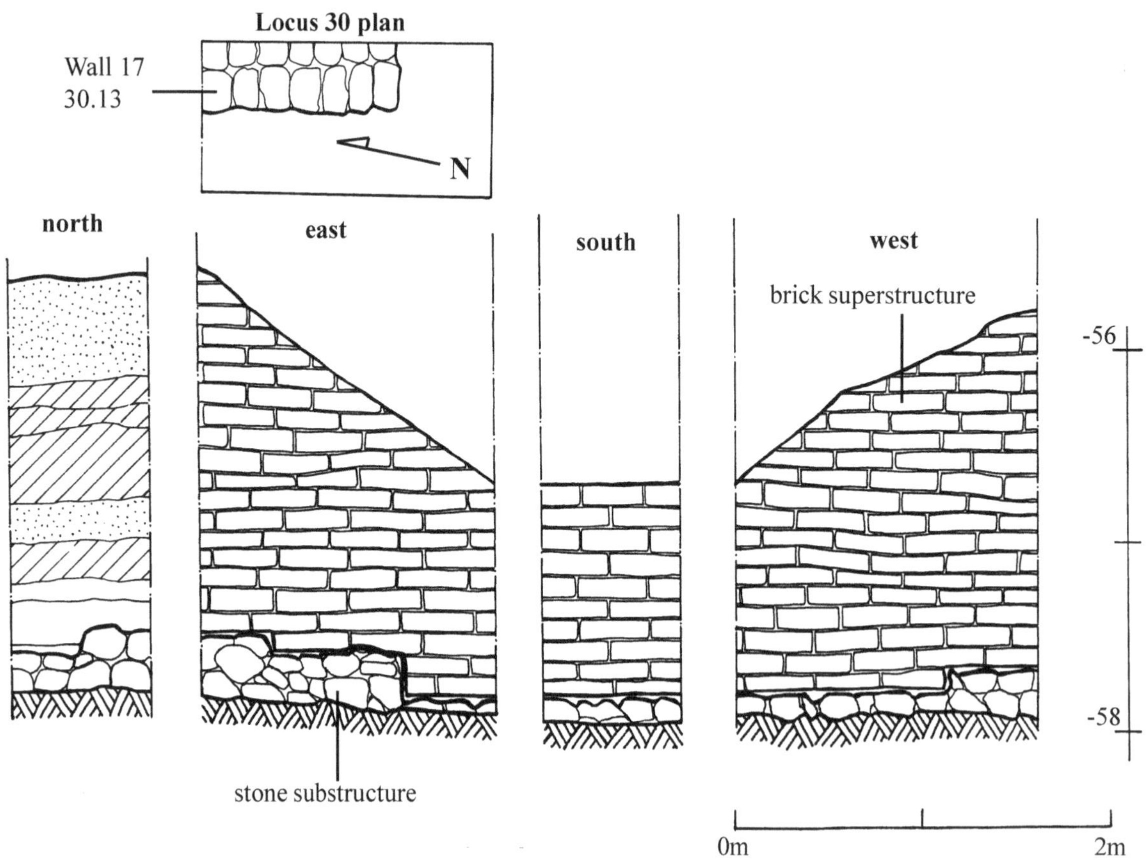

Figure 24a: Pella, Area XXVIIIC, Locus 10 Plan

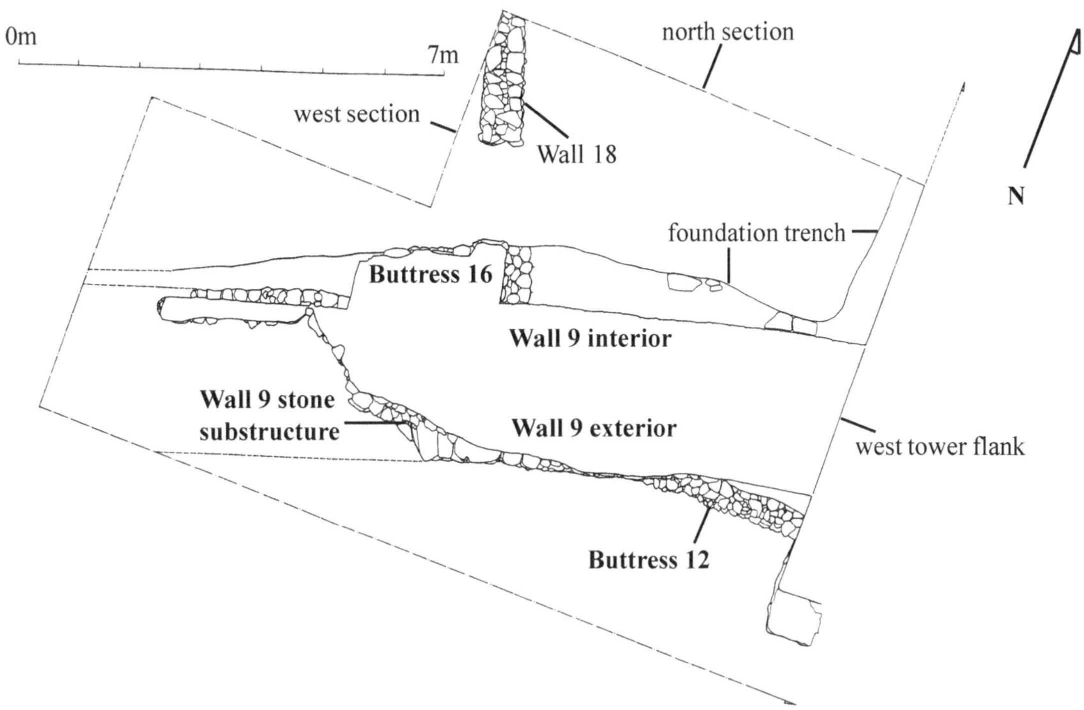

Figure 24b: Pella, Area XXVIIIC, Wall 9 Interior

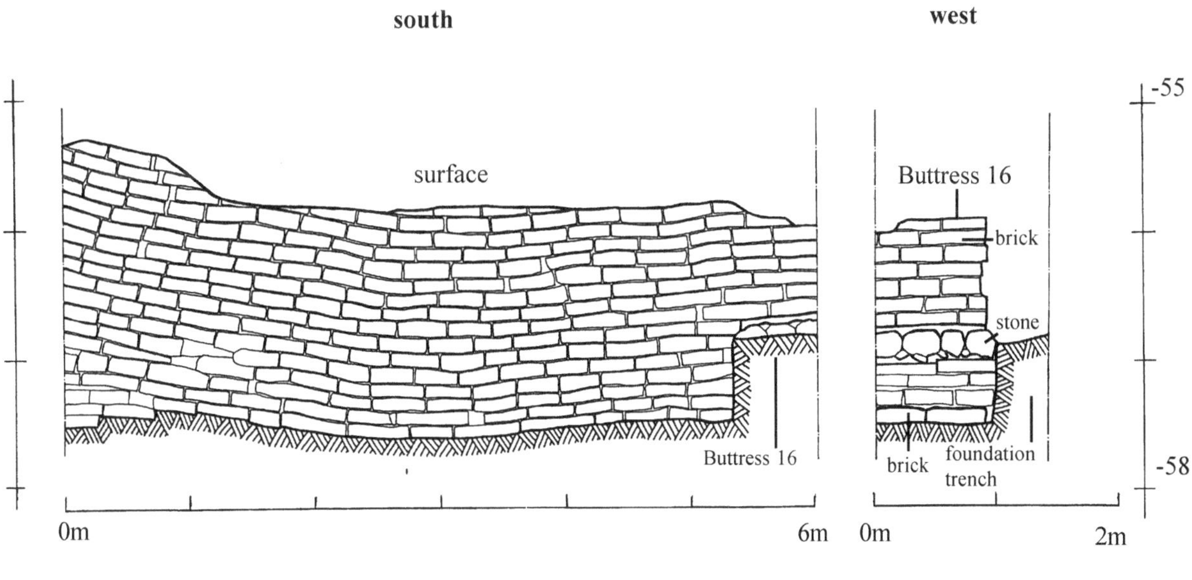

Figure 25a: Pella, Area XXVIIIC, Locus 20 Plan

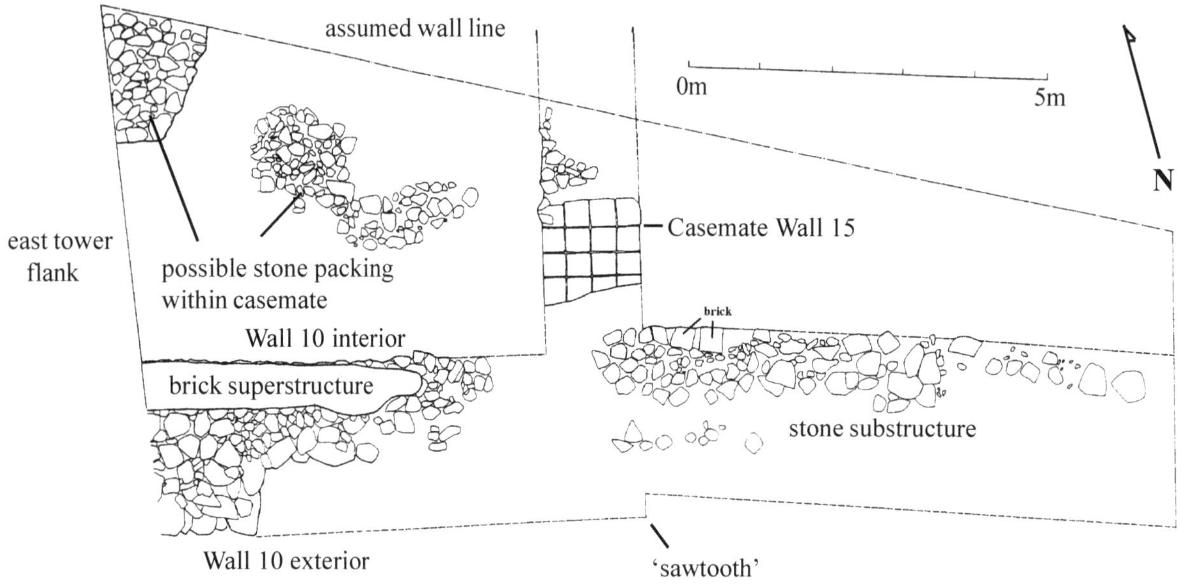

Figure 25b: Pella, Area XXVIIIC, Locus 20, Wall 10 Interior and West Tower Flank

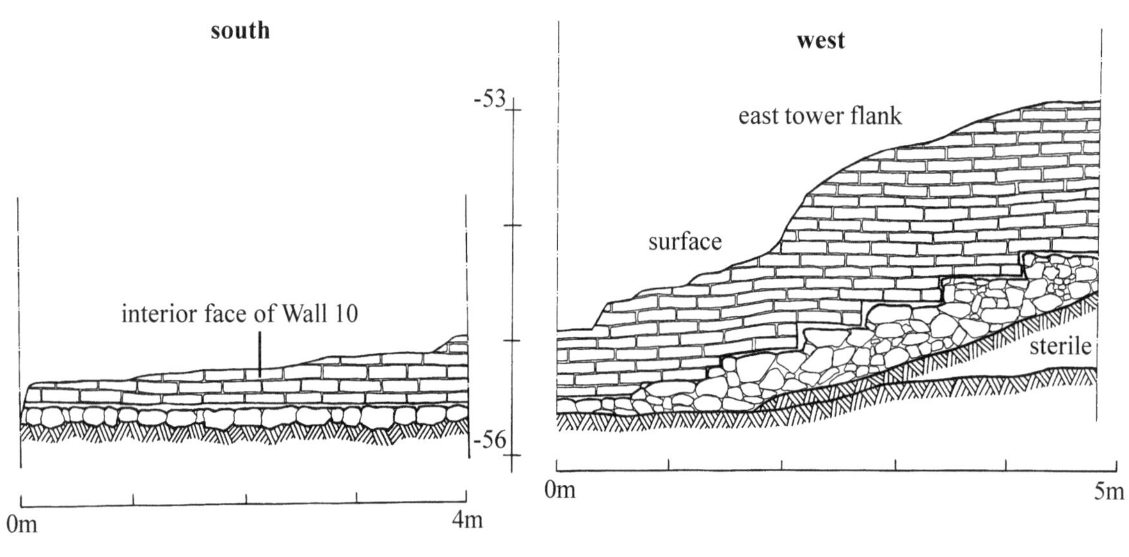

Figure 26: Pella, Area XXVIIIC, Locus 10, Phase E, Pottery and Comparative Ceramic Analysis

Period	Parallel	Variation	Total	%
EBIV				
MBI	20	29	49	31.6
MBII	24	53	77	49.7
MBIII	16	13	29	18.7
Total	60	95	155	

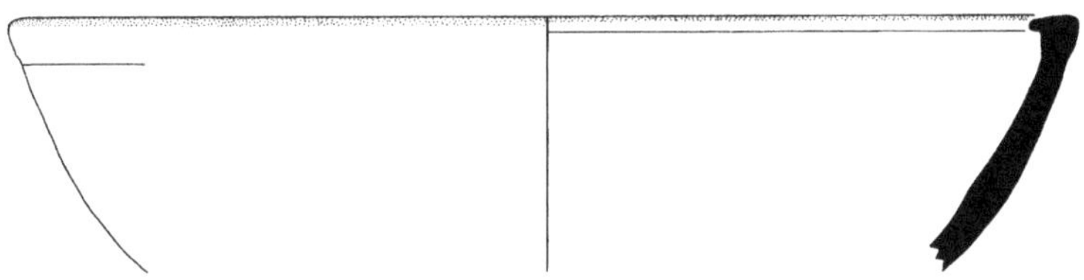

10.3 17145

10.3 17146

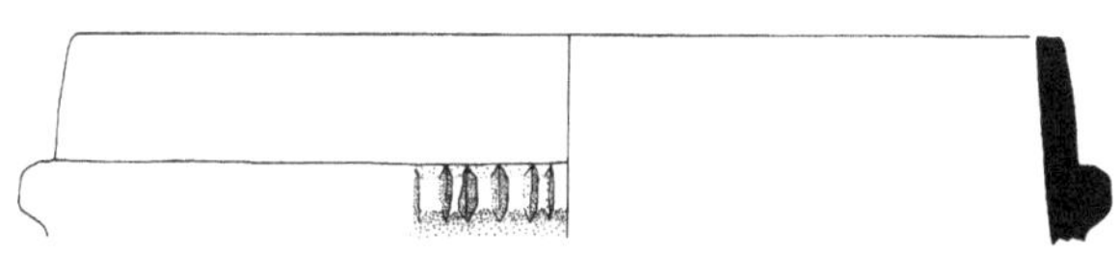

10.15 17190

10.15 17191

10.15 17192

10.15 17196

10.15 17195

Figure 27: Pella, Area XXVIIIC, Locus 10, Phase E, Pottery and Comparative Ceramic Analysis (cont.)

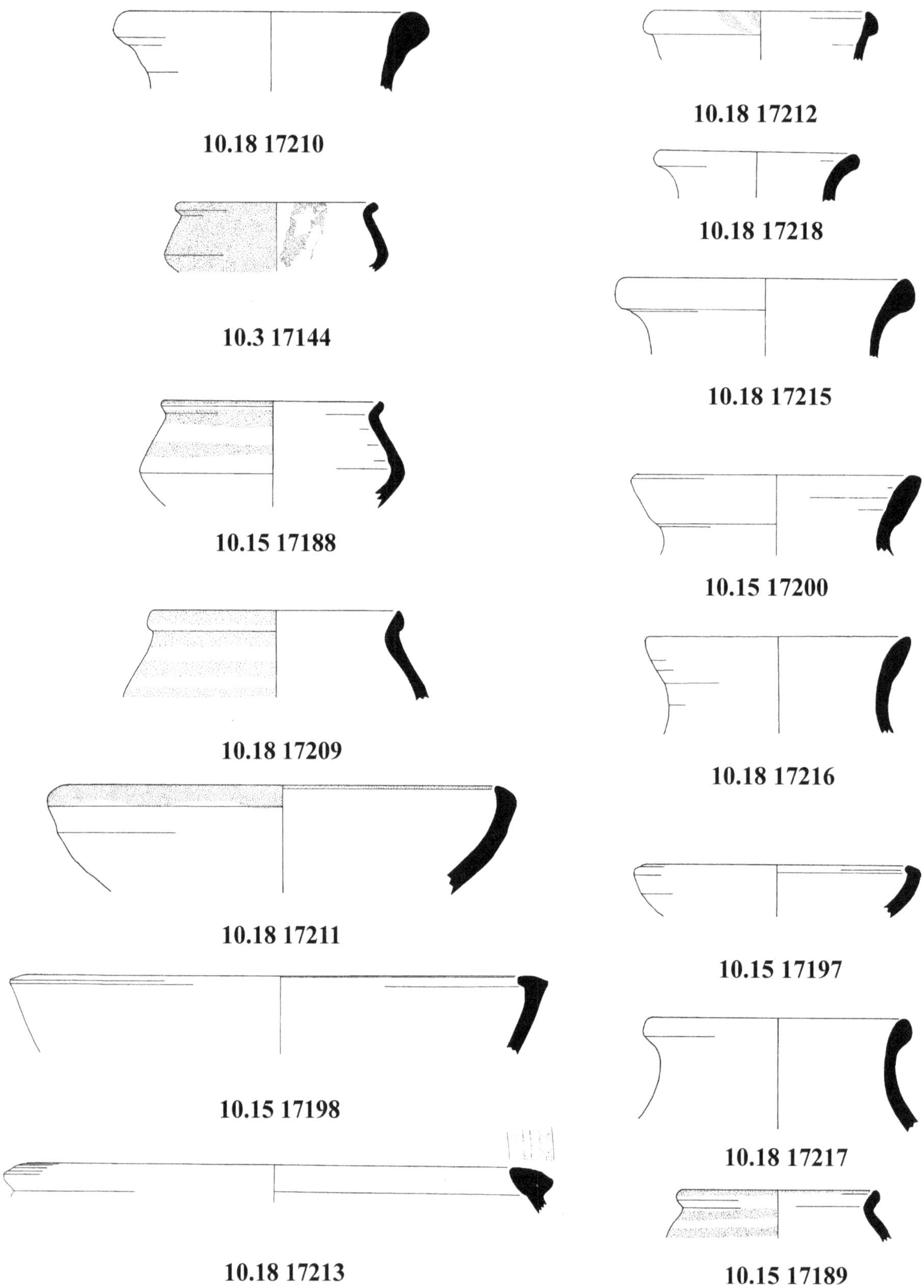

Figure 28: Rukeis, Site Plan

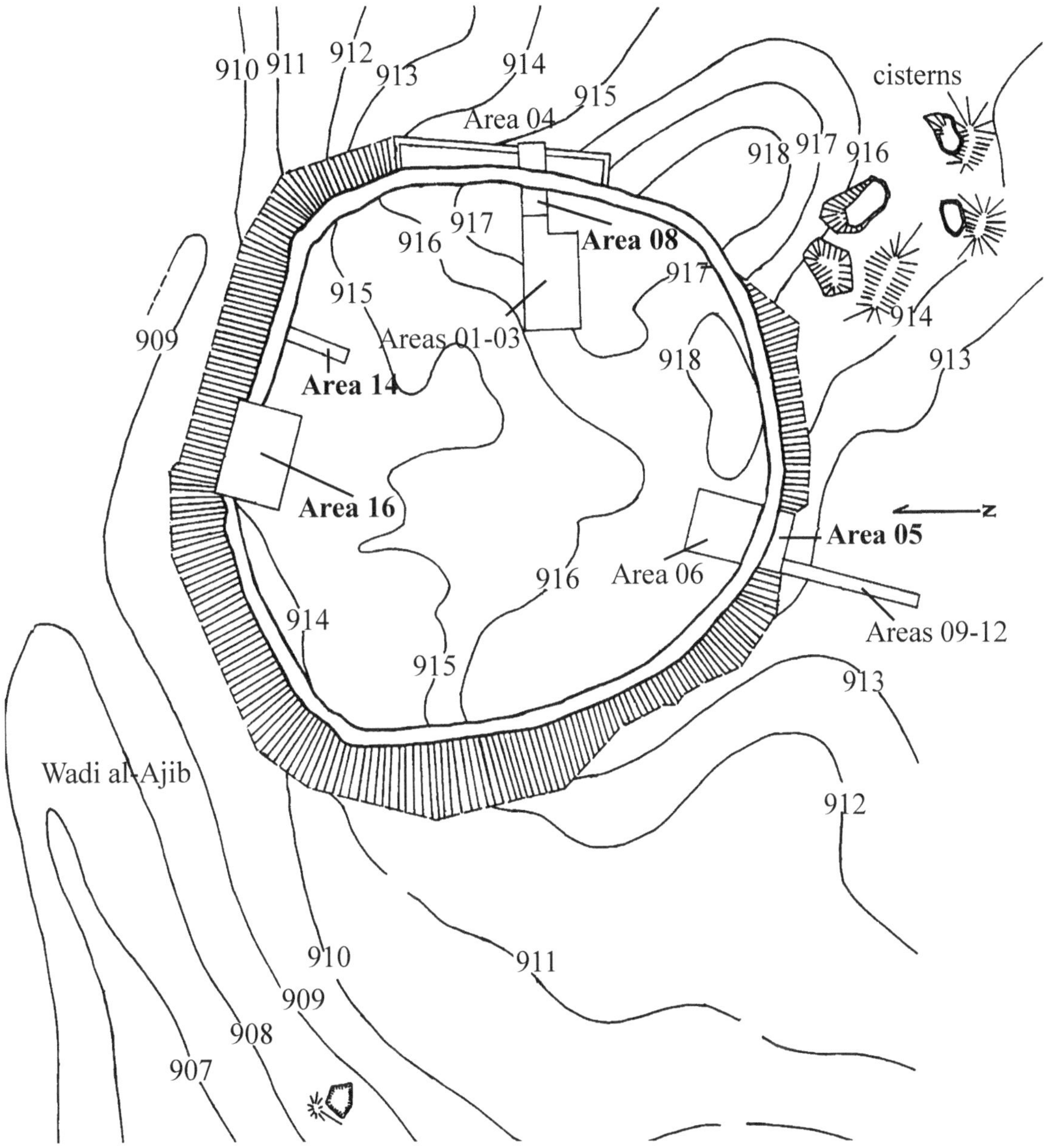

Figure 29a: Rukeis, Site Plan

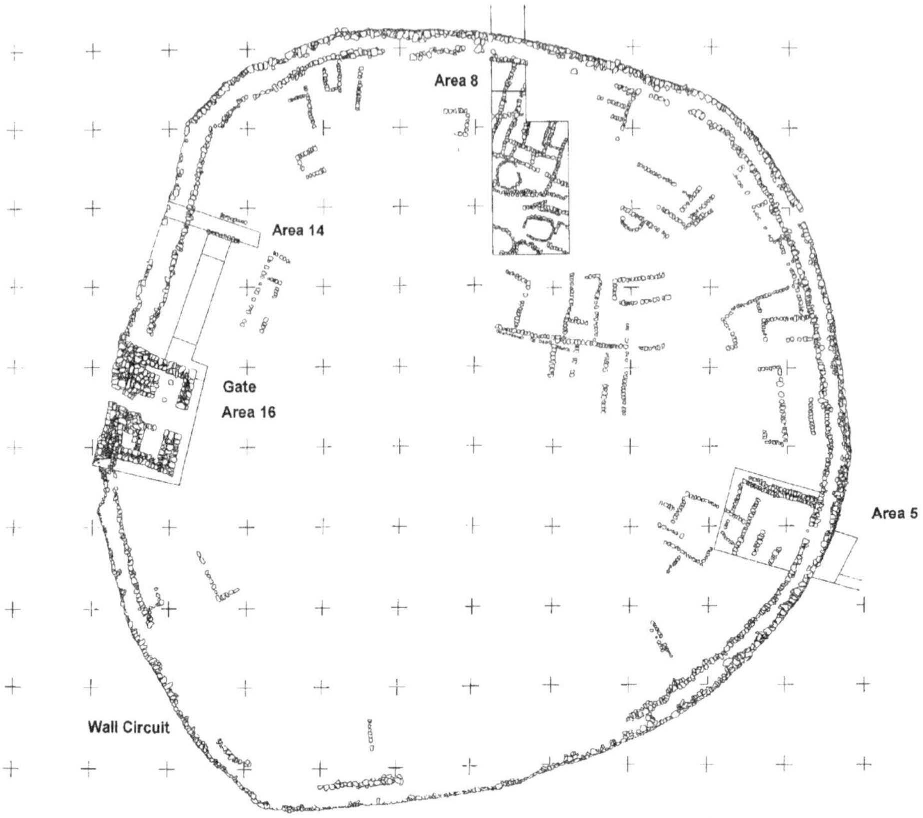

Figure 29b: Rukeis, Aerial Site Photograph

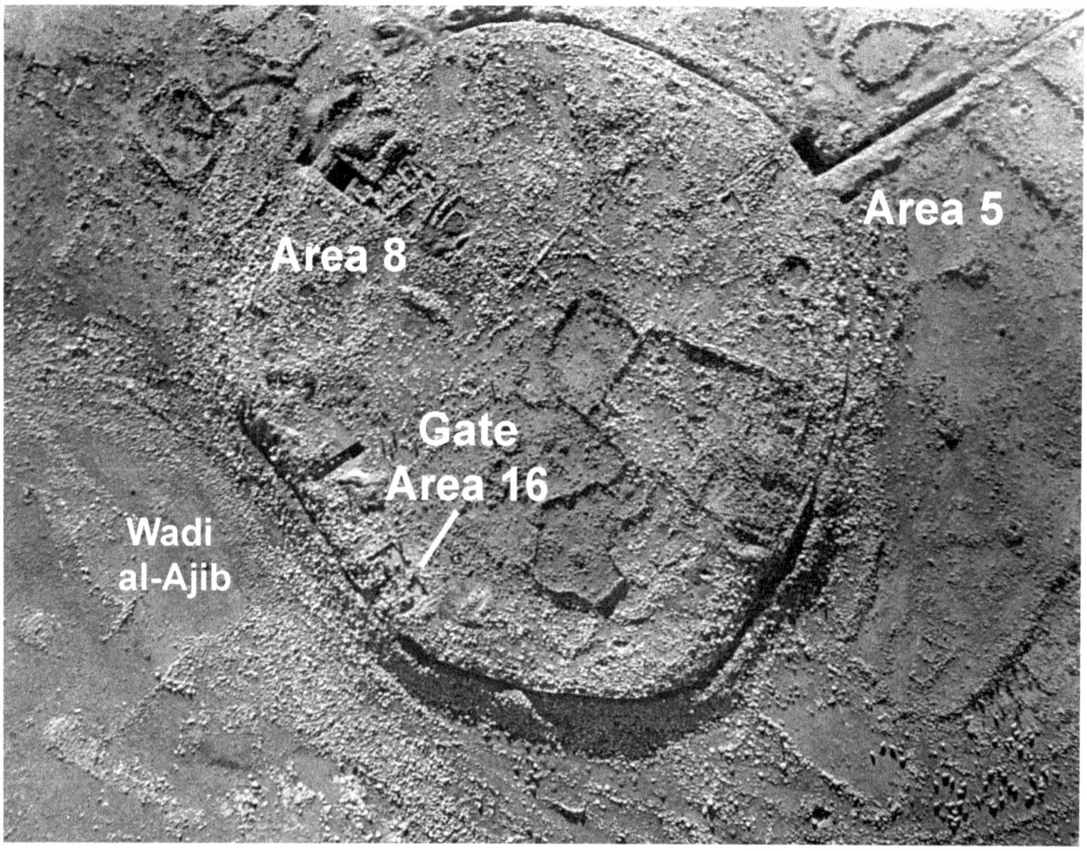

Figure 30: Rukeis, Area 05, West Section, Stratigraphy and Phasing

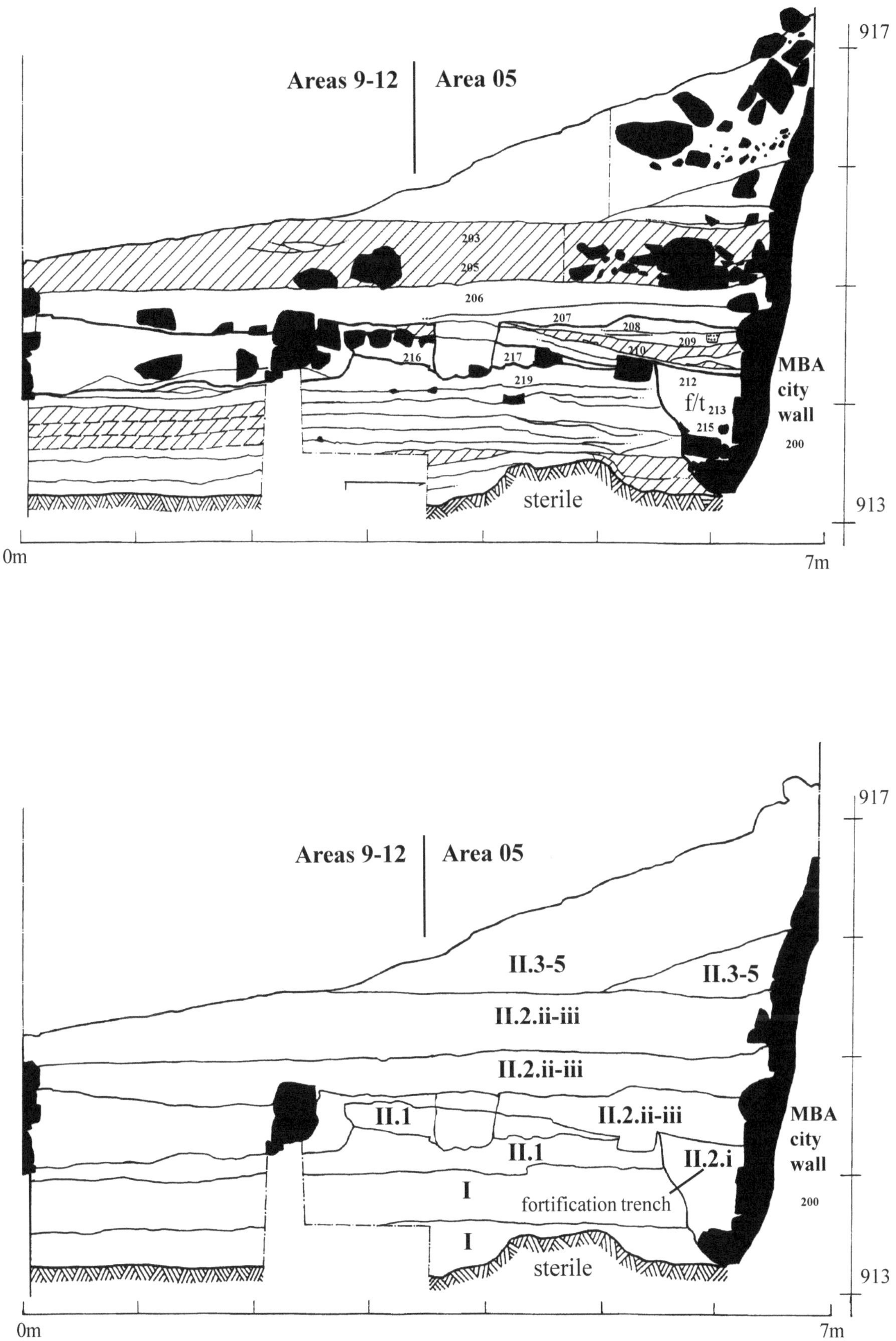

Figure 31: Rukeis, Area 05, South Section, Stratigraphy and Phasing

Area 05

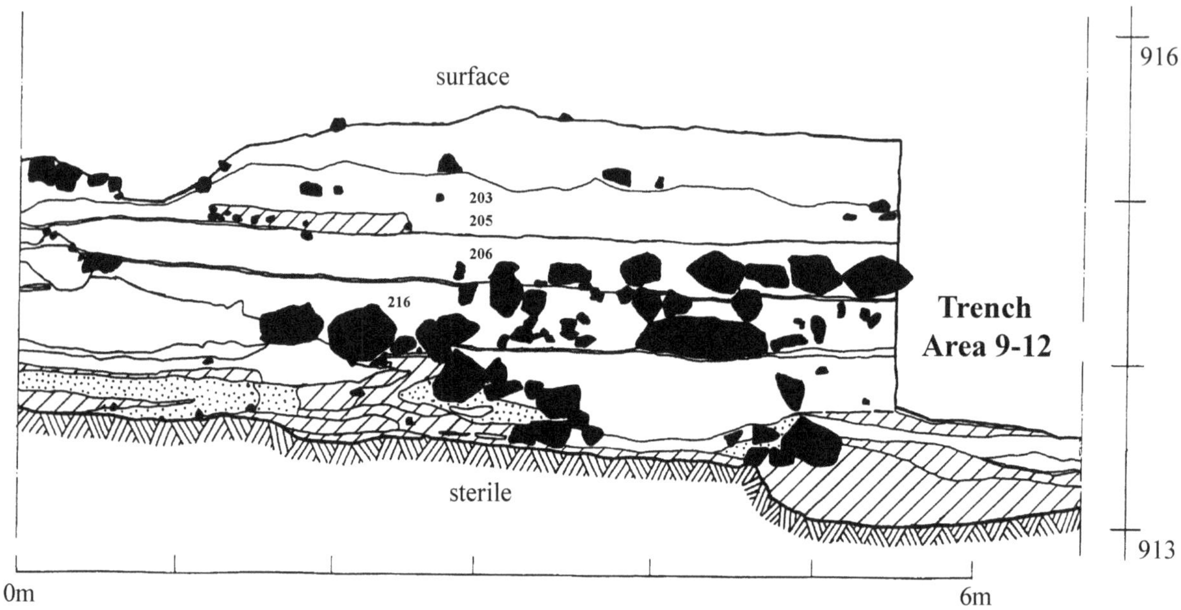

Area 05

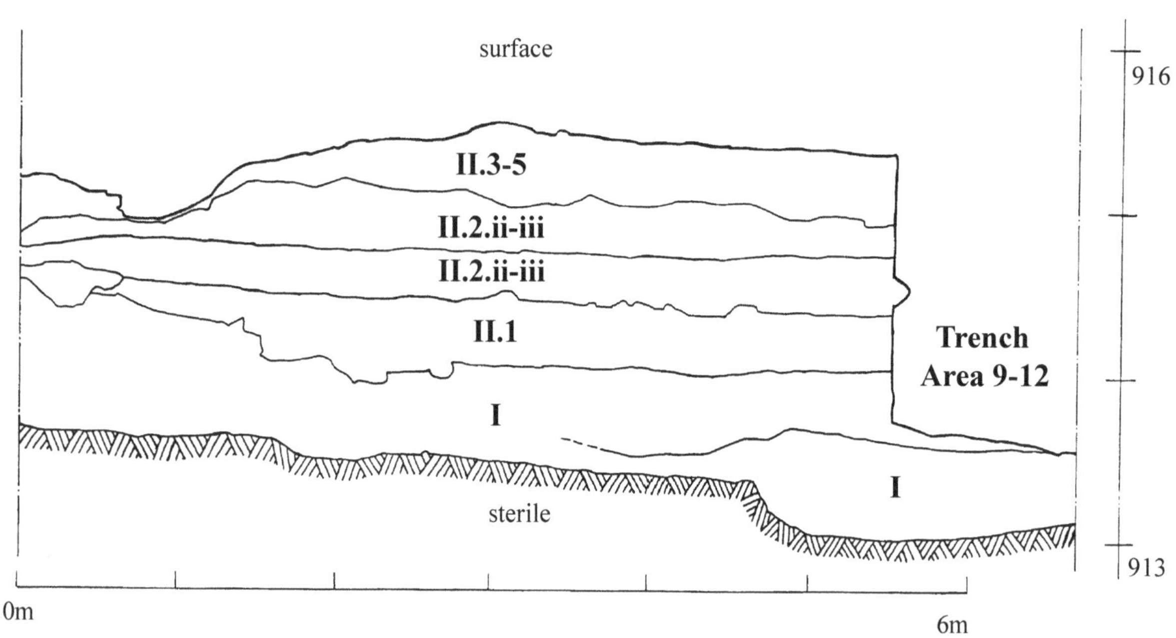

Figure 32: Rukeis, Area 05, East Section, Stratigraphy and Phasing

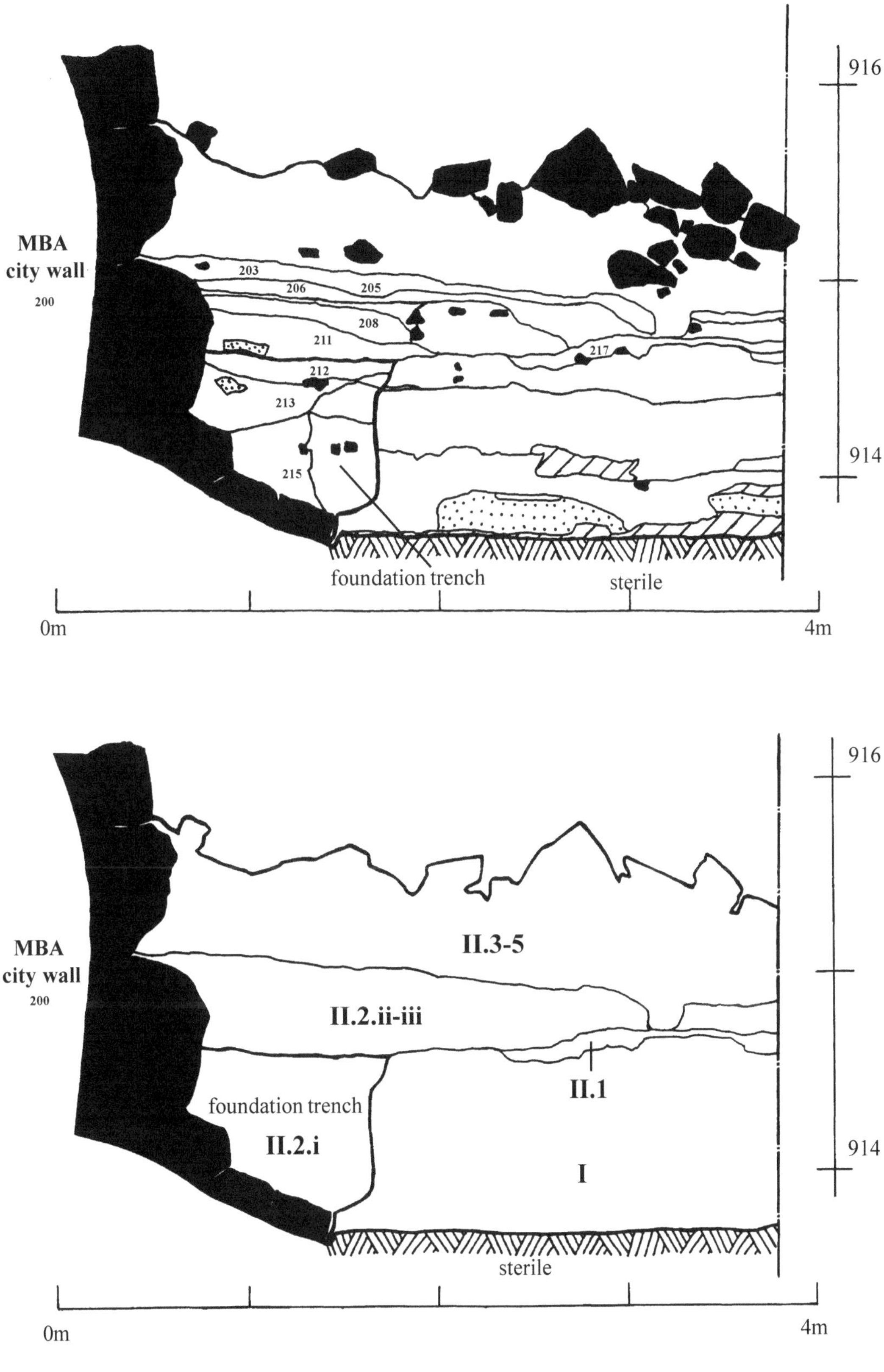

Figure 33: Rukeis, Area 05, Phase II.1, Pottery and Comparative Ceramic Analysis

Period	Parallel	Variation	Total	%
EBIV	2		2	2.7
MBI	8	17	25	33.8
MBII	18	16	34	46
MBIII		13	13	17.6
Total	28	46	74	

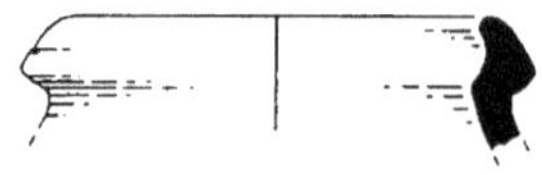

216 1641

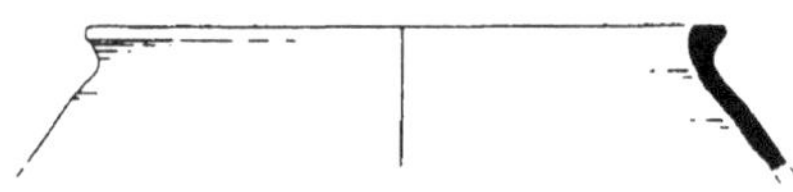

216 1897

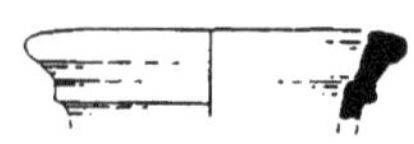

216 1899

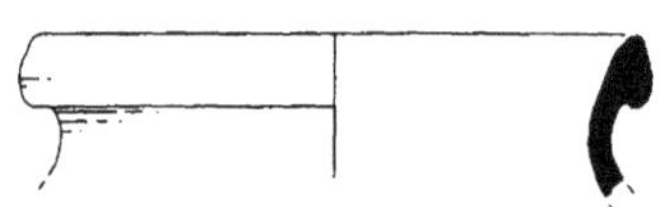

217 1485

219 1866

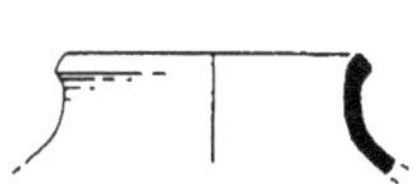

217 1850

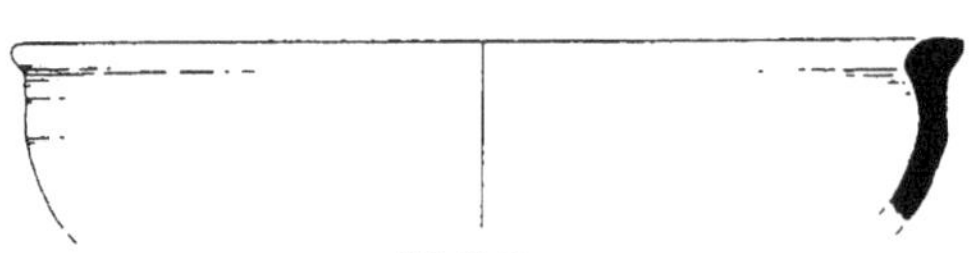

216 2

Figure 34: Rukeis, Area 05, Phase II.2.i, Pottery and Comparative Ceramic Analysis

Period	Parallel	Variation	Total	%
EBIV	1	1	2	3.3
MBI	1	8	9	25
MBII	8	18	26	43.3
MBIII		15	15	28.3
Total	10	42	52	

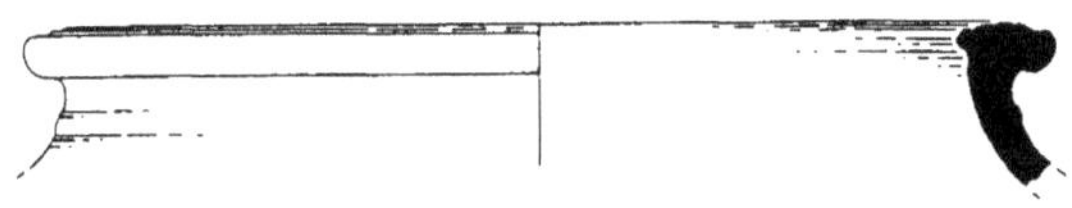

212 1286

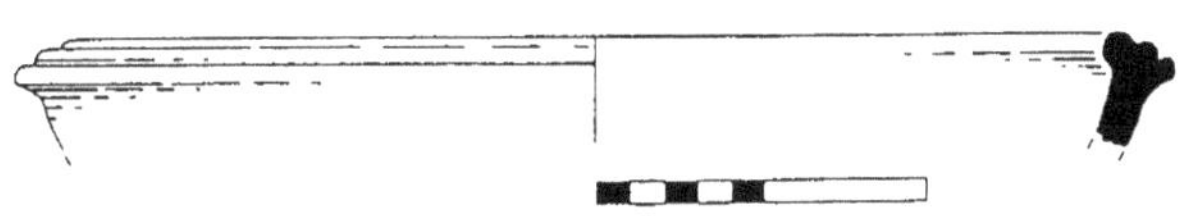

212 1759

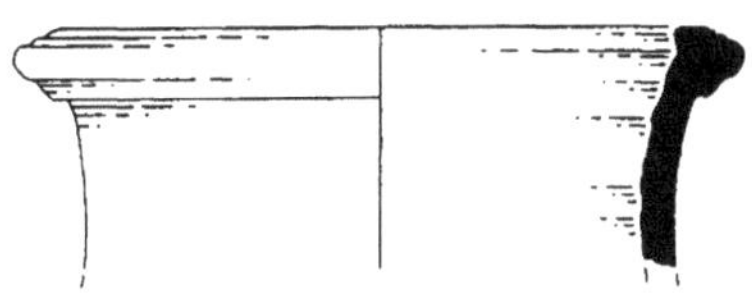

213 312

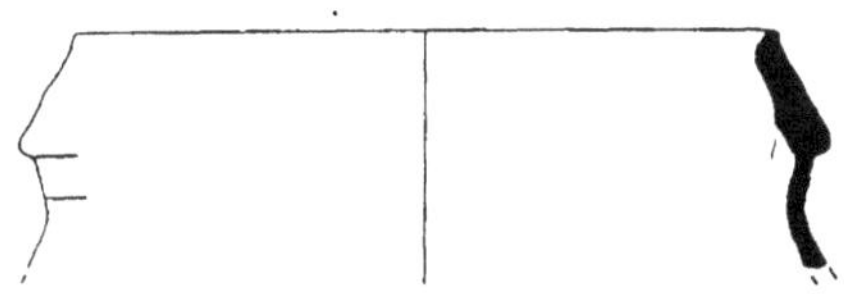

215 1786

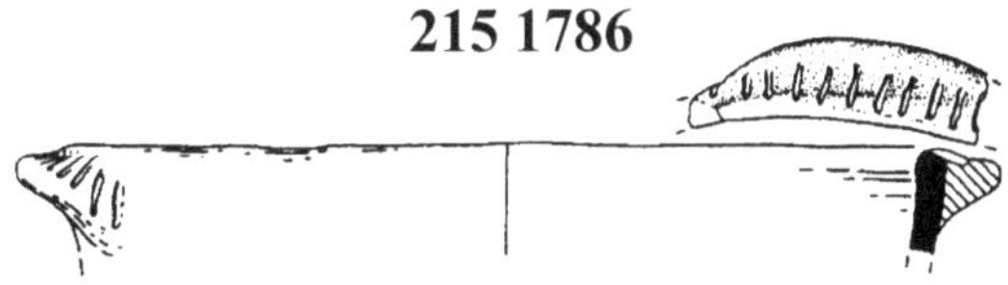

215 1033

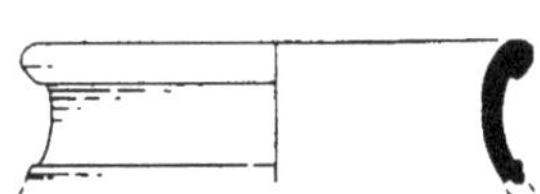

215 1787

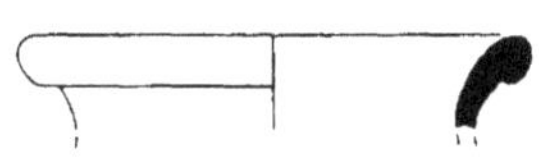

215 1791

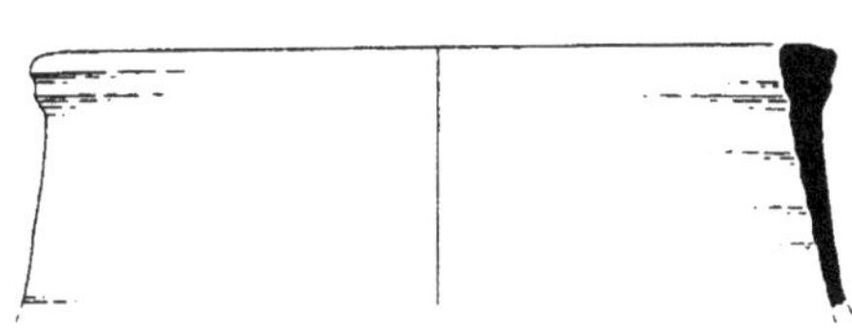

215 1788

Figure 35: Rukeis, Area 05, Phase II.2.ii, Pottery and Comparative Ceramic Analysis

Period	Parallel	Variation	Total	%
EBIV	8	1	9	6.6
MBI	17	27	44	32.1
MBII	30	26	56	40.9
MBIII	4	24	28	20.4
Total	59	78	137	

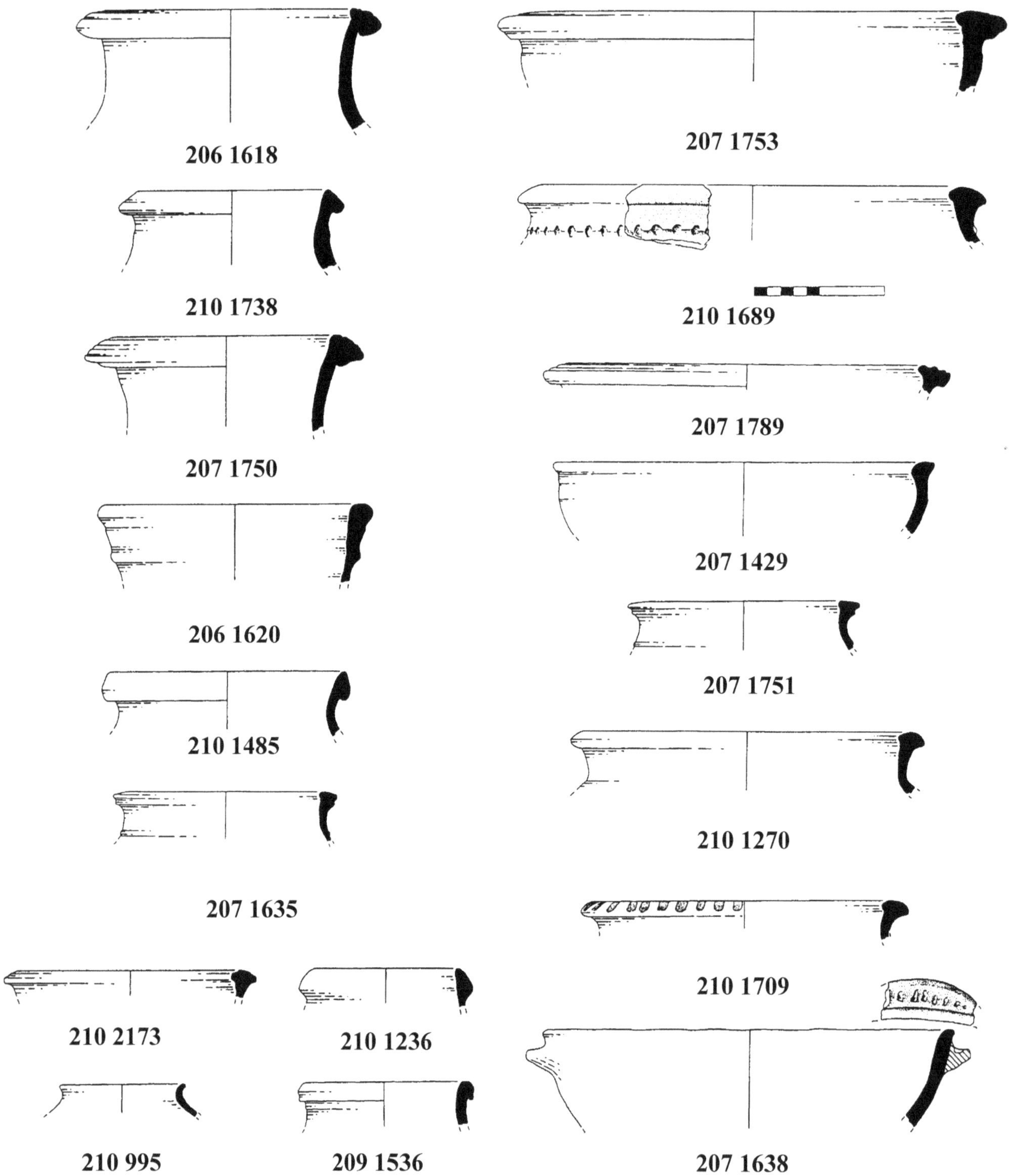

Figure 36: Rukeis, Area 08, Plan, North Section, Stratigraphy and Phasing

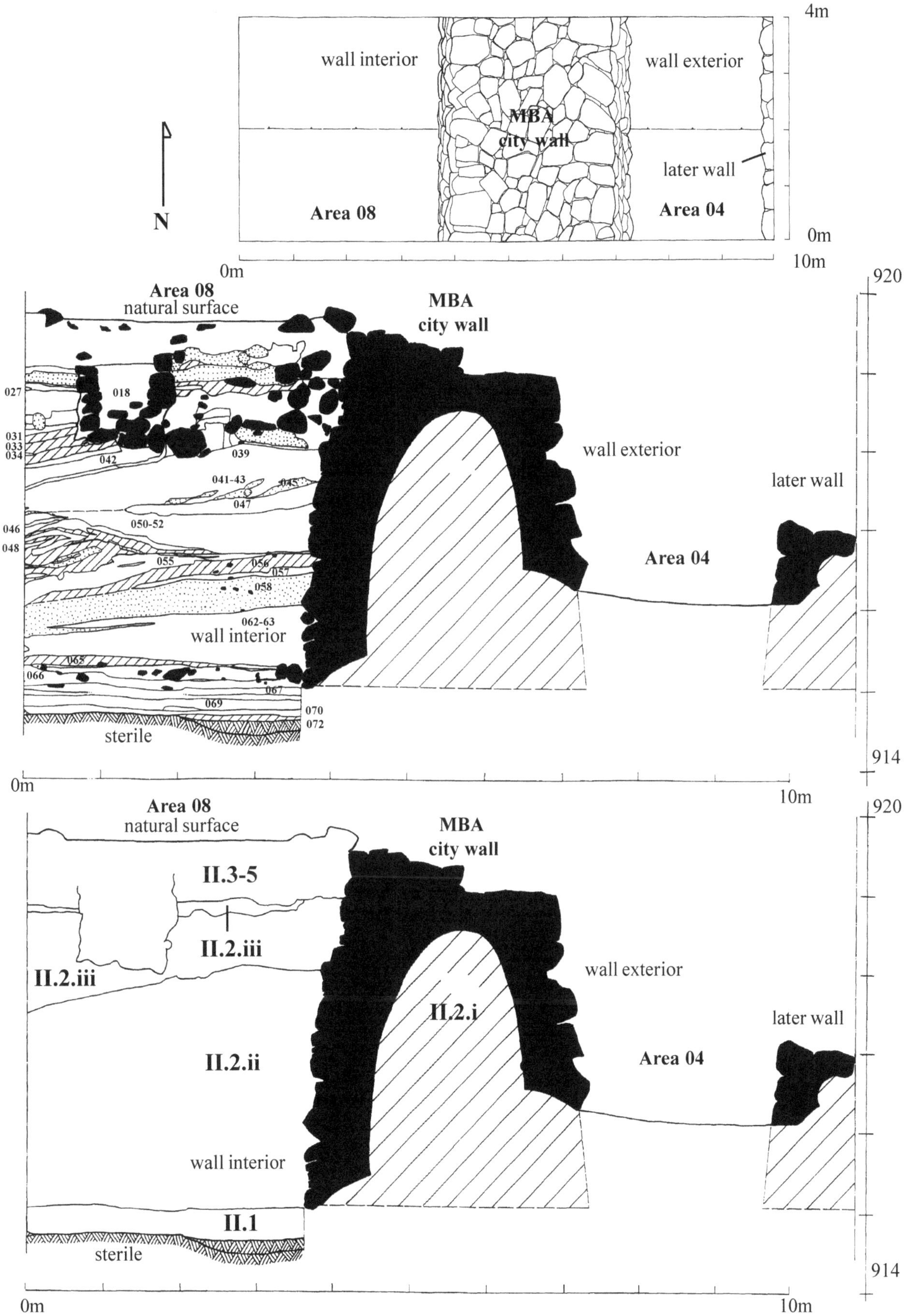

Figure 37: Rukeis, Area 08, South and West Sections, Stratigraphy and Phasing

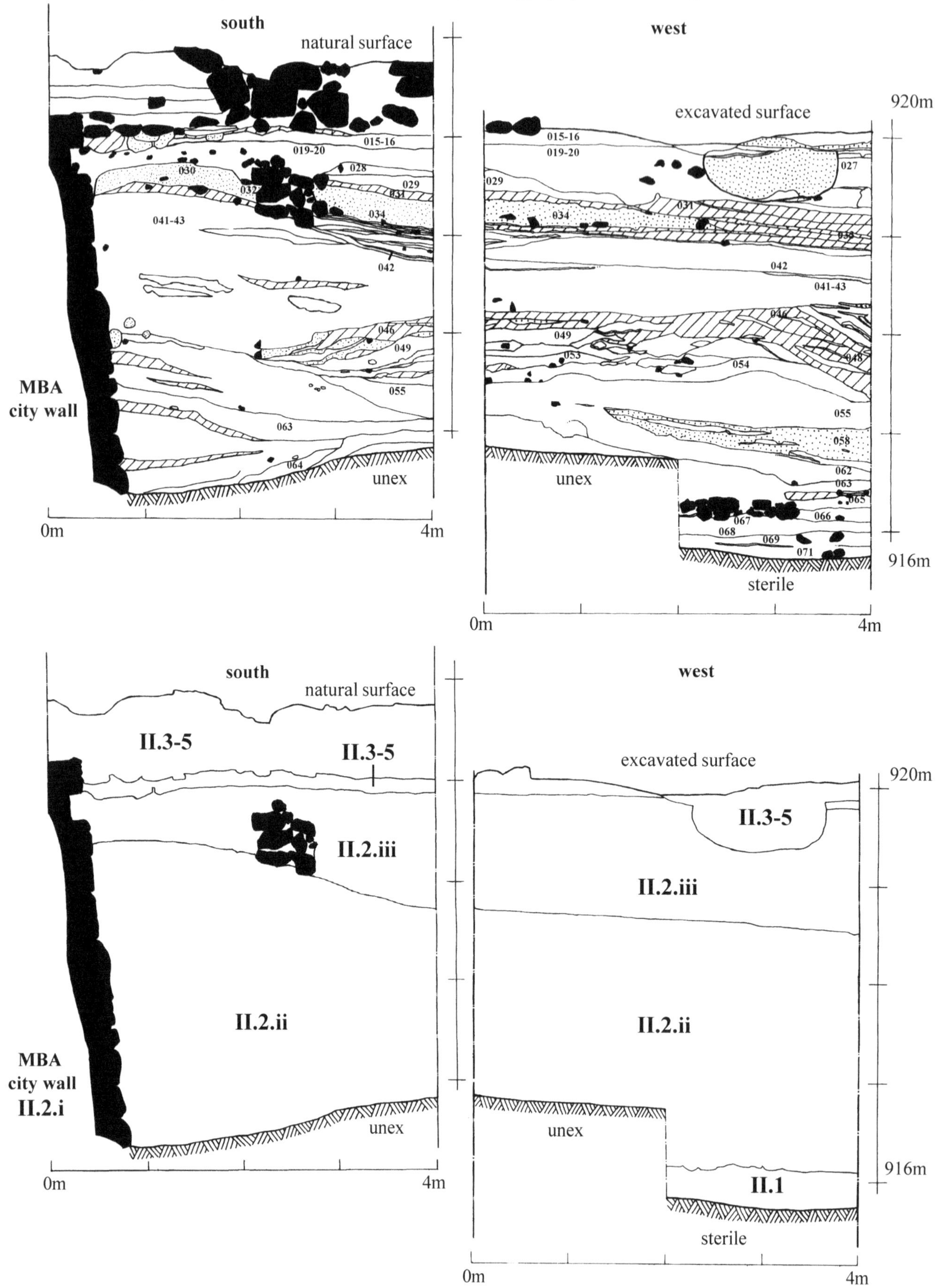

Figure 38: Rukeis, Area 08, Phase II.1, Pottery and Comparative Ceramic Analysis

Period	Parallel	Variation	Total	%
EBIV	5	4	9	9.5
MBI	6	11	17	17.9
MBII	17	24	41	43.2
MBIII	1	27	28	29.5
Total	29	66	95	

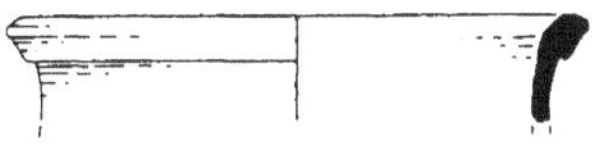

067 2158

067 2132

067 2154

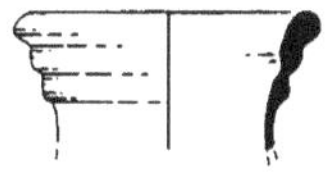

068 2164

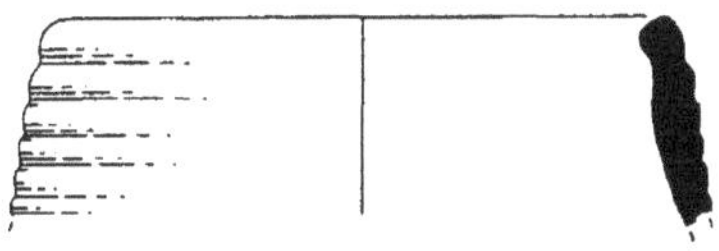

067 2131

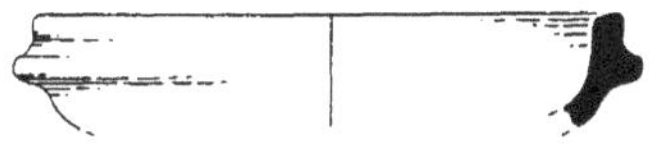

067 2159

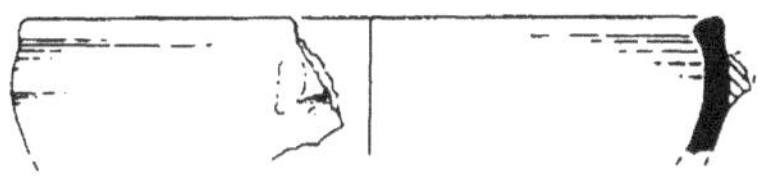

067 2128

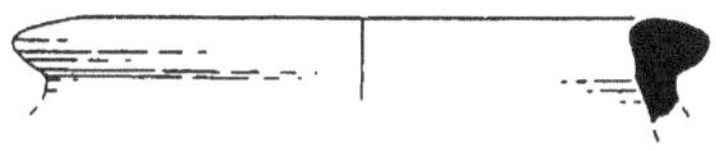

067 2155

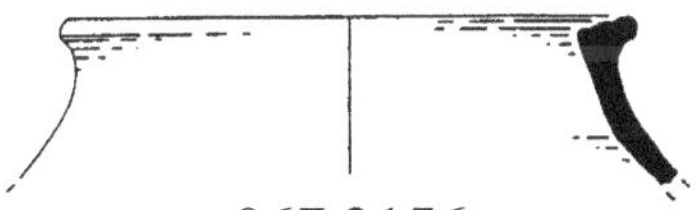

067 2156

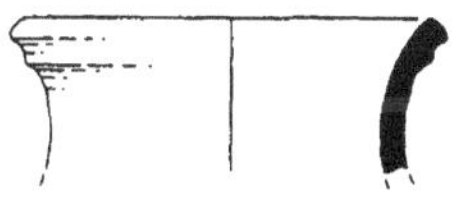

067 2127

068 2166

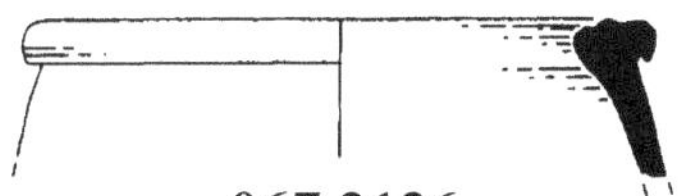

067 2126

Figure 39: Rukeis, Area 08, Phase II.2.ii, Pottery and Comparative Ceramic Analysis

Period	Parallel	Variation	Total	%
EBIV	9	16	25	8.1
MBI	34	32	66	21.3
MBII	47	78	125	40.3
MBIII	10	84	94	30.3
Total	100	210	310	

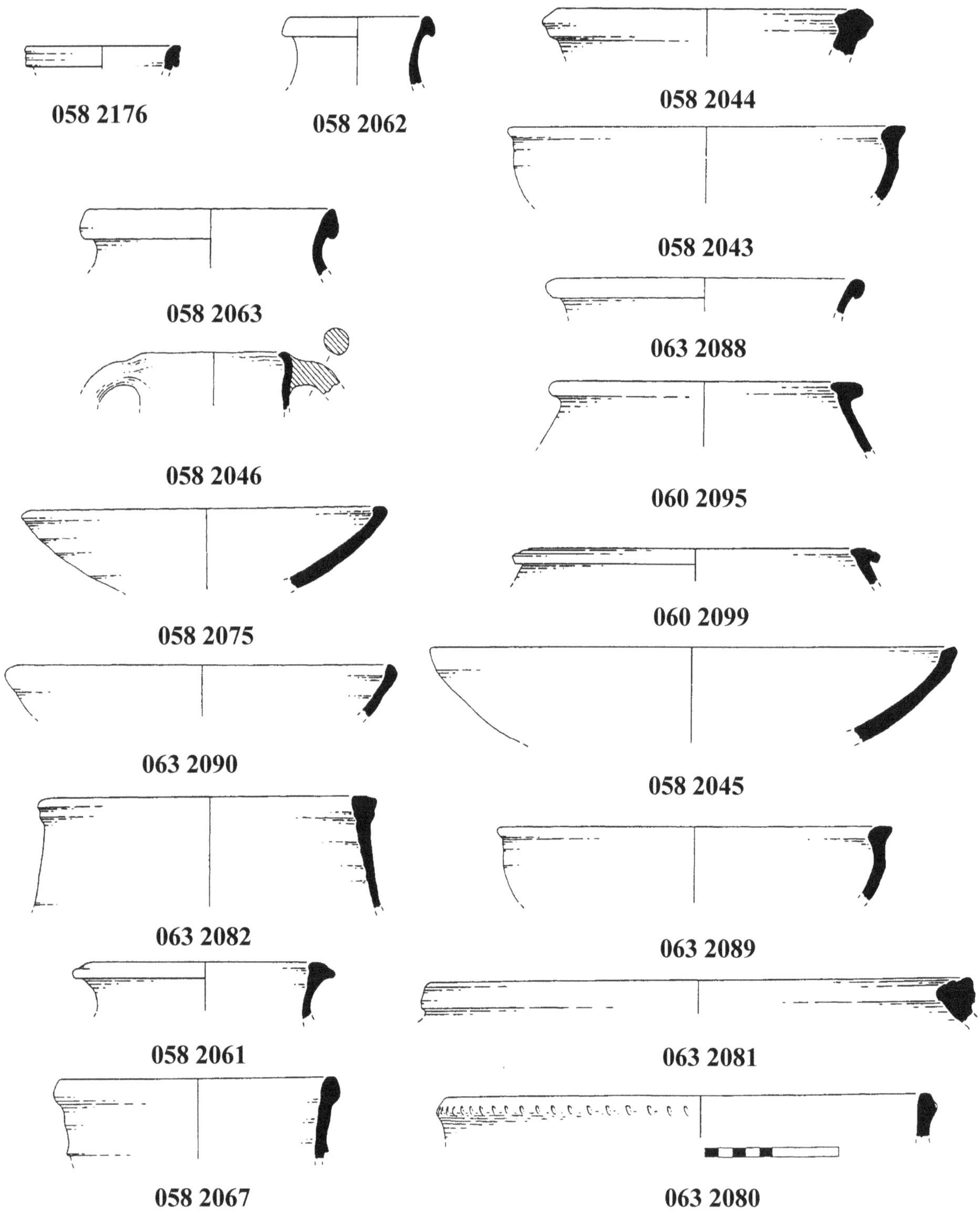

Figure 40: Rukeis, Area 08, Phase II.2.ii, Pottery and Comparative Ceramic Analysis (cont.)

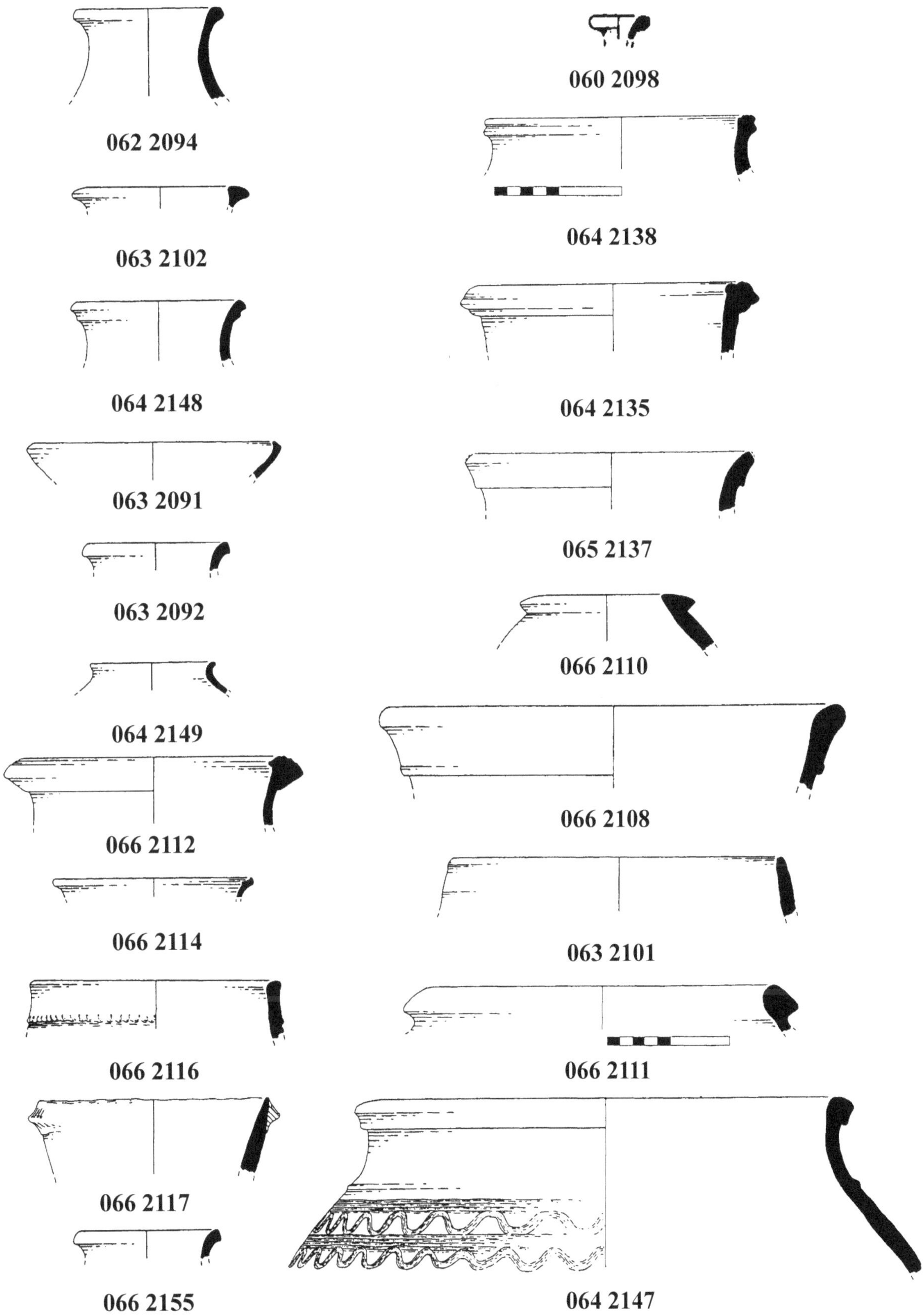

Figure 41: Rukeis, Area 14, Plan, West Section, Stratigraphy and Phasing

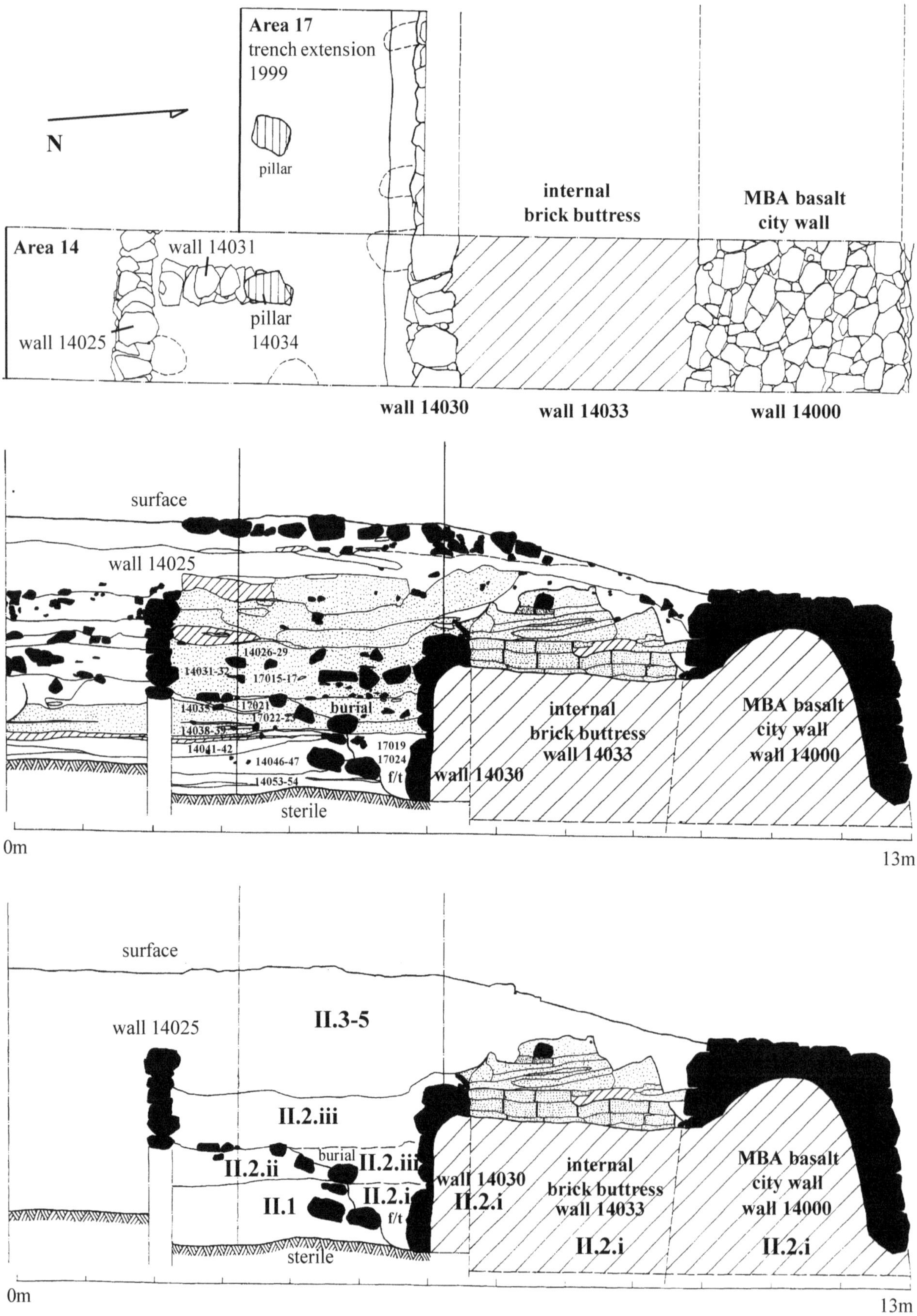

Figure 42a: Rukeis, Area 14, Phase II.2.i, Pottery and Comparative Ceramic Analysis

Period	Parallel	Variation	Total	%
EBIV				
MBI	1	4	5	22.7
MBII	1	9	10	45.4
MBIII		7	7	31.8
Total	2	20	22	

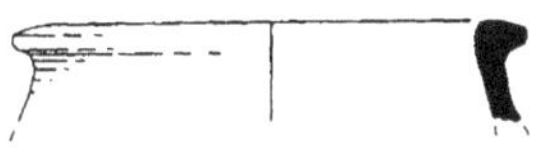

17024 9917

17024 9922

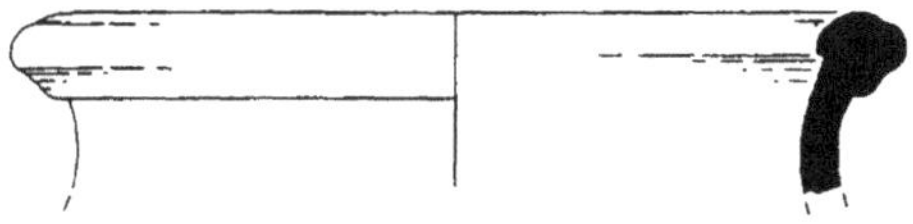

17024 9926

Figure 42b: Rukeis, Area 14, Phase II.2.ii, Pottery and Comparative Ceramic Analysis

Period	Parallel	Variation	Total	%
EBIV	2	2	4	3.6
MBI	15	13	28	25
MBII	19	38	57	50.9
MBIII		23	23	20.5
Total	36	76	112	

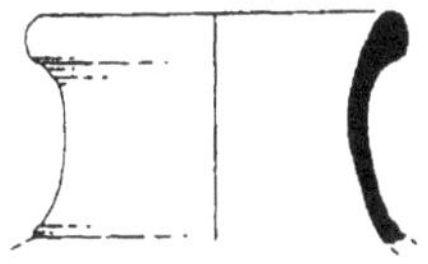

14035 2191

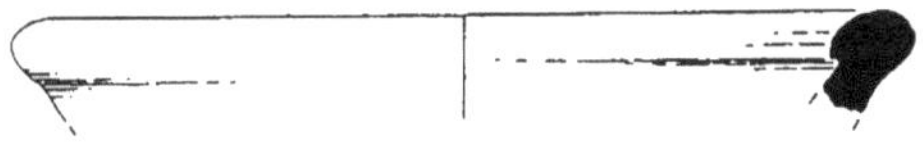

17021 9913

14035 2182

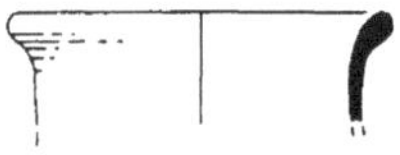

17023 9904

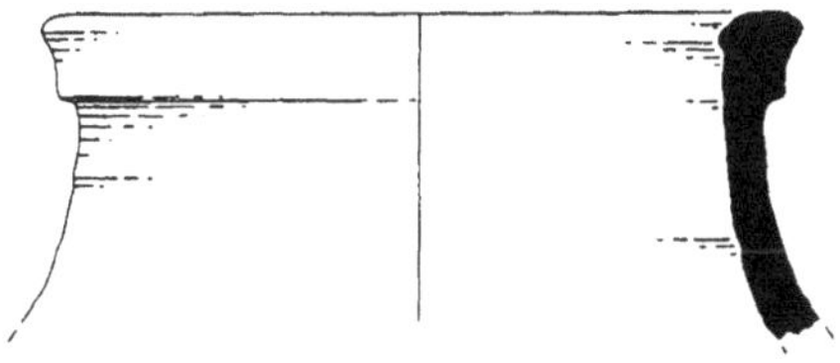

14035 2211

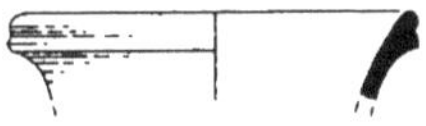

17023 9905

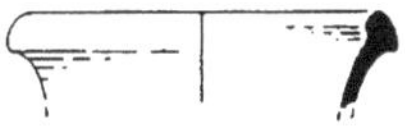

14035 2188

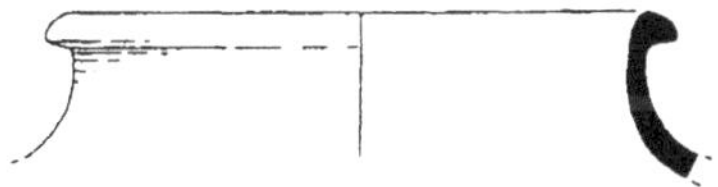

17022 9903

14035 2214

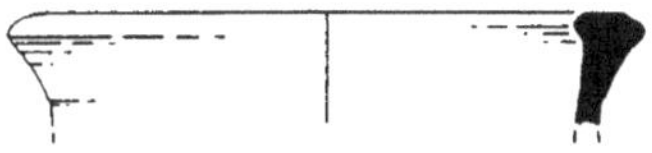

17023 9906

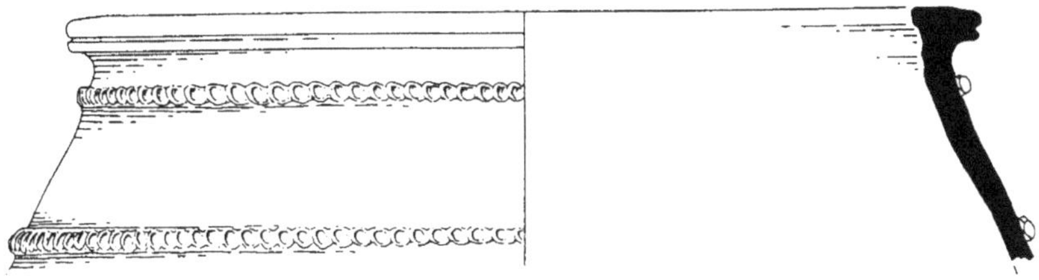

14035 2181

17022 9907

Figure 43: Rukeis, Area 14, Phase II.2.iii, Pottery and Comparative Ceramic Analysis

Period	Parallel	Variation	Total	%
EBIV				
MBI	17	12	29	33.3
MBII	14	25	39	44.8
MBIII	1	18	19	21.8
Total	32	55	87	

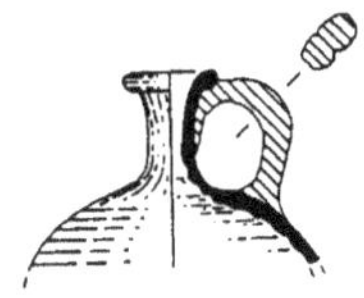

14044 2223

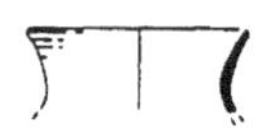

17017 9911

14037 2197

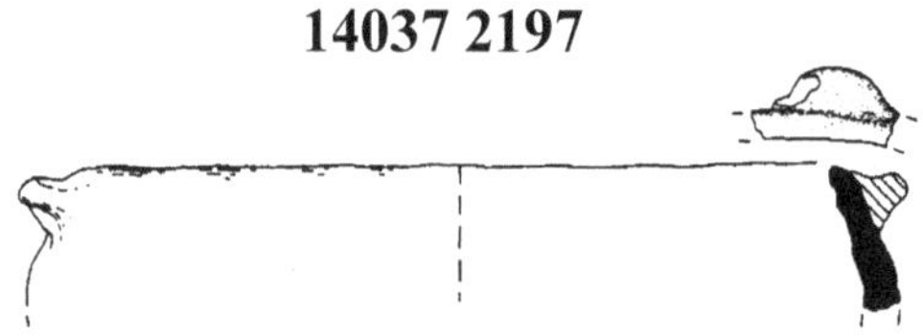

14043 2209

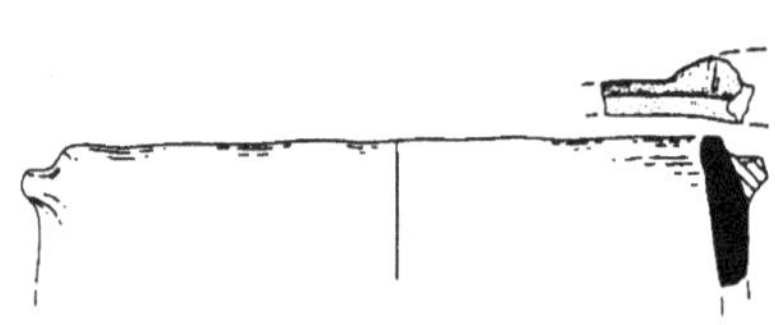

14043 2199

14029 2190

Figure 44a: Rukeis, Area 16, Aerial Photograph

Figure 44b: Rukeis, Area 16, Gate Plan

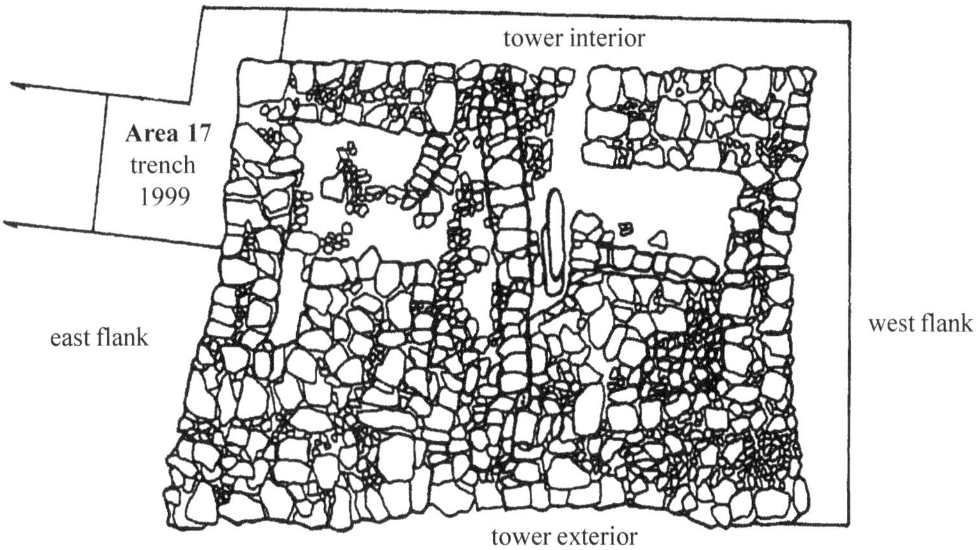

Figure 44c Rukeis, Area 16, Gate Features

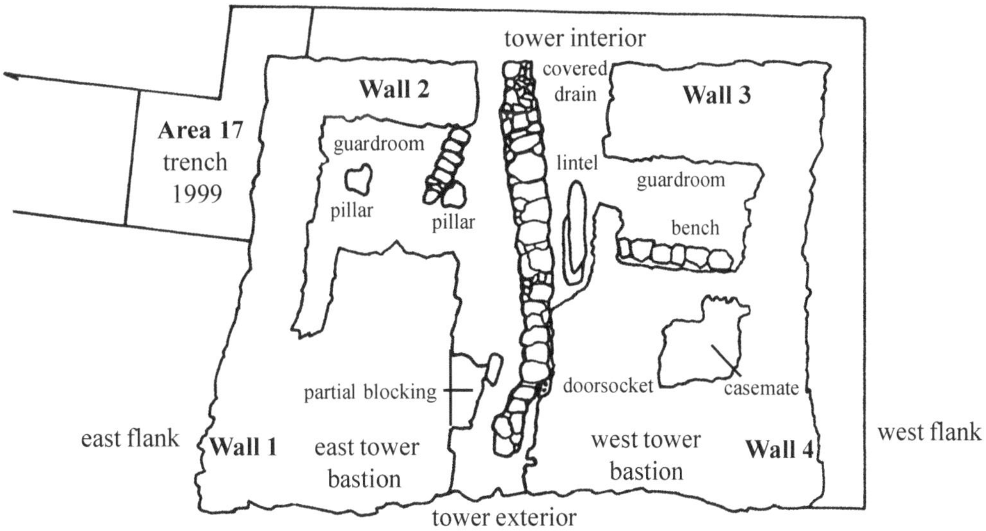

Figure 45a: Rukeis, Areas 14, 16 and 17

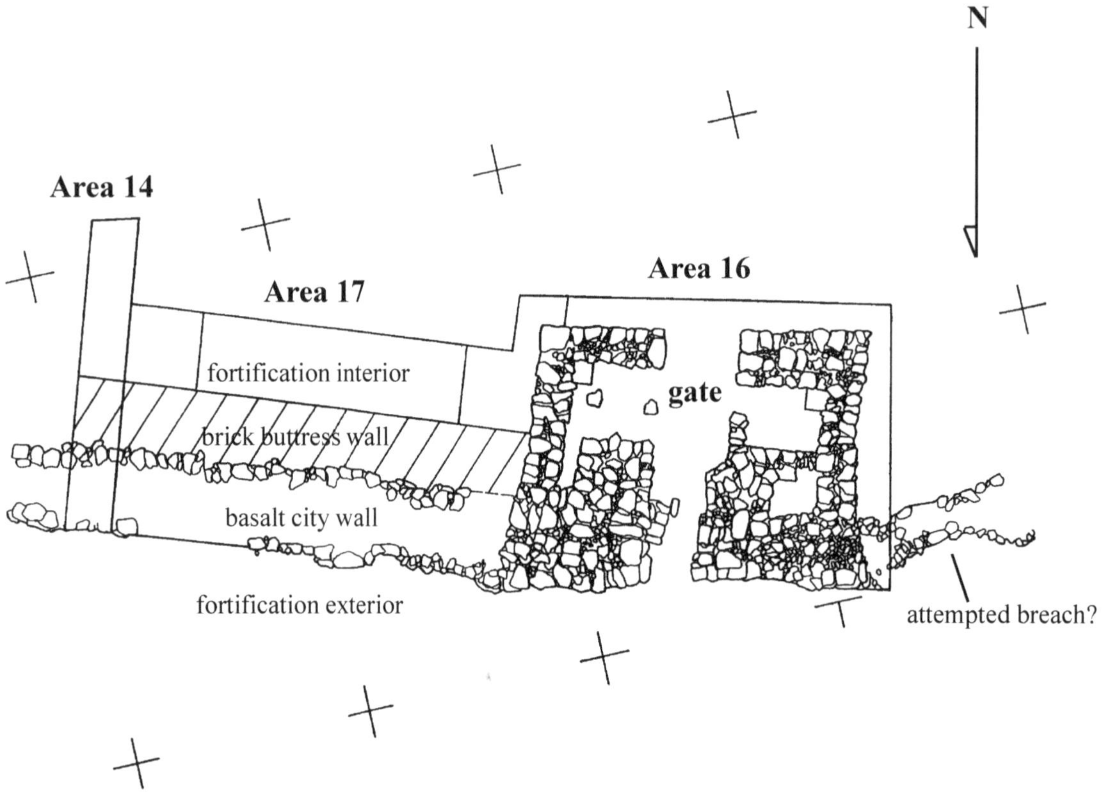

Figure 45b: Rukeis, Areas 14, 16 and 17

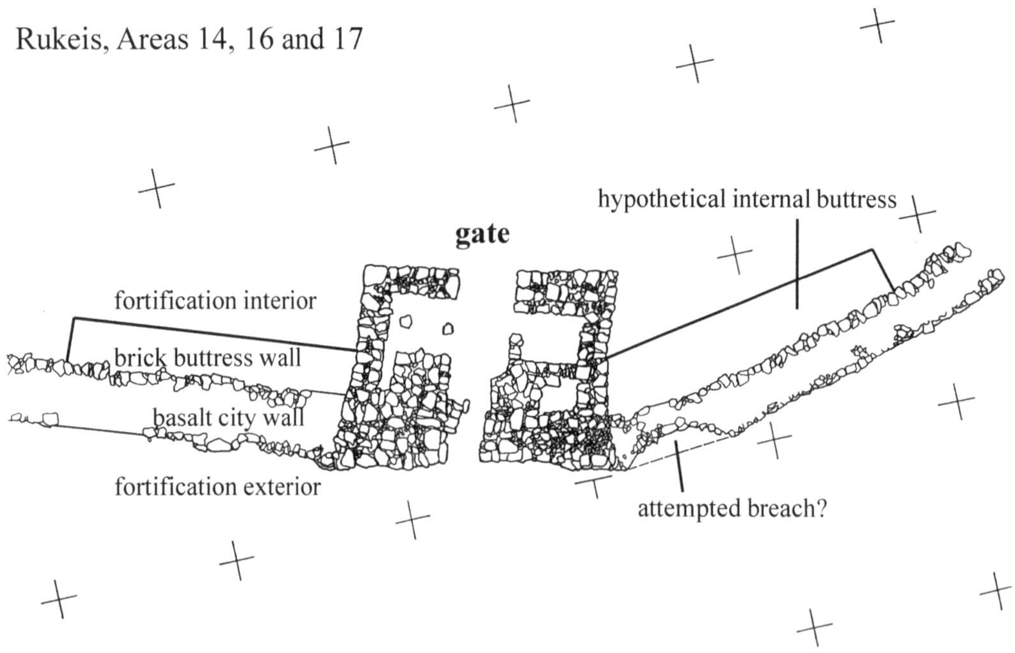

Figure 46: Rukeis, Area 16, Pottery and Comparative Ceramic Analysis

Period	Parallel	Variation	Total	%
EBIV				
MBI	2	4	6	12.5
MBII	9	22	31	64.6
MBIII		11	11	22.9
Total	11	37	48	

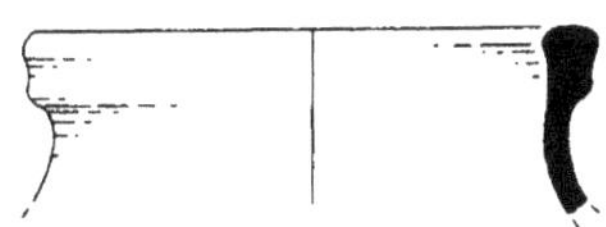

17006 9941

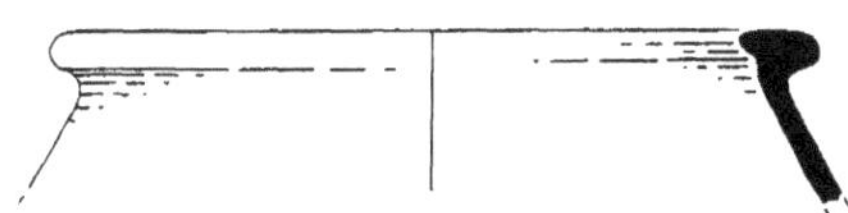

17006 9949

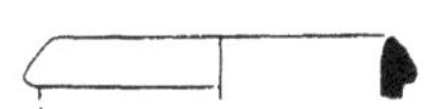

17006 9943

17007 9953

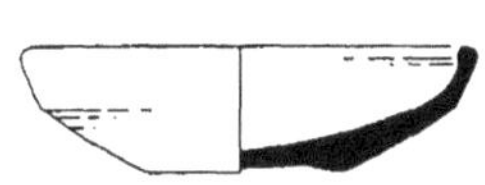

17007 9952

17006 9948

17009 9940

Figure 47a: Example of Casemate Wall

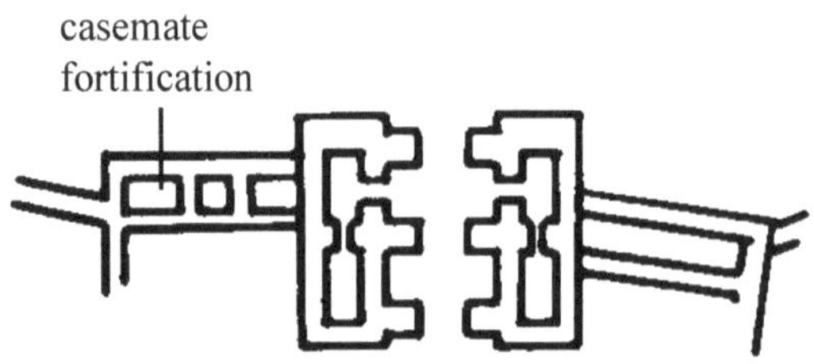

Hazor. Area K
After Wright 1985a. Fig 88.2

Figure 47b: Comparison of Pella and Megiddo Buttresses

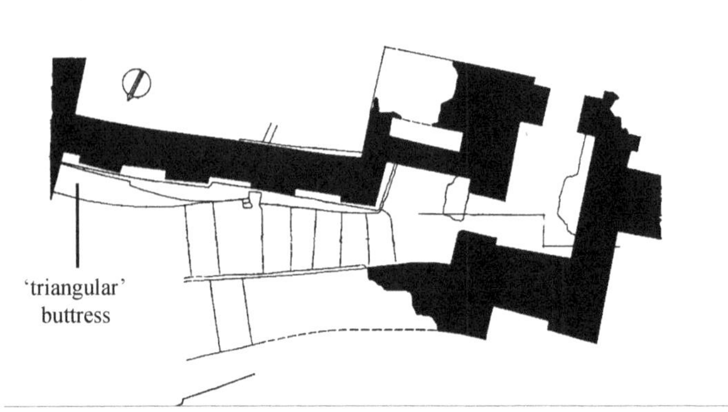

Megiddo. Possible 'triangular' buttress
in Area AA
After Loud 1948:7

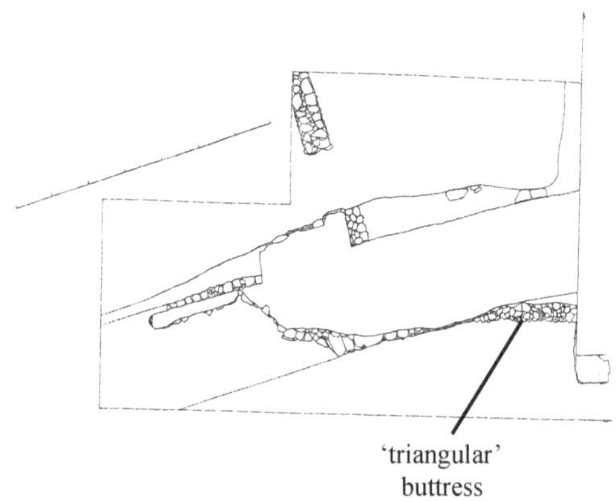

Pella. Area XXVIIIC.
'Triangular' Buttress 12

Figure 47c Example of Buttress-Niche Wall

Tell el-Far'ah North
MBII city wall
After Mallet 1987 plan 14

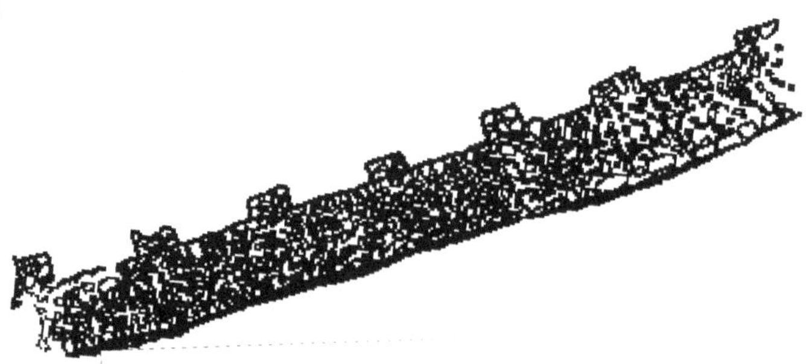

Figure 47d Stepped-In Foundation Construction

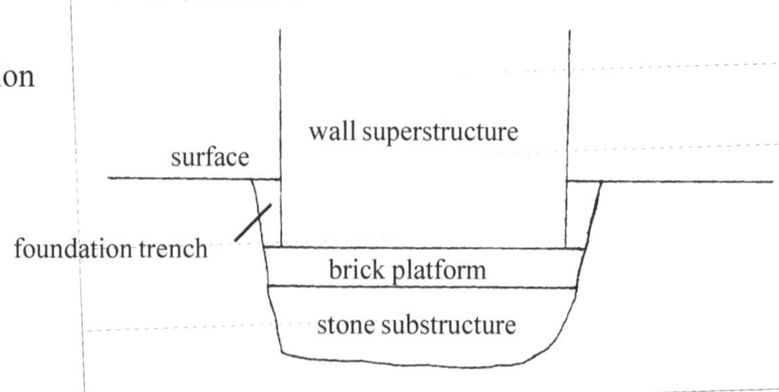

Figure 48a-g: Comparative Table of MBA Gates

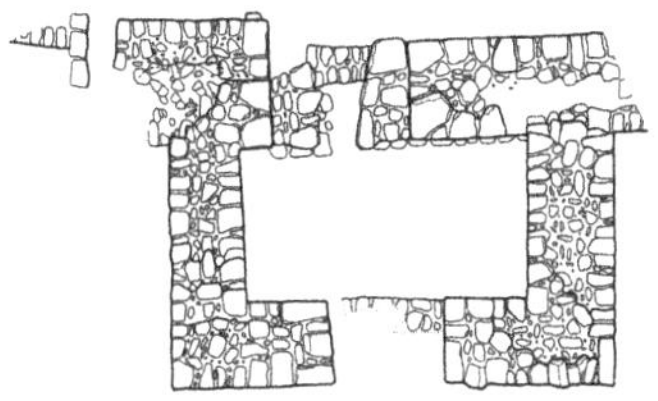

a. Tell el-Far'ah North
After Mallet 1987. Plan IV

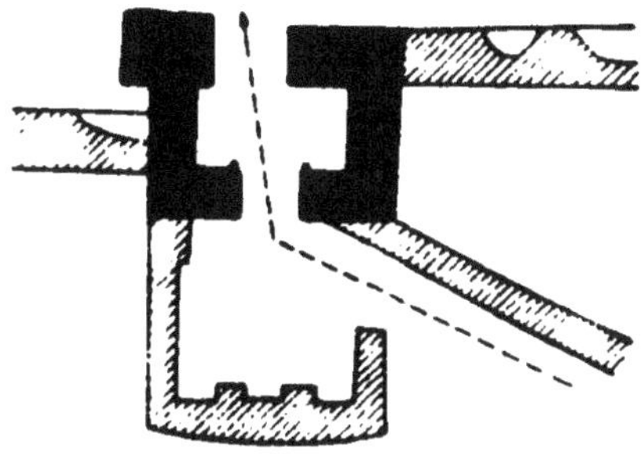

b. Tell el-Far'ah North
After Wright 1985a. Fig 94:17

c. Ashdod
After Wright 1985a. Fig 94:6

d. Shechem
After Wright 1985a. Fig 94:3

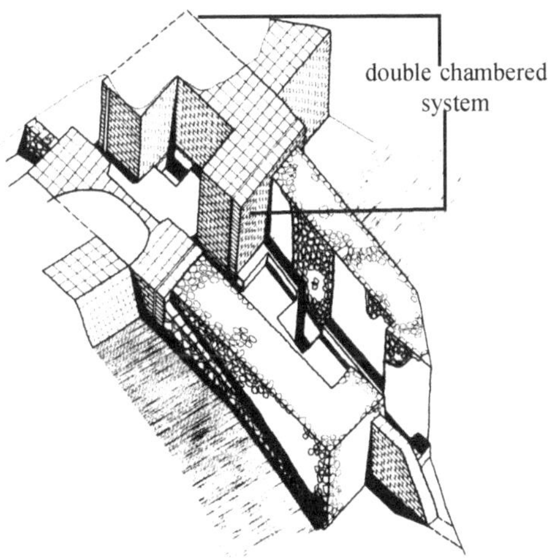

e. Tell Akko
After Dothan and Raban 1990:38

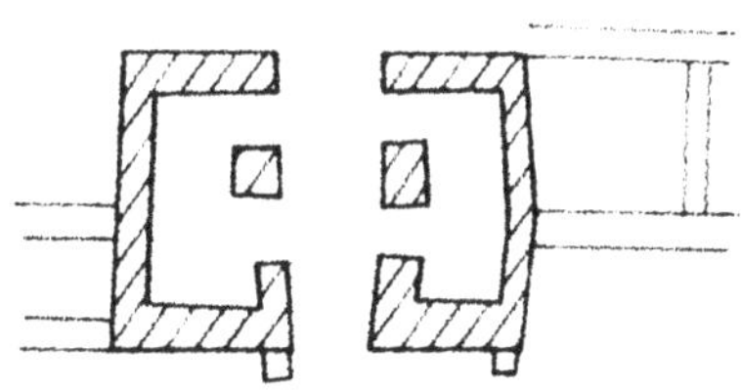

f. Khirbet Iskander
After Richard 1990. Fig 23

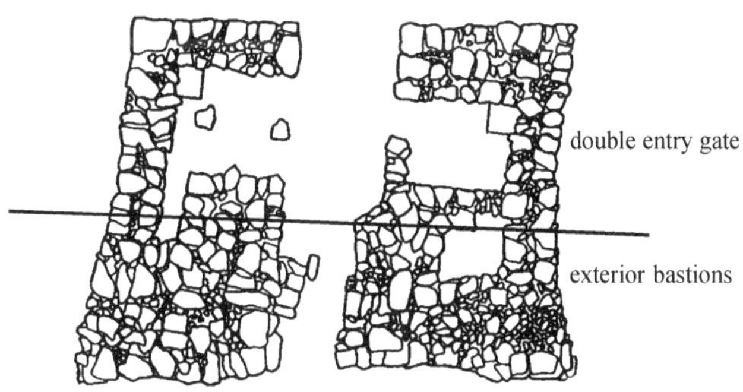

g. Rukeis

Figure 49a-c: Comparative Bronze Age Gate Elevations

a. Tell Dan Area K

After Biran *et al.* 1996 figure 1.61

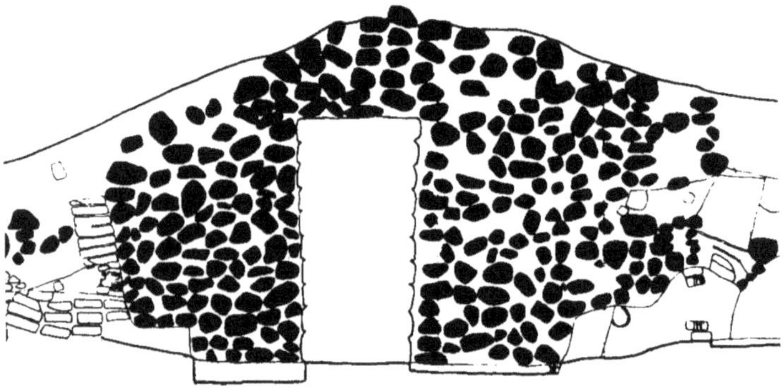

b. Bab edh-Dhra

After Schaub and Rast 1984 figure 4

c. Alalakh

After Woolley 1955 figure 55

Figure 50a: Exterior-Wall Rampart

After Wright 1985a. Fig 91c

Figure 50b: Two-Wall Rampart

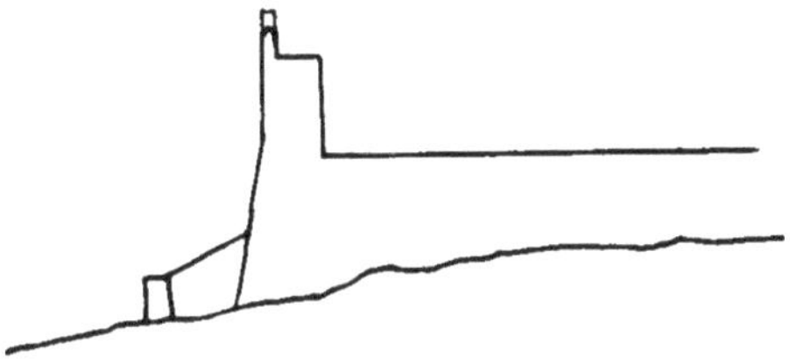

After Wright 1985a. Fig 77c

Figure 50c: Freestanding Earthen Rampart

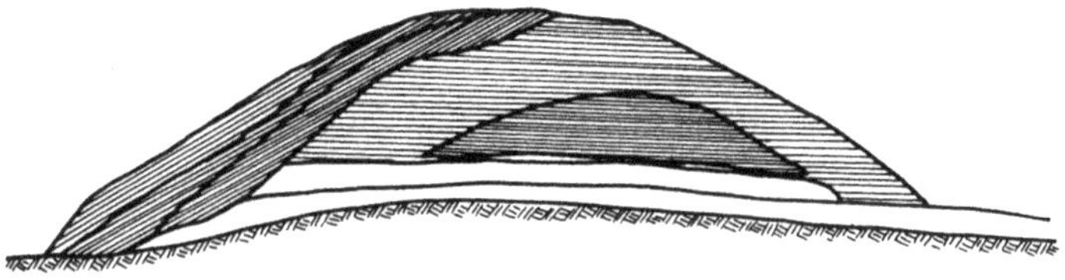

After Mazar 1992. Fig 6.9

Figure 50d: Core-Wall Rampart

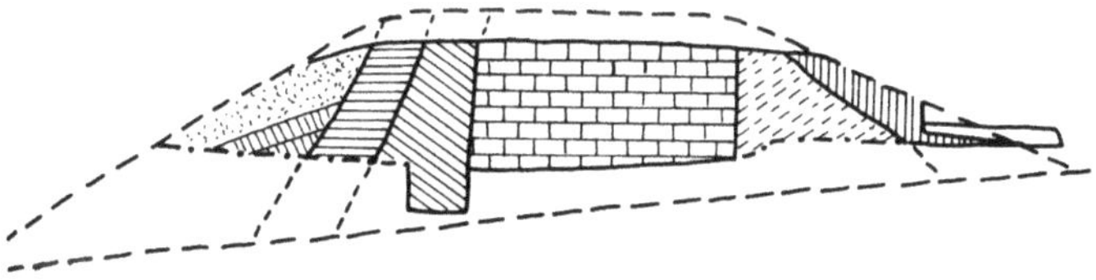

After Mazar 1992. Fig 6.9

Figure 50e: Crowning-Wall Rampart

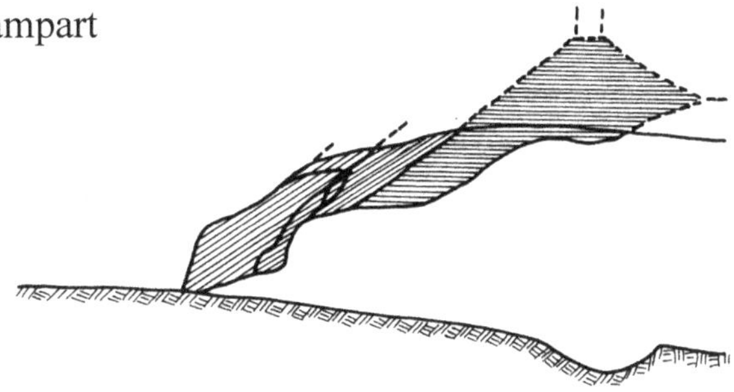

After Mazar 1992. Fig 6.9

Figure 51: Evolution of Rampart Types at EBA Mari

After Margueron 2000. Fig 2

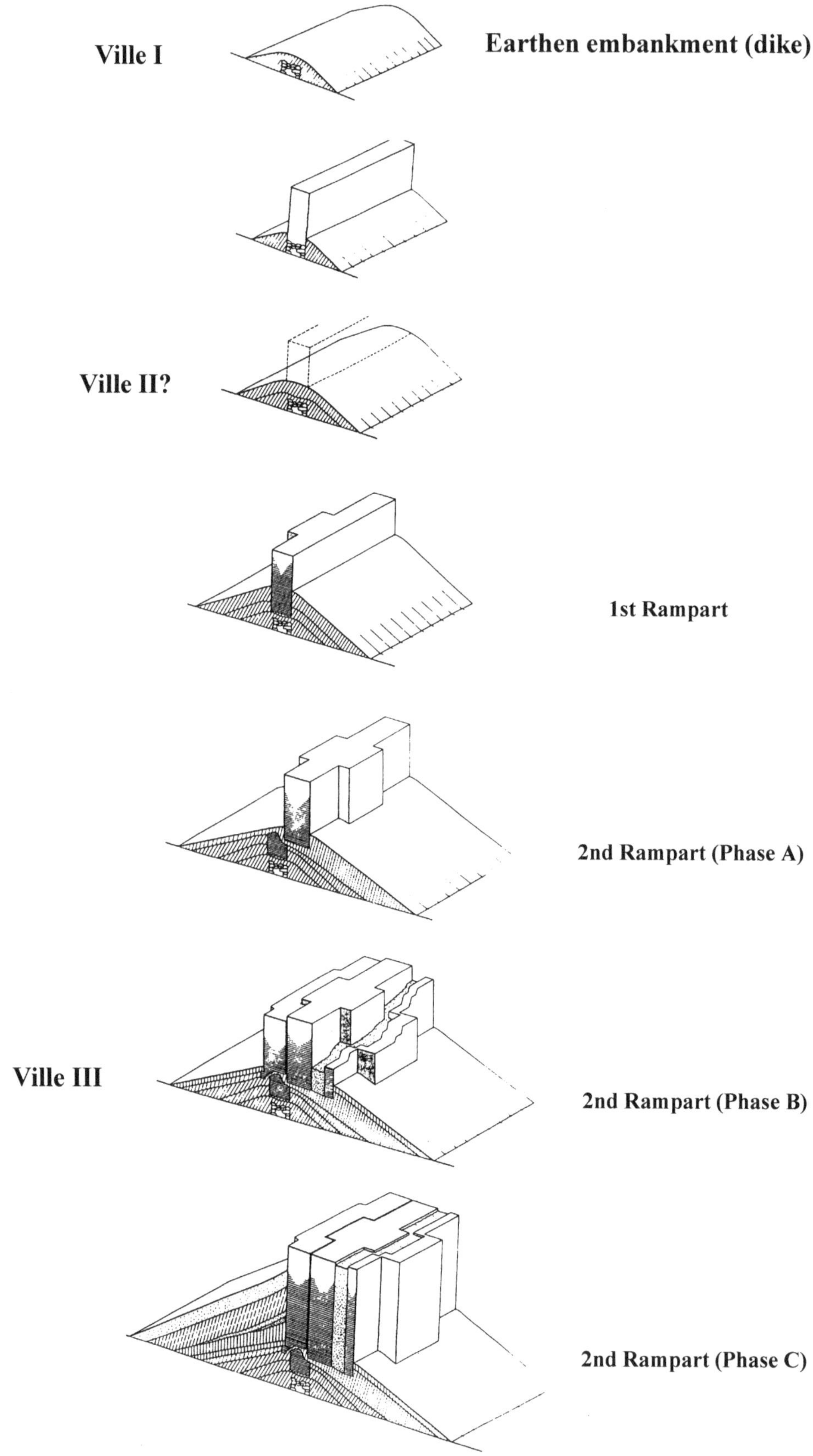

Figure 52: MBI Fortified Sites in the Southern Levant

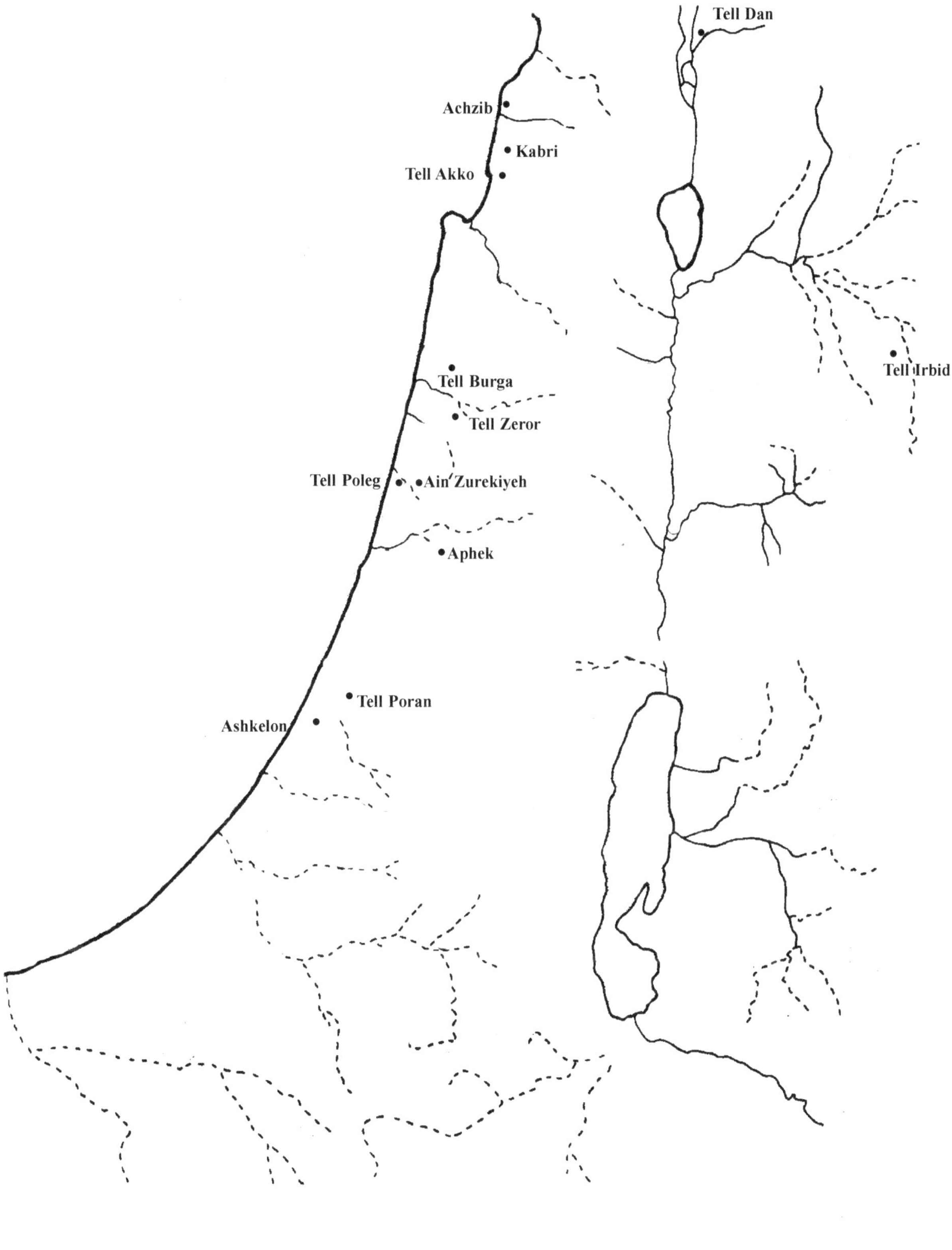

Figure 53: MBI and MBI-II Transition Fortified Sites in the Southern Levant

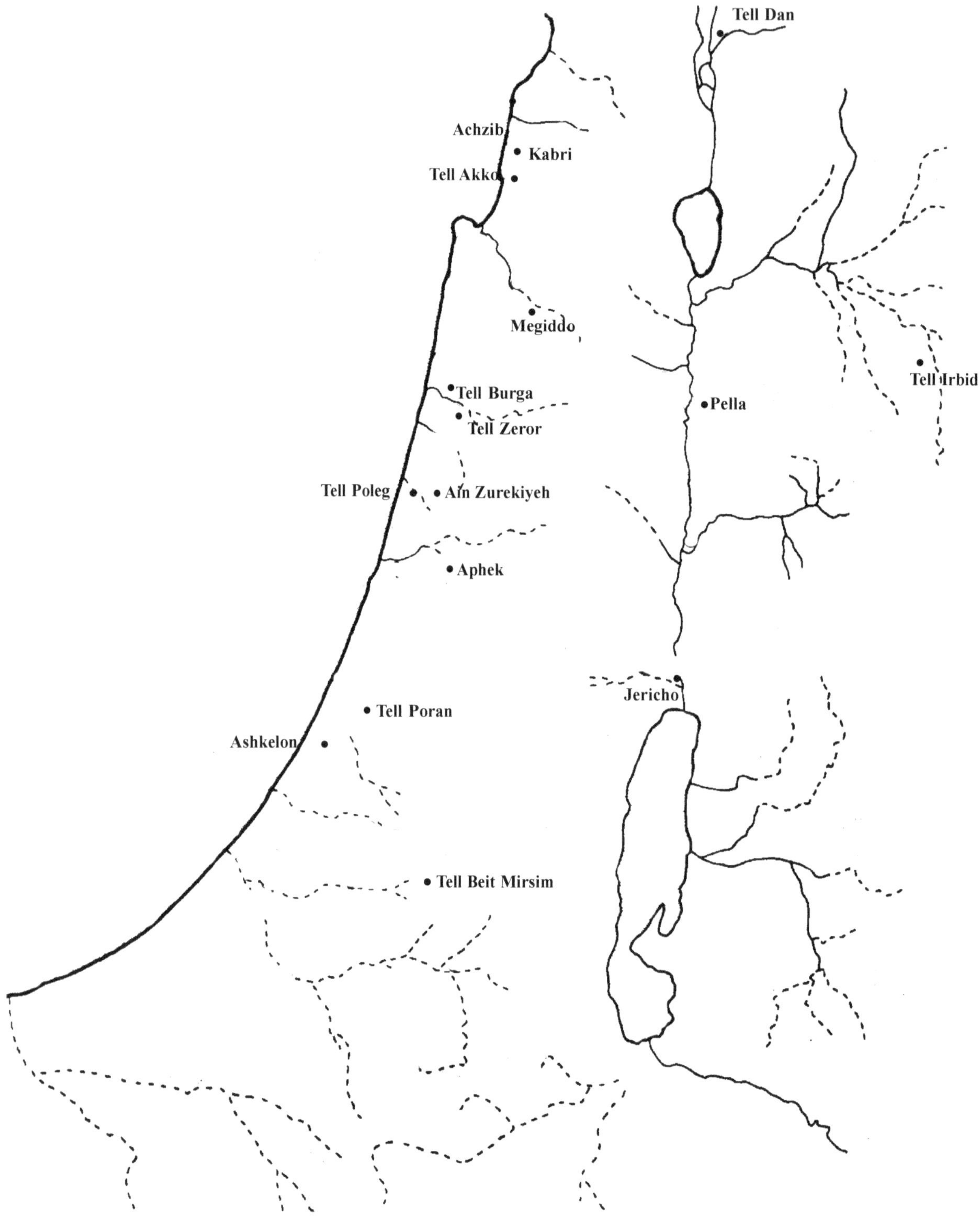

Figure 54: MBI, MBII and MBIII Fortified Sites in the Southern Levant

Bibliography

Abu al-Soof, B.
1968 Tell es-Sawwan excavations (Fourth Season). *Sumer* XXIV, 3-16

Abu Dayyah, A., Greene, J., Hassan, I. and Suleiman, E.
1991 Archaeological Survey of Greater Amman. *ADAJ* 35, 361-95

Aharoni, Y.
1973 *Excavations at Tell Beersheba, 1969-1971 Seasons*, Tel Aviv

Aharoni, Y., Fritz, A. and Kempinski, A.
1974 Excavations at Tel Masos, Preliminary Report on the First Season, 1972. *Tel Aviv* 1, 64-74
1975 Excavations at Tel Masos, Second Season, 1974. *Tel Aviv* 2, 97-124

Albright, W. F.
1922 Palestine in the Earliest Historical Period. *JPOS* 2, 110-139
1923 The Danish Excavations at Shiloh. *BASOR* 9, 10-11
1926 The Excavations at Tell Beit Mirsim 1. *BASOR* 23, 3-6
1928 The Second Campaign at Tell Beit Mirsim. *BASOR* 31, 4-10
1930 The Third Campaign at Tell Beit Mirsim. *BASOR* 39, 1-8
1932a The Fourth Joint Campaign Of Excavation At Tell Beit Mirsim. *BASOR* 47, 6-14
1932b The Excavation of Tell Beit Mirsim in Palestine, Volume 1 : The Pottery of the First Three Campaigns. *AASOR* XII
1933 The Excavation of Tell Beit Mirsim, vol 1A : The Bronze Age Pottery of the Fourth Campaign. *AASOR* XIII, 55-128
1934 The Kyle Memorial Excavation at Bethel. *BASOR* 56, 2-14
1938a The Excavation Of Tell Beit Mirsim II, The Bronze Age. *AASOR* XVII, 1-61
1938b The Chronology of a South Palestinian City, Tell el-Ajjul. *AJSLL* 55, 337-359
1940 New Light on the History of Western Asia in the Second Millenium B.C. *BASOR* 78, 23-30
1941 New Egyptian Data on Palestine in the Patriachal Age. *BASOR* 81, 16-20
1952a A Note on the Chronology of the Second Millenium B.C. *BASOR* 126, 24-25
1952b Further Observations on the Chronology of the Early Second Millenium B.C. *BASOR* 127, 27-30
1956 Stratigraphic Confirmation of the 'Low' Mesopotamian Chronology. *BASOR* 144, 26-29
1957 Further Observations on the Chronology of Alalakh. *BASOR* 146, 26-33
1962 The Chronology of Middle Bronze I (EB-MB). *BASOR* 168, 36-42
1973 The Historical Framework of Palestine Archaeology Between 2100 and 1600 B.C. *BASOR* 209, 12-18

Albright, W.F. and Sellers, O.R.
1931 The First Campaign of Excavation at Beth-Zur. *BASOR* 43, 2-11

Alkim, U.D.
1964 Tilmen Huyuk. *AS* XIV, 23-24

Alp, S.
1959 Karahoyuk. *AS* IX, 30

Amiran, R.
1960 The Pottery of the MBI in Palestine. *IEJ* 10.4, 204-225
1965 A Preliminary Note on the Synchronisms Between the Early Bronze Strata of Arad and the First Dynasty. *BASOR* 179, 30-34
1969a A Second Note on the Synchronism Between Early Bronze Arad and the First Dynasty. *BASOR* 195, 50-54
1969b *Ancient Pottery of the Holy Land*, New Brunswick.
1978 *Early Arad I 1962-1966*, Jerusalem

Amiran, R. and Eitan, A.
1963a Tell Nagila. *IEJ* 13.2, 144-145
1963b Tell Nagila. *IEJ* 13.4, 334
1965 A Canaanite-Hyksos City at Tell Nagila. *Archaeology* 18, 113-123

Amiran, R. and Ilan, O.
1996 *Early Arad II, The Chalcolithic and Early Bronze IB Settlements and the Early Bronze II City : Architecture and Town Planning, 1971-1978, 1980-1984*, Jerusalem

Anderson, W.P.
1988 *Sarepta I : The Late Bronze and Iron Age Strata of Area II.Y*, Beirut

Astrom, P.
1992 Implications of an Ultra Low Chronology. *SICAC* III, 19-21

Beck, P.
1975 The Pottery of the MBIIA at Tel Aphek. *Tel Aviv* 2, 45-85
1985 The MBIIA Pottery From Aphek 1972-1984 : First Summary. *Tel Aviv* 12.1, 181-203

Beek, G.W. van
1992 The Rammed Earth Revetment at Tell Jemmah. *EI* 23, 4-9
1997 Tell Jemmah, in Meyers, E.M., (ed) The Oxford Encyclopedia of Archaeology in the Near East, Oxford.

Behm-Blancke, M.R.
1980 Hassek-Huyuk, 1978-9. *AS* XXX, 215

Ben-Tor, A.
1991 New Light on the Relations Between Egypt and Southern Palestine During the Early Bronze Age. *BASOR* 281, 3-10

Ben-Tor, A., Bonfil, R., Garfinkel, Y., Greenberg, R., Maeir, A.M. and Mazar, A.
1997 *Hazor V*, Jerusalem

Ben-Tor, D.
1994 The Historical Implications of Middle Kingdom Scarabs Found in Palestine Bearing Private Names and Titles of Officials. *BASOR* 294, 7-22
1997 The Relations Between Egypt and Palestine in the Middle Kingdom as Reflected by Contemporary Canaanite Scarabs. *IEJ* 47.3-4, 162-189

Betts, A.V.G.
1991 *Excavations at Jawa 1972-1986*, Edinburgh

Betts, A.V.G. and Helms, S.W.
1989 A Water Harvesting and Storage System at Ibn al-Ghazzi in Eastern Jordan. Levant 21, 3-11

Betts, A.V.G., Eames, S., Schroder, M. and al-Qader al-Husan, A.
1995 Archaeological Survey of the Wadi al-Ajib, al-Mafraq. ADAJ XXXIX,149-168

Betts. A.V.G., Eames, S., Hulka, S., Schroder, M., Rust, J. and McLaren, P.B.
1996 Studies of Bronze Age Occupation in the Wadi al-Ajib, Southern Hauran. *Levant* XXVIII, 27-40

Bienkowski, P.
1989 The Division of MBIIB-C in Palestine. *Levant* XXI, 169-179

Bietak, M.
1987a The Middle Bronze Age of the Levant - A New Approach to Relative and Absolute Chronology. *ICAC* III, 78-123
1987b 'The Impact of Egypt on Canaan During the Middle Bronze Age'. in Rainey, A.F. (ed), *Egypt, Israel, Sinai: Archaeological and Historical Relationships in the Biblical Period*, Tel Aviv, 41-56
1991 Egypt and Canaan During the Middle Bronze Age. *BASOR* 281, 27-72
1997 'Avaris, Capital of the Hyksos Kingdom: New Results of Excavations'. in Oren, E.D. (ed), *The Hyksos: New Historical and Archaeological Perspectives*, Philadelphia:87-140.

Bietak, M. and Kleinsgutl, D.
2000 The Synchronization of Civilizations in the Eastern Mediterranean in the 2nd Millennium BC. *ICANE* I, 99-107

Bikai, P.
1978 *The Pottery of* Tyre, Warminster

Bittel, K.
1957 Boghazkoy. *AS* VII, 21
1958 Boghazkoy. *AS* VIII, 24

Biran, A.
1980 Two Discoveries at Tel Dan. *IEJ* 30.1-2, 174-195
1984 The Triple-Arched Gate of Laish at Tel Dan. *IEJ* 34.1, 1-19
1994 *Biblical Dan*, Jerusalem

Biran, A., Ilan, D. and Greenberg, R.
1996 *Dan I*, Jerusalem

Bourke, S.J.
1991 *The Transition From the Middle To the Late Bronze Age in the Northern Levant : The Evidence From Tell Nebi Mend*, Unpublished Doctoral Dissertation, University of London
1997 Pre-Classical Pella in Jordan : A Conspectus of Ten Years Work (1985-1995). *PEQ* 129, 94-115

Bourke, S.J. and Sparks, R.T.
1995 The DAJ Excavations at Pella in Jordan 1963/64. in Bourke, S.J. and Descoeudres, J-P. (eds), *Trade, Contact and the Movement of Peoples in the Eastern Mediterranean : Studies in Honour of J. Basil Hennessy*, 149-167, Sydney

Bourke, S.J., Sparks, R.T., Sowada, K.N. and Mairs, L.D.
1994 Preliminary Report on the University of Sydney's Fourteenth Season of Excavations at Pella (Tabaqat Fahl) in 1992. *ADAJ* 38, 81-126

Bourke, S.J., Sparks, R.T., Sowada, K.N., McLaren, P.B. and Mairs, L.D.
1998 Preliminary Report on the University of Sydney's Sixteenth and Seventeenth Seasons of Excavations at Pella (Tabaqat Fahl) in 1994/1995. *ADAJ* XLII, 179-212

Boling, R.G.
1989 Site Survey in the el-Umeiri region. *Madaba Plains Project* I, 98-188
Berrien Springs

Braemer, F.
1984 Prospections Archeologiques Dans Le Hawran (Syrie). *Syria* LXI, 219-250
1987 Two Campaigns of Excavations on the Ancient Tell of Jerash. *ADAJ* XXXI, 525-529
1988 Prospections Archeologiques Dans Le Hawran. *Syria* LXV, 99-137
1991 Tell Zheir : Village d'altitude dans le Jebel el'Arab (Syrie). *Paleorient* 17/1, 143-152
1993 Prospections Archeologiques Dans Le Hawran (Syrie) III. *Syria* LXX,117-170
1994 Khirbet el-Umbachi. *AJA* 98,124-125

Braun, E.
1989 Me'ona. *IEJ* 39.1-2, 96-97

Buccellati, G.
1988 The Kingdom and Period of Khana. *BASOR* 270, 43-62

Buccellati, G. and Kelly-Buccellati, M.
1978 The Terqa Archaeological Project. First Preliminary Report. *AAAS* XXVII, 71-96
1983 Terqa : The First Eight Seasons. *AAAS* XXXIII.2, 47-67

Buhl, M.L. and Holm-Nielsen, S.
1969 *Shiloh, The Danish Excavations at Tall Sailun, Palestine in 1926, 1929, 1932 and 1963*, Copenhagen

Bull, R.J., Callaway, J.A., Campbell, E.F., Ross, J.F. and Wright, G.E.
1965 The Fifth Campaign at Balatah (Shechem). *BASOR* 180, 7-41

Bull, R.J. and Campbell, E.F.
1968 The Sixth Campaign at Balatah (Shechem). *BASOR* 190, 2-40

Bunimovitz, S.
1992 The Middle Bronze Age Fortifications in Palestine as a Social Phenomenon. *Tel Aviv* 19.2, 221-234

Burney, C.
1961 Excavations at Yanik Tepe, North-west Iran. *Iraq* XXIII, 138-153

Cahill, J.M., Lipton, G., Tarler, D.
1988 Tell el-Hammah, 1988. *IEJ* 38.3, 194

Callaway, J.A.
1972
The Early Bronze Age Sanctuary at Ai (Et-Tell), London
1978 New Perspectives on the EBIII in Canaan. in Moorey, P.R.S. and Parr, P.
Archaeology in the Levant : Essays for Kathleen Kenyon, 46-58
1980 *The EBA Citadel and Lower City at Ai, (et-Tell)*, Cambridge, Mass

Campbell, E.F., Ross, J.F. and Toombs, L.E.
1971 The Eighth Campaign at Balateh (Shechem). *BASOR* 204, 2-17

Caneva, I.
2000 Mersin-Yumuktepe from 7th through 5th Millennium BC: From a Rural Settlement to a Fortified Town. *ICANE* I, 165-170

Carroll, W.D.
1925 Bitter and it's Archaeological Remains. *AASOR* V, 77-104

Caskey, J.L.
1964 Greece, Crete and the Aegean Islands in the Bronze Age. *CAH* I, XXVIa

Chavalas, M.
1996 Terqa and the Kingdom of Khana. *BA* 59.2, 90-103

Clark, D.R.
1994 The Iron I Western Defense System at Tell el-Umeiri, Jordan. *BA* 57.3, 138-149

Cole, D.P.
1978 Tell Halif (Lahav),1977. *IEJ* 28.1-2, 119
1984 *Shechem I : The Middle Bronze IIB Pottery*, Winona Lake

Coogan, M.D.
1984 Numeira, 1981. *BASOR* 255, 75-82

Cresson, B. and Gophna, R.
1979 Tell Dalit, 1978. *IEJ* 29.2, 122

Culican, W. and McClellan, T.L.
1983 El-Qitar 1982-1983 : A Preliminary Report. *AAAS* XXXIII, 289-297

Dahlberg, B.T. and O'Connell, K.G.
1989 *Tell el-Hesi IV*, Winona Lake

Dalley, S.
1977 Old Babylonian Trade in Textiles at Tell al Rimah. *Iraq* XXXIX.2, 155-160

Daviau, P.M.M.
1996 The Fifth Season of Excavations at Tell Jawa (1994), a Preliminary Report. *ADAJ* XL, 83-99
2000 Survey and Excavation in Northern Moab. *ICANNE* I, 279-292

Delougaz, P.
1938 A Short Investigation of the Temple at Al-Ubaid. *Iraq* V, 1-11

Delougaz, P. and Lloyd, S.
1942 *Pre-Sargonid Temples in the Diyala Region*, Oriental Institute Press 58, Chicago

Detlef Jericke, V. von
1992 Tell es-Seba, Stratum V. *ZDPV* 108.2, 122-148

Dever, W.G.
1973a The EBIV-MBI Horizon in TransJordan and Southern Palestine. *BASOR* 210, 37-62
1973b Tower 5017 at Gezer : A Rejoinder. *IEJ* 23.1, 23-26
1976 The Beginning of the Middle Bronze Age in Syria-Palestine. in Cross, F.M., Lemke, W.E. and Miller, P.D. *Magnelia Dei : The Mighty Acts of God. Essays on the Bible*

and Archaeology in Memory of G. Ernest. Wright, New York, 3-38
1980 New Vista's on the EBIV ('MBI') Horizon in Syria-Palestine. *BASOR* 237, 35-64
1985 Relations Between Syria-Palestine and Egypt in the Hyksos Period. in Tubb, J. (ed) *Palestine in the Bronze and Iron Ages*, Institute of Archaeology, London.
1986, *Gezer IV*, Jerusalem
1987 Palestine in the Middle Bronze Age: The Zenith of Urban Canaanite Civilisation. *BA* 50, 148-77
1990
Hyksos, Egyptian Destructions and the End of the Palestinian MBA. *Levant* XXII, 75-81
1991 Tell el-Dab'a and Levantine Middle Bronze Age Chronology : A Rejoinder to Manfred Bietak. *BASOR* 281, 73-80
1992 The Chronology of Syria-Palestine in the Second Millenium B.C.E : A Review of Current Issues. *BASOR* 288, 1-26
1997a Settlement Patterns and Chronology of Palestine in the Middle Bronze Age. in Oren, E.D. (ed) *The Hyksos: New Historical and Archaeological Perspectives*, Philadelphia, 285-302
1997b Beth Shemesh. in Meyers, E.M., (ed) *The Oxford Encyclopedia of Archaeology in the Near East*, Oxford
1997c Kabri. in Meyers, E.M., (ed) The Oxford Encyclopedia of Archaeology in the Near East, Oxford

Dever, W.G., Bullard, R.G., Cole, D.P. and Seger, J.D.
1974 *Gezer II*, Jerusalem

Dever, W.G., Lance, H.D. and Wright. G.E.
1970 *Gezer I*, Jerusalem

Dornemann, R.H.
1979 Tell Hadidi : A Millenium of Bronze Age City Occupation. *AASOR* 44, 113-152.
1983 *The Archaeology of the TransJordan in the Bronze and Iron Ages*, Milwaukee
1987 Some Observations on the Geographical Extent of Cultural Areas in Syria and the Tran Jordan. *SHAJ* III, 275-277
1990 Preliminary Comments on the PotteryTraditions at Tell Nimrin. *ADAJ* 34, 153-81

Dothan, M.
1965 The Fortress at Kadesh-Barnea. *IEJ* 15.3, 134-151
1969 Tell Ashdod. *IEJ* 19.4, 243-244
1971 Tell Ashdod. *IEJ* 21.2-3, 175
1973a Accho. *IEJ* 23.4, 257-258
1973b The Foundation of Tel Mor and of Ashdod. *IEJ* 23.1, 1-17
1974 Accho. *IEJ* 24.3-4, 276-277
1975 Accho. *IEJ* 25.2-3, 164
1976 Akko, Interim Excavation Report, First Season, 1973/4. *BASOR* 224, 1-48
1981 Akko, 1980. *IEJ* 31.1-2, 111-112

Dothan, M., Conrad, D.
1978 Akko, 1978. *IEJ* 28.4,264-266
1979 Akko, 1979. *IEJ* 29.3-4,227
1983 Akko, 1982. *IEJ* 33.1-2,113
1984 Akko, 1983. *IEJ* 34.4,190-191

Dothan, M., Raban, A.
1990 The Sea Gate of Ancient Akko. *BA* 43, 35-39

Doumet, C.
1986 *Les Fouilles de Tell el Ghassil de 1972 'a 1974*, Unpublished Doctoral Dissertation, Sorbonne, Paris

Dubertret, L. and Dunand, M.
1955 Les Gisements Ossiferes de Khirbet el-Umbachi et de Hebariye. *AAAS* 5.

Duru, R.
1981 Kurucay Hoyuk, 1980. *AS* XXXI, 196

Eames, S.
2001 *Between the Desert and the Sown : Settlement in the Hauran in the Middle Bronze Age*, Unpublished Doctoral Dissertation, University of Sydney

Edwards, P.C., Bourke, S.J., Da Costa, K.A., Tidmarsh, J.C., Walmsley, A.G. and Watson, P.M.
1990 Preliminary Report on the University of Sydney's Tenth Season of Excavations at Pella in Jordan. *ADAJ* 34, 57-93

Egami, N.
1983 The Archaeological Researches in Idlib Prefecture. *AAAS* 1983, 75-82

el-Eisawi, D.M.
1985 Vegetation in Jordan. *SHAJ* II, 45-57

Emery, W.B.
1958 A Preliminary Report on the Excavations at Buhen. *Kush* VII, 7-14
1959 A Preliminary Report on the Excavations of the Egyptian Exploration Society at Buhen, 1958-1959. *Kush* VIII, 7-10
1960 A Preliminary Report on the Excavations of the Egyptian Exploration Society at Buhen, 1959-1960. *Kush* IX, 81-86

Esse, D.L.
1991 *Trade, Subsistence and Social Change In Early Bronze Age Palestine*,
SAOC 50, University Of Chicago Press, Chicago

Falconer, S.E. and Magness-Gardiner, B.
1983 1982 Excavations of the Tell el-Hayyat Project. *ADAJ* XXVII, 87-104
1984 Preliminary Report of the First Season of the Tell el-Hayyat Project. *BASOR* 255, 49-74

Finet, A.
1977 Bilan provisoire des fouilles belges du Tell Kannas, *AASOR* 44, 79-98

Finkelstein, I.
1985 Excavations at Shiloh 1981-1984 : Preliminary Report. *Tel Aviv* 12.2, 123-180
1989a Further Observations on the Socio-Demographic Structure of the Intermediate Bronze Age. *Levant* 21, 129-137
1989b The Land of Ephraim Survey, 1980-1987. Preliminary Report. *Tel Aviv* 15-16.2, 117-183
1991 The Central Hill Country in the Intermediate Bronze Age. *IEJ* 41.1-3, 19-45
1992 Middle Bronze Age Fortifications : A Reflection of Social Organisation and Political Formations. *Tel Aviv* 19.2, 201-220

Finkelstein, I., Bunimovitz, S. and Lederman, Z.
1993 *Shiloh*, Jerusalem

Finkelstein, I. and Gophna, R.
1993 Settlement, Demographic and Economic Patterns in the Highlands of Palestine in the Chalcolithic and Early Bronze Periods and the Beginning of Urbanism. *BASOR* 289, 1-22

Finkelstein, I. and Perevolotsky, A.
1990 Processes of Sedentaization and Nomadism in the History of Sinai and the Negev. *BASOR* 279, 67-88

Finkelstein, I., Ussishkin, D. and Halpern, B.
2000 *Megiddo III*, Tel Aviv

Fischer, P.M.
2000 A Synthesis of Ten Campaigns at Tell Abu al-Kharas, Jordan Valley: The Early, Middle, Late Bronze and Iron Ages. *ICANNE* I:447-470

Flanagan, J. W. and McCreery, D. W.
1990 First Preliminary Report on the 1989 Tell Nimrin Project. *ADAJ* 34,131-149

Flanagan, J. W., McCreery, D. W. and Yassine, K. N.
1992 Preliminary Report of the 1990 Excavation at Tell Nimrin. *ADAJ* 36, 89-107
1994 Preliminary Report on the 1993 Season at Tell Nimrin. *ADAJ* 38, 205-226

Foster, B.R.
1977 Commercial Activity in Sargonic Mesopotamia. *Iraq* XXXIX.1, 31-44

Franken, H. J.
1961 The Excavations at Deir 'Alla in Jordan, Second Season. *VT* 11, 361-372

Franken, H.J. and Ibrahim, M.
1978 Two Seasons of Excavations at Tell Deir Alla, 1976-1978. *ADAJ* XXII, 57-80
1989 Deir Alla. in Homes-Fredericq, D. and Hennessy, J.B. (eds) *The Archaeology of Jordan* 2, Leuven, 201-5

Free, J.P.
1958 The Fifth Season at Dothan. *BASOR* 152, 10-18

Fritz, V. and Kempinski, A.
1983 *Ergebinesse Der Ausgrabungen Auf Der Hirbet El-Msas (Tell Masos). Teil I.* Wiesbaden

Fugmann, E.
1958 *Hama : Fouilles et Recherches, 1931-1938, II. L'Architecture des Periodes pre-Hellenistiques,* Copenhagen.

Funk, R. W.
1958 The 1957 Campaign at Beth-Zur. *BASOR* 150, 8-20

Funk, R.W. and Richardson, H.N.
1958 The 1958 Sounding at Pella. *BA* 21, 82-96

Garstang, J.
1932 Jericho : City and Necropolis. *AAAL* XIX, 1-2:3-22

Gates, M-H.
1987 Alalakh and Chronology Again. In ICAC 1, 60-86

Geraty, L.T., Herr, L.G. and LaBianca, O.S.
1987 The Madaba Plains Project. A Preliminary Report on the First Season at Tell el-Umeiri and Vicinity. *ADAJ* XXXI, 187-199

Geraty, L.T., Herr, L.G, LaBianca, O.S. and Younker, R.W.
1989 *Madaba Plains Project 1*, Berrien Springs

Gerkan, A.V. von
1935 Die Stadtmauer von Palmyra. *Berytus* II, 25-33

Gerstenblith, P.
1980 A Reassessment of the Beginning of the Middle Bronze Age in Syria-Palestine. *BASOR* 237, 65-84
1983 The Levant at the Beginning of the Middle Bronze Age. *ASOR* Dissertation Series Number 5

Geva, S.
1983 *Tell* Jerishe, Qedem 15
Giannessi, D.
2000 Tell Afis During the Late Chalcolithic Period. The late Chalcolithic Levels: Architecture and Pottery. *ICANE* I, 555-560

Gittlen, B. M.1981 The Cultural and Chronological Implications of the Cypro-Palestinian Trade During the Late Bronze Age. *BASOR* 241, 49-60

Glueck, N.
1934 Explorations in Eastern Palestine I. *AASOR* XIV
1935 Explorations in Eastern Palestine II. *AASOR* XV
1939a Explorations in Eastern Palestine III. *AASOR* XIX
1939b The Earliest History of Jerash. *BASOR* 75, 22-29
1951 Explorations in Eastern Palestine IV. *AASOR* XXV-XXVIII

Goedicke, H.
1992 The Chronology of the Thera Santorini Explosion. *SICAC* III, 57-62

Goetze, A.
1952 The Date of the Hittite Raid on Babylon. *BASOR* 127, 16-20
1957 Alalakh and Hittite Chronology. *BASOR* 146, 20-25

Gonen, R.
1984 Urban Canaan in the Late Bronze Period. *BASOR* 253, 61-74

Gophna, R.
1978 Archaeological Survey of the Central Coastal Plain, 1977. *Tel Aviv* 5.3-4, 136-147
1992 EBA Fortification Wall and MBA Rampart at Tel Poran. *Tel Aviv* 19.2, 267-273

Gophna, R. and Ayalon, E.
1982 A Fortified MBIIA Site at Ain Zurekiyeh in the Poleg Basin. *Tel Aviv* 9.1, 69-78

Gophna, R., Liphschitz, N. and Lev-Yadun, S.
1986 Man's Impact on the Natural Vegetation of the Central Coastal Plain of Israel During the Chalcolithic Period and the Bronze age. *Tel Aviv* 13,71-84

Gophna, R. and Portugali, J.
1988 Settlement and Demographic Processes in Israel's Coastal Plain from the Chalcolithic to the Middle Bronze Age. *BASOR* 269, 11-28

Graber, O.
1965 Qasr al-Hayr al-Sharqi. *AAAS* XV, 107-122

Greene, J.A. and 'Amr, K.
1992 Deep Sounding on the Lower Terrace of the Amman Citadel : Final Report. *ADAJ* XXXVI, 113-144

Gregori, B.
1986 Three Entrance City Gates of the MBA in Syria and Palestine. *Levant* 18, 83-103

Guardata, F.B.
1990 Tell Touqan, une ville paleosyrienne. *AAAS* XL, 64
1996 Tell Touqan, une ville paleosyrienne. *AAAS* XLII, 64-72

Gurney, O.R.
1966 Anatolia c.1600-1380 B.C. *CAH* Vol II. XVa

Haiman, M.,
1996 Early Bronze IV Settlement Patterns of the Negev and Sinai Deserts : View From Small Marginal Temporary Sites. *BASOR* 303, 1-32

Hammond, P.C.
1960 Excavations at Petra in 1959. *BASOR* 159, 26-31

Harif, A.
1979 Common Architectural Features at Alalakh, Megiddo and Shechem. *Levant* XI, 162-166

Harlan, J.R.
1985 The Early Bronze Age Environment of the Southern Ghor and the Moab Plateau. *SHAJ* II, 125-129

Harrison, T. P.
1993 Economics With an Entrepreneurial Spirit : Early Bronze Trade with Late PreDynastic Egypt. *BA* 56.2, 81-93
1997 Shifting Patterns of Settlement in the Highlands of Central Jordan During the Early Bronze Age. *BASOR* 306, 1-38

Hauptmann, A. and Weisgerber, G.
1992 Periods of Ore Exploitation and Metal Production in the Area of Feinan, Wadi Arabah, Jordan. *SHAJ* IV, 61-66

Heinz, M.
1992 *Tell Atchana / Alalakh. Die Schichten VII-XVII.* Wiesbaden

Helms, S.W.
1973 Excavations at Jawa 1973, A Preliminary Report. *ADAJ* 18, 41-44
1975 Jawa 1973 : A Preliminary Report. *Levant* VII, 20-38
1976a The Early Bronze Age Gate at Ras en-Naqira. *ZDPV* 92, 1-9
1976b Jawa Excavations 1974 - A Preliminary Report. *Levant* VIII, 1-35
1977a Early Bronze Age Fortifications at Tell Dothan. *Levant* IX,101-114
1977b Jawa Excavations 1975 : Third Preliminary Report. *Levant* IX, 21-35
1987a A Question of Economic Control During the Proto-Historical Era of Palestine / Tran Jordan. *SHAJ* III, 41-51
1987b Jawa, Tell Um Hammad and the EBI/Late Chalcolithic Landscape. *Levant* XIX, 49-81
1989 Jawa at the Beginning of the MBA. *Levant* XXI, 141-168

Hennessy, J.B.
1967a Preliminary Report on a First Season of Excavations at Teleilat Ghassul. *Levant* 1, 1-24

1967b *The Foreign Relations Of Palestine During the Early Bronze Age*. London

Hennessy, J.B., McNicoll, A.W., Hanbury-Tenison, J.W., Watson, P.M., Randle, L. and Walmsley, A.G.
1983 Preliminary Report on the Fourth Season of Excavations at Pella, 1982. *ADAJ* XXVII, 325-362

Herr, L.G., Geraty, L.T., LaBianca, O.S. and Younker, R.W.
1991a *Madaba Plains Project 2*, Andrews University Press, Berrien Springs, MI
1991b Madaba Plains Project : The 1989 Excavations at Tell el-Umeiri and Vicinity. *ADAJ* XXXV, 151-194

Herr, L.G., Clark, D.R., Geraty, L.T. and LaBianca, O.S.
1999 Madaba Plains Project: Excavations at Tell el-Umeiri 1998. *ADAJ* XLIII, 99-114

Herzog, Z.
1986 *Das Stadttor in Israel und in den Nachbalandern*,
1997 *Archaeology of the City : Urban Planning in Ancient Israel and its Social Implications*, Tel Aviv

Hoffmeier, J. K.
1989 Reconsidering Egypts Part in the Termination of the MBA in Palestine. *Levant* XXI, 181-193
1990 Some Thoughts on W.G. Dever's 'Hyksos, Egyptian Destructions, and the End of the Palestinian MBA'. *Levant* XXII, 83-89
1991 James Weinstein's 'Egypt and the MBIIC/LBIA Transition' : A Rejoinder. *Levant* XXIII, 117-124

Holland, T.A.
1976 Preliminary Report on Excavations at Tell es-Sweyhat. *Levant* VIII, 36-70
1977 Preliminary Report on Excavations at Tell es-Sweyhat, 1975. *Levant* IX, 36-65

Huber, J.
1987 Astronomical Evidence for the Long and Against the Middle and Short Chronologies. ICAC 1, 5-17

Ibach, R.D.
1987 *Hesban 5 : Archaeological Survey of the Hesban Region*, Berrien Springs

Ibrahim, M. M.
1975 Third Season of Excavations at Sahab, 1975. *ADAJ* XX, 69-83
1987 Sahab and it's Foreign Relations. *SHAJ* III, 73-81
1989 Sahab. In Homes-Frederiq, D. and Hennessy, J.B. (eds) *The Archaeology of Jordan*, Leuven
1997 Sahab. In Meyers, E.M. (ed) *The Oxford Encyclopedia of Archaeology in the Near East*, Oxford

Ibrahim, M. and Mittman, S.
1997 Khirbet ez-Zeraqun. In Meyers, E.M. (ed) *The Oxford Encyclopedia of Archaeology in the Near East*, Oxford

Jesus, P. de
1978 Metal Resources In Ancient Anatolia. *AS* XXVIII, 97-102

Jacobs, P.F.
1984 Tell Halif, 1983. *IEJ* 34.4, 198-199

Jacoby, R.
1991 The Representation and Identification of Cities on Assyrian Reliefs. *IEJ* 41.1-3, 112-129

Joffe, A.
1997 Tell el-Far'ah North. In Meyers, E.M. (ed) *The Oxford Encyclopedia of Archaeology in the Near East*, Oxford

Kafafi, Z.A. and Knauf E.A.
1989 Dahr el Medineh. In Homes-Frederiq, D. and Hennessy, J.B. (eds) *The Archaeology of Jordan*, 198-200

Kaplan, J.
1964 Jaffa. *IEJ* 14.4, 286
1967 Yavneh Yam and Ashdod Yam. *IEJ* 17.4, 269
1969 Yavneh Yam. *IEJ* 19.2, 120-121
1975 Further Aspects of the MBAII Fortifications in Palestine. *ZDPV* 91, 1-17

Kelly-Buccellati, M.
1990 Three Seasons of Excavation at Tell Mozan. In Eichler, S. *Tall al-Hamidiya 2*, 119-132

Kelm, G.L. and Mazar, A.
1979 Tel Batash (Timnah), 1979. *IEJ* 29.3-4, 242
1982 Three Seasons of Excavations at Tel Batash - Biblical Timnah. *BASOR* 248, 1-36
1985 Tel Batash (Timnah), 1984. *IEJ* 35.2-3, 201
1986 Tel Batash (Timnah), 1985. *IEJ* 36.1-2, 107-108
1987 Tel Batash (Timnah), 1986. *IEJ* 37.1, 59-60

Kelm, G.L.
1991 Tel Batash (Timnah) Excavations : Third Preliminary Report, 1984-1989. *BASOR* Supp. 27, 47-68

Kelso, J. L.
1958 The Third Campaign at Bethel. *BASOR* 151,3-8
1961 The Fourth Campaign at Bethel. *BASOR* 164, 5-18

Kempinski, A.
1978 *The Rise of an Urban Culture*, Israel Ethnographic Society Studies 4, Jerusalem
1983 Early Bronze Age Urbanism in Palestine : Some Topics in a Debate. *IEJ* 33.3-4, 235-241
1987 in Katsenstayn, H. (ed) *The Architecture of Ancient Israel, From the Prehistoric to the Persian Periods; in Memory of I. Dunayevsky*. Jerusalem

1989 *Megiddo, A City State and Rotal Centre in North Israel*, Munich
1992 The Middle Bronze Age in Northern Israel, Local and External Synchronisms. In SICAC, 69-73
1997 The Hyksos: A View From Northern Canaan and Syria. in Oren, E.D. (ed) *The Hyksos: New Historical and Archaeological Perspectives*, Philadelphia, 327-334

Kempinski, A., Miron, E.
1987 Kabri 1986-7. *IEJ* 37.2-3, 175-177

Kenyon, K.M.
1951 Excavations at Jericho, 1951. *PEQ* 83, 101-138
1952 Excavations at Jericho. *PEQ* 84, 62-71
1953 Excavations at Jericho, 1953. *PEQ* 85, 81-96
1962 Excavations in Jerusalem. *PEQ*, 76-87
1963 Excavations in Jerusalem, 1962, *PEQ,* 8-21
1966a Excavations in Jerusalem, 1965. *PEQ*, 73-88
1966b *Amorites and Canaanites*, London
1967 Excavations in Jerusalem, 1966. *PEQ*, 65-73
1968 Excavations in Jerusalem, 1967. *PEQ*, 97-109
1969 The Middle and Late Bronze Age Strata at Megiddo. *Levant* I, 25-60
1973 Palestine in the Middle Bronze age. In Edwards, I.E.S., Gadd, C.J., Hammond, N.G.L. and Sollberger, E. (eds) *CAH* 3.2, Part 1, 77-116
1979 *Archaeology in the Holy Land*, New York

Kenyon, K.M., Bottero, J. and Posener, G.
1971 Syria and Palestine c.2160-1780 BC. in *CAH* 1, Part 2, 532-594

Kenyon, K.M. and Holland, T.A.
1981 *Excavations at Jericho III. The Architecture and Stratigraphy of the Tell*, Oxford
1982 *Excavations at Jericho IV. The Pottery Type Series and Other Finds*, London
1983 *Excavations at Jericho V. The Pottery Phases of the Tell and Other Finds*, London

Kitchen, K.A.
1967 Byblos, Egypt and Mari in the Early Second Millenium B.C. *Orientalia* 36.1, 39-54
1987a The Basics of Egyptian Chronology in Relation to the Bronze Age. *ICAC* 1, 37-55
1987b Supplementary Notes on 'The Basics of Egyptian Chronology. *ICAC* 3, 152-159

Klengal, H.
1990 The Region of Idlib in the Bronze Age : A Historical Outline According to Cuneiform Sources. *AAAS* XL, 93-99

Kochavi, M.
1967 Tell Malhata. *IEJ* 17.4, 272
1975 The First Two Seasons of Excavation at Aphek Antipatris. *Tel Aviv* 2, 17-44
1994 Mithiam Leviah. *IEJ* 44.1-2, 141
1997 Aphek. In Meyers, E.M. (ed) *The Oxford Encyclopedia of Archaeology in the Near East*, Oxford

Kochavi, M., Beck, P. and Gophna, R.
1979 Aphek-Antipatris, Tel Poleg, Tel Zaror and Tel Burga; Four Fortified Sites of the MBIIA in the Sharon Plain. *ZPDV* 95.2, 121-165

Kochavi, M., Beck, P. and Yadin, E.
2000 *Aphek-Antipatris I*, Tel Aviv.

Korfmann, M.
1977 Demirchihuyuk, 1976. *AS* XXVII, 37-38
1978 Demirchihuyuk, 1977. *AS* XXVIII, 16
1979 Demirchihuyuk, 1978. *AS* XXIX, 192

Kramer, S.N.
1977 Commerce and Trade : Gleanings From Sumerian Literature. *Iraq* XXXIX.1, 59-66

Langdon, S.
1935 Excavations at Kish and Barghuthiat 1933. *Iraq* I.II, 113-124

Lapp, N.
1976 Casemate Walls in Palestine and Tell el-Ful. *BASOR* 223, 25-42

Lapp, P.W.
1964 The 1963 Excavations at Ta'annek. *BASOR* 173, 4-44
1966 *The Dhahr Mirzbaneh Tombs*, New Haven
1967 The 1966 Excavations at Ta'annek. *BASOR* 179, 2-38
1969 The 1968 Excavations at Ta'annek. *BASOR* 195, 2-49
1970 Palestine in the Early Bronze Age. in Sanders, J.A. (ed) *Near Eastern Archaeology in the Twentieth Century : Essays in Honor of Nelson Glueck*, New York, 101-131

Larsen, M. T.
1976 *The Old Assyrian City State and it's Colonies.* Copenhagen

Lawrence, A.W.
1965 Ancient Egyptian Fortifications. *JEA* 51, 71

Lehmann, G., Lampreche, R., Kerner, S. and Bernbeck, R.
1991 The 1990 Excavations at Abu Snesleh. *ADAJ* 35, 41-65

Leonard, A.
1992 The Jordan Valley Survey, 1953 : Some Unpublished Soundings Conducted by James Mellaart. *AASOR* 50

Lenzen, C.S.,
1997 Tell Irbid. In Meyers, E.M. (ed) *The Oxford Encyclopedia of Archaeology in the Near East*, Oxford

Lenzen, C.S., Gordon, R.L. and McQuitty, A.M.
1985 Excavations at Tell Irbid and Beit Ras, 1985. *ADAJ* XXIX, 151-160

Lenzen, C.S. and McQuitty, A.
1984 Tel Irbid. *ADAJ* XXVIII, 297
1989 Irbid (Tell). In Homes-Fredericq, D., Hennessy, J.B., (eds)
The Archaeology of Jordan 2, 298-300

Levy, T. E., Alon, D., Smith, P., Yekutieli, Y., Rowan, Y., Goldberg, P., Porat, N., van Den Brink, E. C. M., Witten, A. J., Golden, J., Grigson, C., Kansa, E., Dawson, L., Holl, A., Moreno, J. and Kersel, M.
1997 Egyptian-Canaanite Interaction at Nahal Tillah, Israel (ca. 4500-3000 B.C.E). An Interim Report on the 1994-1995 Excavations. *BASOR* 307, 1-52

Levy, T. E., van Den Brink, E. C. M., Goren, Y. and Alon, D.
1995 New Light on King Narmer and the ProtoDynastic Egyptian Presence in Canaan. *BA* 58.1, 26-36

Liere, W.J. van
1963 Capitals and Citadels of Bronze-Iron Age Syria in Their Relationship to Land and Water. *AAAS* 13, 109-122

Lloyd, S.
1940 Iraq Government Soundings at Sinjar. *Iraq* VII, 13-21

Lloyd, S., Mellaart, J.
1958 Beycesultan Excavations : Fourth Preliminary Report, 1957. *AS* VIII, 93-126

Loon, M. van
1968 First Results of the 1967 Excavations at Tell Selenkahiye. *AAAS* XVIII.1-2, 21-36
1970 Korucutepe. *AS* XX, 10
1973 First Results of the 1972 Excavations at Tell Selenkahiye. *AAAS* XXIII, 145-158
1977 First Results of the 1974 and 1975 Excavations at Selenkahiye Near Meskene, Syria. *AAAS* XXVII, 165-186
1979 1974 and 1975 Preliminary Results of the Excavations at Selenkahiye Near Meskene, Syria. *AASOR* 44, 97-112

Loon, M. van, and Meijer, D.
1983 Hammam et-Turkman on the Balikh : First Results. *AAAS* XXXIII, 131-135

Loon, M. van, Meijer, D. and Akkermans, P.
1987 Hammam et-Turkman and Sabi Abyad on the Balikh. The University of Amsterdam's 1986 Excavations. *AAAS* XXXVII, 309-324

Long, G.
1957
The Bioclimatology and Vegetation of East Jordan.
Rome.

Loud, G.
1948 *Megiddo II*, Chicago

Mare, H.
1991 The 1988 Season of Excavation at Abila of the Decapolis. *ADAJ* 35, 103-221

Ma'ayeh, F.
1960 Recent Archaeological Discoveries in Jordan. *ADAJ* 4/5, 115

Mabry, J.
1989 Investigations at Tell el-Handaquq, Jordan (1987-88). *ADAJ* XXXIII, 59-95
1996 Early Town Development and Water Management in the Jordan Valley : Investigations at Tell el-Handaquq North. *ASOR* 53, 115-154

Mabry, J. and Palumbo, G.
1992 Environmental, Economic and Political Constraints on Ancient Settlement Patterns in the Wadi al-Yabis Region. *SHAJ* III, 67-72

MacAlister, R.A.S.
1911 *The Excavation of Gezer, Volume I*, London
1912 *The Excavation of Gezer, Volumes II-III*, London

MacAlister, R.A.S. and Duncan, J.G.
1926 Excavations on the Hill of Ophel, Jerusalem, 1923-1925. *APEF* 4

MacDonald, B.
1988 *The Wadi el-Hasa Archaeological Survey, 1979-1983, West-Central Jordan*, Waterloo, Ontario
1992 *The Southern Ghors and Northeast Arabah Archaeological Survey*,
Sheffield

MacKenzie, D.L.
1913 Excavations at Ain Shems. *APEF* 2-3

Magness-Gardiner, B.
1997 The Middle Bronze Age in Tran Jordan. in Oren, E. (ed) *The Hyksos: New Historical and Archaeological Perspectives*, Philadelphia, 303-326

Maisler, B., Stekelis, M. and Avi-Yonah, M.
1952 The Excavations at Beth Yerah (Khirbet el-Kerak) 1944-46. *IEJ* 2.3, 165-173

Malamat, A.
1971 Syro-Palestinian Destinations in a Mari Tin Inventory. *IEJ* 21.1, 31-38
1992 Mari and Hazor : The Implication for the Middle Bronze Age Chronology. In SICAC III, 121-123

Mallet, J.
1973 *Tell el-Far'ah. L'installation du Moyen Bronze anterieure au rempart*,
Paris
1987 *Tell el-Far'ah II,1. Le Bronze* Moyen, Paris

Mallowan, M.E.L.
1946 Excavations in the Balih Valley, 1938. *Iraq* VIII, 111-159
1950 The Excavations at Nimrud, 1949-50. *Iraq XII.2*, 159

Marchetti, N., Nigro, L. and Taha, H.
2000 Preliminary Report on the Second Season of Excavations of the Italian-Palestinian Expedition at Tell es-Sultan/Jericho, 1998. *ICANNE* I, 869-902

Marfoe, L.
1979a The Integrative Transformation : Patterns of Socio-political Organization in Southern Syria. *BASOR* 234, 1-42
1979b *Tell Kamid el-Loz, 1972-73. Preliminary Type Series Catalogue. North Slope and Palace Areas*, Unpublished manuscript, University of Chicago

Margueron, J.
1983 Mari. *Syria* XXXIII, 281-293
1997 Mari. in Meyers, E.M. (ed) *The Oxford Encyclopedia of Archaeology in the Near East*, Oxford
2000 Mari: Derniers Developpments des Recherches Conduits Sur le Tell Hariri'. *ICANE* I, 909-928

Matthiae, P.
1967 Mission Archeologique de l'Universite' de Rome a Tell Mardikh. *AAAS* XVII.1-2, 25-44
1968 Mission Archeologique de l'Universite' de Rome a Tell Mardikh (1966). *AAAS* XVIII.1-2, 5-20
1970 Mission Archeologique A Tell Mardikh. *AAAS* XX, 60-68
1982 The Mature Early Syrian Culture of Ebla and the Development of Early Bronze Civilisation of Jordan. *SHAJ* I, 77-91
1984 New Discoveries at Ebla. *BA* 47.1, 18-35
1987 The Destruction of Ebla Royal Palace : Interconnections Between Syria, Mesopotamia and Egypt in the Late EBIVA. ICAC 3, 163-169
1990 A New Monumental Temple of Middle Bronze II at Ebla and the Unity of the Architectural Tradition of Syria-Palestine. *AAAS* XL, 111-119
1997a Ebla. in Meyers, E.M. (ed) *The Oxford Encyclopedia of Archaeology in the Near East*, Oxford
1997b Ebla and Syria in the Middle Bronze Age. in Oren, E.D. (ed) *The Hyksos: New Historical and Archaeological Perspectives*, Philadelphia, 379-414
2000 Monuments seculaires et religieux dans la Ville Basse et fortifications d'Ebla au BMI-II: fouilles a Tell Mardikh, 1991-1997. *ICANE* I, 1029-1042

Mazar, A.
1992 *Archaeology of the Land of the Bible. 10 000-586 B.C.E.* New York
1997*Timnah (Tel Batash)*. Qedem 37

Mazar, B.
1968 The Middle Bronze Age in Palestine. *IEJ* 18.2, 65-97

Mazar, B., Biran, A., Dothan, M. and Dunyevsky, I.
1964 Ein Gev : Excavations in 1961. *IEJ* 14.1-2, 1-49

McClellan, T. L.
1986 A Syrian Fortress of the Bronze Age : el-Qitar. *NGR* 2(4), 418-440

McClellan, T. L. and Porter, A.
1995 Jawa and North Syria. *SHAJ* V, 49-65

McGovern, P.E.
1989 Baq'ah Valley Project. in Homes-Fredericq, D., Hennessy, J.B. (eds)
The Archaeology of Jordan 2, 25-44

McGovern, P. E., Bourriau, J., Harbottle, G. and Allen, S. J.
1994 The Archaeological Origin and Significance of the Dolphin Vase as Determined by Neutron Activation Analysis. *BASOR* 296, 31-44

McGovern, P.E. and Harbottle, G.
1997 Hyksos Trade Connections Between Tell el-Dab'a (Avaris) and the Levant: A Neutron Activation Study of the Canaanite Jar. in Oren, E. (ed) *The Hyksos : New Historical and Archaeological Perspectives*, Philadelphia, 141-158

McLaren, P.B.
2001a Comparative Ceramic Analysis as a Dating Mechanism in Marginal Zones - A Case Study. *SHAJ* VII, 247-258
2001b Problems Associated With Using Comparative Ceramic Analysis in Marginal Zones - The Jordanian Hauran as a Case Study. in Hopkins, L. and Parker, A. (eds) *The Archaeology of the Near East - An Australian Perspective*, Sydney University Archaeological Methods Series 7, 37-48

McNicoll, A.W., Ball, W., Bassett, S., Edwards, P.C., Macumber, P.G., Petocz, D., Potts, T.F., Randle, L., Villiers, L. and Watson, P.M.
1984 Preliminary Report on the University of Sydney's Fifth Season of Excavations at Pella in Jordan. *ADAJ* XXVIII, 55-86

McNicoll, A.W., Edwards, P.C., Hosking, J., Macumber, P.G., Walmsley, A.G. and Watson, P.M.
1986 Preliminary Report on the University of Sydney's Seventh Season of Excavations at Pella (Tabaqat Fahl) in 1985. *ADAJ* XXX, 155-198

McNicoll, A.W., Edwards, P.C., Hanbury-Tenison, J., Hennessy, J.B., Potts, T.F., Smith, R.H., Walmsley, A. and Watson, P.
1992 Pella in Jordan 2, Sydney

McNicoll, A.W., Hennessy, J.B., Walmsley, A.G. and Potts, T.F.
1982 Preliminary Report on the Third Season of Excavations at Pella, 1981. *ADAJ* XXVI, 343-364

McNicoll, A.W., Smith, R.H. and Hennessy, J.B.
1982 Pella in Jordan 1, Canberra

Mee, C.
1978 Aegean Trade and Settlement in Anatolia in the Second Millenium B.C. *AS* XXVIII, 121-156

Mellaart, J.
1960 Excavations at Hacilar, Third Preliminary Report, 1959. *AS* X, 83-104

Mellink, M.
1966 Karatas-Semayuk. *AS* XVI, 36

Mesnil du Buisson. de
1935 *Le Site Archeologique de Mishrife-Qatna*, Paris

Meyer, J.W.
1988 Halawa 1986-1987. Les Fouilles sur le tell Halawa A. *AAAS* XXXVIII, 242-249

Meyer, J. W. and Orthmann, W.
1983 Halawa 1980-1982. *AAAS* XXXIII, 93-110

Miller, J.M. (ed)
1991 *Archaeological Survey of the Kerak Plateau*, Atlanta

Miroschedji, P. de
1990 The Early Bronze Age Fortifications at Tel Yarmuth - An Interim Report. *EI* 21, 48-61

Mittman, S.
1970 *Beitrage zur Siedlungs- und Territorial Geschichte des Nordlichen Ostjordanlandes*, Wiesbaden

Muhly. J.D.
1977 The Copper Ox-Hide Ingots and the Bronze Age Metals Trade. *Iraq* XXXIX.1, 73-82

Munn-Rankin, J.M.
1956 Diplomacy in Western Asia in the Early 2nd Millenium B.C. *Iraq* XVII, 68-110

Najjar, M.
1992 The Jordan Valley (East Bank) During the Middle Bronze Age in the Light of New Excavations. *SHAJ* III, 149-153

Najjar, M., Hauptmann, A., Weisgerber, G. and Bachmann, H. G.
1995 The Early Bronze Age at Faynan / Wadi Arabah : A Period of New Technology in Copper Production. *SHAJ* V, 519-521

Nigro, L.
2000 Coordinating the MBI Pottery Horizon of Syria and Palestine. *ICANNE* I, 1187-1212

Northedge, A.
1983 The Fortifications of Qal'at 'Amman (Amman Citadel) :Preliminary Report. *ADAJ* XXVII, 436-460

Northedge, A., Wilkinson, T.J. and Falkner, R.
1990 Survey and Excavations at Samarra, 1989. *Iraq* LII, 121-148

Oates, D.
1967 The Excavations at Tell al-Rimah, 1966. *Iraq* XXIX.2, 70-96
1983 Tell Brak : The Most Recent Discoveries. *AAAS* XXXIII, 83-92

O'Connor, D.
1983 Review of W.A. Ward, Studies on Scarab Seals I : Pre-12th Dynasty Scarab Amulets. *Chronique d'Egypte* 58, 163-172
1985 The Chronology of Scarabs of the Middle Kingdom and the Second Intermediate Period. *JSSEA* 15, 1-41
1997 The Hyksos Period in Egypt. in Oren, E.D. (ed) *The Hyksos: New Historical and Archaeological Perspectives*, Philadelphia, 45-68

Ohata, K.
1970 *Tel Zeror III*, Tokyo

Oren, E.D.
1973 The Early Bronze IV Period in Palestine and it's Cultural and Chronological Setting. *BASOR* 210, 20-37
1997a Introduction and Overview: Structure and Goals. in Oren, E.D. (ed) *The Hyksos: New Historical and Archaeological Perspectives*, Philadelphia, xix-xxvi
1997b The Kingdom of Sharuhen and the Hyksos Kingdom. in Oren, E.D. (ed) *The Hyksos: New Historical and Archaeological Perspectives*, Philadelphia, 253-284

Oren, E. and Yekutieli, Y.
1996 The MBA Defence System at Tel Haror. *EI* 25, 15-26

Orthmann, W.
1975 Bericht uber die dritte Grabunges kampagne 1973 in Mumbaqat. *AAAS* XXV, 129-154

Ory, J.
1938 Excavations at Ras el-'Ain II. *QDAP* 6, 99-120

Osten, H.H. von der
1956 *Svenska Syrien Expeditionen, 1952-53, Die Grabung von Tell Es-Salihiyeh*,
Lund

Ozguc, T.
1960 Kultepe. *AS* X, 24-25
2000 Kanish-Nesa, The Earliest International Trade Center of the Near East. *ICANNE* I, 1247-1258

Paley, S., Porath, Y. and Stieglitz, R.R.
1984 The Emeq Hefer Archaeological Research Project, 1984. *IEJ* 34.4, 277

Palmieri, A.
1981 Excavations at Arslantepe (Malatya). *AS* XXXI, 101-120
1984 Arslantepe, 1983. *AS* XXXIV:208

Parker, S.T.
1982 Preliminary Report on the 1980 Season of the Central '*Limes Arabicus*' Project. *BASOR* 247, 1-26

Parr, P.
1968 The Origin of the Rampart Fortifications of MBA Palestine and Syria. *ZDPV* 84, 18-45
1983 The Tell Nebi Mend Project. *AAAS* XXXIII, 99-117
1997 Qadesh. in Meyers, E.M. (ed) *The Oxford Encyclopedia of Archaeology in the Near East*, Oxford

Peyronel, L.
2000 The Middle Bronze II Fortress V at Tell Mardikh-Ebla (Syria). Preliminary Analysis of Architectural Contexts and Archaeological Materials. *ICANE* I, 1353-1378

Pennels, E.
1983 Middle Bronze Age Earthworks : A Contemporary Engineering Evaluation. *BA* 46.1, 57-62

Petrie, W.M.F.
1891 *Tel el-Hesy*, London
1898 *Deshashe*, London
1906 *Hyksos and Israelite Cities*, British School of Archaeology in Egypt, Vol 12, London
1928 *Gerar* London
1930 *Beth-Pelet I (Tel Fara)*, Vienna
1931 *Ancient Gaza I (Ajjul)*, London
1932 *Ancient Gaza II*, London
1933 *Ancient Gaza III*, London

Petrie, W.M.F., Mackay, E.J.H. and Murray, M.A.
1952 *City Of Shepherd Kings And Ancient Gaza V*, London

Pezard, M.
1931 *Qadesh. Mission Archeologique A Tell Nebi Mend, 1921-1922*, Paris

Pfalzner, P.
1987 The Excavations at Tell Bderi 1986. *AAAS* XXXVII, 292-308

Philip, G.
1989 *Metal Weapons of the Early and Middle Bronze Ages in Syria-Palestine*, BAR 526, Oxford.
1995 The Same But Different : A Comparison of Middle Bronze Age Metalwork From Jericho and Tell ed-Dab'a. *SHAJ* V, 523-530

Plat Taylor, J. du, Seton Williams, M.V. and Waechter, J.
1950 Excavations at Sacke Gozu. *Iraq* XII.2, 53-138

Porada, E.
1966 Les Cylindres de la Jarre Montet. *Syria* 43, 243-258

Potts, T.F.
1985 A Bronze Age Ivory-decorated Box from Pella (Pahel) and Its Foreign Relations. *SHAJ* III, 59-71

Potts, T.F., Bourke, S.J., Edwards, P.C., Richards, F. and Wightman, G.J.
1988 Preliminary Report on the Eighth and Ninth Season of Excavations at the University of Sydney at Pella (Tabaqat Fahl), 1986 and 1987. *ADAJ* 32, 115-149

Potts, T.F., Colledge, S.M. and Edwards, P.C.
1985 Preliminary Report on a Sixth Season of Excavations by the University of Sydney at Pella in Jordan (1983/1984). *ADAJ* XXIX, 181-210

Prag, K.
1974 The Intermediate EB-MB Age : An Interpretation of the Evidence From TransJordan, Syria and Lebanon. *Levant* VI, 69-107
1985 Ancient and Modern Pastoral Migration in the Levant. *Levant* XVII, 81-89
1992 Bronze Age Settlement in the South Jordan Valley : Archaeology, Environment and Ethnology. *SHAJ* III, 155-160

Prausnitz, M. W.
1963 Achzib. *IEJ* 13.4, 337
1975 The Planning of the MBA Town at Achzib and it's Defences. *IEJ* 25, 202-210

Prausnitz, M.W. and Kempinski, A.
1977 Kabri. *IEJ* 27.2-3, 166

Pumpelly, R.
1908 *Explorations in Turkestan. Prehistoric Civilization of Anau, Volume 1*, Washington

Raban, A.
1987 The City Walls of Stratons Towers : Some New Archaeological Data. *BASOR* 268, 71-88

Rainey, A. F.
1994 Remarks on Donald Redfords 'Egypt, Canaan, and Israel in Ancient Times'. *BASOR* 295, 81-84

Rast, W.E.
1995 Building on Marl : The Case of Bab adh-Dhra. *SHAJ* V, 123-128

Rast, W.T. and Schaub, R.T.
1978 Bab edh-Dhra, 1975. *ASOR* 43, 1-60
1980 Preliminary Report of the 1979 Expedition to the Dead Sea Plain, Jordan. *BASOR* 240, 21-62
1981 The Southeastern Dead Sea Plain Expedition : An Interim Report of the 1977 Season. *ASOR* 46

Reade, J.E.
1968 Tell Taya (1967) : Summary Report. *Iraq* XXX.2, 234-264

Redford, D.
1982 Contact Between Egypt and Jordan in the New Kingdom : Some Comments on Sources. *SHAJ* I, 115-119

Reisner, G.A.
1960 The Egyptian Forts from Halfa to Semna. *Kush* VIII, 11-24

Renfrew, C.
1972 *The Emergence of Civilisation : The Cyclades and the Aegean in the Third Millenium B.*C, London

Richard, S.
1980 Toward a Consensus of Opinion on the End of the Early Bronze Age in Palestine-TransJordan. *BASOR* 237, 5-34
1983 Report on the 1982 Season of Excavations at Khirbet Iskander. *ADAJ* XXVII, 45-53
1987 Questions of Nomadic Incursions at the End of the Third Millenium B.C. *SHAJ* III, 241-246
1990 The 1987 Expedition to Khirbet Iskander and it's Vicinity : Fourth Preliminary Report. *BASOR* Supp 26, 33-58
2001 Khirbet Iskander. *AJA* 105.3, 441

Richard, S. and Boraas, R.
1984 Preliminary Report of the 1981-1982 Seasons of the Expedition to Khirbet Iskander and it's Vicinity. *BASOR* 254, 63-87
1988 The Early Bronze IV Fortified Site of Khirbet Iskander, Jordan : Third Preliminary Report, 1984 Season. *BASOR* Supp 25, 107-130

Rose, D. G. and Toombs, L.E.
1978 Four Seasons of Excavation at Tell el-Hesi : a Preliminary Report. *ASOR* 43, 109-150

Ross, J. F.
1979 Early Bronze Age Structures at Tell el-Hesi. *BASOR* 236, 11-22

Ryholt, K.S.B.
1998 Hotepibre, a Supposed Asiatic King in Egypt with Relations to Ebla. *BASOR* 311, 1-6

Saller, S.J. and Bagatti, B.
1949 *The Town of Nebo (Khirbet el-Mekhayyer)*, Jerusalem

Sasson, J. M.
1968 Instances of Mobility Among
Mari Artisans. *BASOR* 190, 46-54

Sauer, J.
1985 Ammon, Moab and Edom. *Biblical Theology Today*, Jerusalem, 206-214
1986 TransJordan in the Bronze and Iron ages : A Critique of Gluecks Synthesis. *BASOR* 263, 1-26

Scandone Matthiae, G.
1990 Egyptianizing Ivory Inlays From Palace P at Ebla. *AAAS* XL, 146-160
1997 The Relations Between Ebla and Egypt. in Oren, E.D. (ed) *The Hyksos: New Historical and Archaeological Perspectives*, Philadelphia, 415-428

Schaub, R. T.
1982 The Origins of the Early Bronze Age Walled Town Culture of Jordan. *SHAJ* I, 67-75
1992 A Reassessment of Nelson Glueck on Settlement on the Jordan Plateau in Early Bronze III and IV. *SHAJ* IV, 161-167

Schaub, R.T. and Rast, W.T.
1984 Preliminary Report of the 1981 Expedition to the Dead Sea Plain, Jordan. *BASOR* 254, 35-60

Schloen, D.
1997 Ashkelon. in Meyers, E.M. (ed) *The Oxford Encyclopedia of Archaeology in the Near East*, Oxford

Schroder, M.
1997 *The Ceramics From the Wadi al-'Ajib, Jordan*, Unpublished doctoral dissertation, University of Sydney

Schwartz, G.M., Curvers, H.H., Dunham, S. and Stuart, B.
2003 A Third Millennium B.C. Elite Tomb and Other New Evidence From Tell Umm el-Marra, Syria. *AJA* 107.3, 325-362

Seeden, H.
1986 Bronze Age Village Occupation at Busra : AUB Excavations on the Northwest Tell, 1983-1984. *Berytus* XXXIV, 11-82

Seeden, H. and Wilson, J.
1984 Busra in the Hawran : AUBs Ethnoarchaeological Projects 1980-1985. *Berytus* XXXII, 19-34

Seger, J.D.
1974 The MBIIC Date of the East Gate at Shechem. *Levant* VI, 117-130
1977 Tell Halif (Lahav),1976. *IEJ* 27.1, 45-46
1983 Investigations at Tel Halif, Israel, 1976-1980. *BASOR* 252, 1-24
1987 Tell Halif 1986. *IEJ* 37.2-3, 192-194

Seger, J.D., Baum, B., Borowski, O., Cole, D.P., Forshey, H., Futato, E., Jacobs, P.F., Laustrup, M., O'Connor Seger, P. and Zeder, M.
1990 The Bronze Age Settlements at Tell Halif : Phase II Excavations, 1983-1987. *BASOR* Supp 26, 1-32

Sellers, O.R., Funk, R.W., McKenzie, J.L., Lapp, P. and Lapp, N.
1968 The 1957 Excavation at Beth Zur'. *AASOR* XXXVIII

Sellin, E.
1904 *Tell Ta'annek*, Vienna
1905 *Eine Nachlese auf dem Tell Ta'annek*, Vienna
1926 Die Ausgrabung von Sichem. *ZDPV* 49, 229-236, 304-320
1927 Die Ausgrabung von Sichem, Fruhjahr und Sommer 1927. *ZDPV* 50, 265-274
1928 Die Masseben des El-Berit in Sichem. *ZDPV* 51, 119-123

Sellin, E. and Watzinger, C.
1913 *Jericho. Die Ergebinesse der Ausgrabungen*, Leipzig

Seton-Williams, M. V.
1980 Tell Keisan (1971-1976), une cite phenicienne en Galilee. in Briend, J. and Humbert, J.P

Sevin, V.
1984 Imikusagi, 1983. *AS* XXXIV, 222

Shea, W. H.
1979 The Conquests of Sharuhen and Megiddo Reconsidered. *IEJ* 29.1, 1-5

Shehadeh, N.
1985 The Climate of Jordan in the Past and the Present. *SHAJ* II, 25-37

Shiloh, Y.
1984a *Excavations at the City of David I, 1978-1982*, Qedem 19, Jerusalem
1984b City of David, Jerusalem. *IEJ* 34, 57-58

Simmons, A. and Kafafi, Z.,
1988 Preliminary Report on the 'Ain Ghazal Archaeological Survey, 1987. *ADAJ* 31, 27-40

Smith, R.H.
1973 *Pella of the Decapolis. The 1967 Season of the College of Wooster Expedition*, Wooster
1987 Trade in the Life of Pella of the Decapolis. *SHAJ* III, 53-55

Stager, L.E.
1991 *Ashkelon Discovered*, Washington D.C
1999 The Fortress Temple at Shechem and the 'House of El, Lord of the Covenant'. in Williams, P.H. and Hiebert, J. (eds) *Realia Dei. Essays in Archaeology and Biblical Interpretation*, Atlanta

Steele, C.S.
1990 Early Bronze Age Socio-Political Organization in South Western Jordan. *ZDPV* 106, 1-33

Stern, E.
1973 Tel Meborach. *IEJ* 23.4, 256
1983 Tel Dor, 1983. *IEJ* 33.3-4, 261
1984 *Tel Mevorakh 2*, Qedem 18, Jerusalem

Stewart, J.R.,
1974 Tell el-Ajjul : The Middle Bronze Age Remains. *Studies in Mediterranean Archaeology*, Volume 38, Goteburg

Strange, J.,
2000 A Proposal for a New Periodisation for the Bronze and Iron Ages in the Southern Levant (Jordan, Palestine and Israel). *ICANE* I, 1571-1578

Strommenger, E.
1979 Ausgrabungen der Deutschen Orient-Gesellschaft in Habuba Kabira. *AASOR* 44, 63-78
1980 *Habuba Kabira, Eine Stadt vor 5000 Jahren*, Mainz

Stronach, D. and Lumsden, S.
1992 UC Berkley's Excavations at Nineveh. *BA* 55.4, 227-233

Stucky, R.A.
1975 Tell el Hajj 1972. *AAAS* XXV, 165-182

Tadmor, M. and Prausnitz, M.
1959 Excavations at Rosh Hanniqra. *Atiqot* II, 72-88

Tefnin, R.
1983 Apercu Sur Neuf Campagnes de Fouilles Belges Aux Tells Abou Danne et Oumm el-Marra (1975-1983). *AAAS* XXXIII, 141-152

Thalmann, J-P.
1978 Tell Arqa : Campagnes 1972-74. *Syria* 55, 88-144
1979 Tell Arqa 1978-1979 : Rapport Provisoire. *BMB* 30, 61-80
1991 L'Age du Bronze 'a Tell 'Arqa. Bilan et Perspectives (1981-1991). *Berytus* XXX, 21-38

Thompson, L.T.
1974 Observations on the Bronze age in Jordan. *ADAJ* 19, 63-70

Toombs, L.E. and Wright, G.E.
1961 The Third Campaign at Balatah (Shechem). *BASOR* 161, 11-54
1963 The Fourth Campaign at Balatah (Shechem). *BASOR* 169, 1-60

Tubb, J.N.
1983 The MBIIA Period In Palestine : Its Relationship With Syria and its Origin. *Levant* XV, 49-62

Tucker, D.J.
1994 Representations of Imgur-Enlil on the Balawat Gates. *Iraq* LVI, 107-116

Tufnell, O.
1956 'Hyksos' Scarabs From Canaan. *AS* VI, 67-74
1958 *Lachish IV*, London
Tufnell, O. and Ward, W.A.
1966 Relations Between Byblos, Egypt and Mesopotamia at the End of the Third Millenium B.C. *Syria* 43, 164-241

Tuman, V.S.
1987 Astronomical Omens From Lunar Eclipses as a Source For Babylonian Chronology. *ICAC* III, 197-206

Ussishkin, D.
1978 *Excavations at Tel Lachish 1973-1977, Preliminary Report*. Tel Aviv University, Institute Of Archaeology, Reprint Series No 3. Tel Aviv
1989 Notes on the Fortifications of the Middle Bronze II Period at Jericho and Shechem. *BASOR* 276, 29-54
1992 Notes on the Middle Bronze Age Fortifications of Hazor. *Tel Aviv* 19.2, 274-281

Vaux, R. de
1955 La Cinquieme Campagne de Fouilles a Tell el-Far'ah, Pres Naplous. *RB* 52, 541-589
1962 Les Fouilles de Tell el-Far'ah. *RB* 59, 212-253
1970 Palestine During the Neolithic and Chalcolithic Periods. in Edwards, I.E.S., Gadd, C.J. and Hammond, N.G.L. *CAH* Vol 1, Pt 1, 499-538
1971 Palestine in the Early Bronze Age. in Edwards, I.E.S., Gadd, C.J. and Hammond, N.G.L. *CAH*, Vol 1, Pt 2, 208-237

Vries, B. de
1985 Urbanisation in the Basalt Region of North Jordan in Late Antiquity : The Case of Umm el-Jimal. *SHAJ* II, 249-256

Veenhof, K.R.
1977 Some Social Effects of Old Assyrian Trade. *Iraq* XXXIX.1, 109-118

Wafler, M.
1990 The Excavations at Tell Hamidi. in Eichler, S. *Tall al Hamidiya 2*, 219-228

Walker, C.
1980 Some Assyrians at Sippar in the Old Babylonian Period. *AS* XXX, 15-22

Walmsley, A.G., Macumber, P.G., Edwards, P.C., Bourke, S.J. and Watson, P.M.
1993 The Eleventh and Twelfth Seasons of Excavation at Pella (Tabaqat Fahl) 1989-1990. *ADAJ* 37, 165-240

Wapnish, P.
1997 Middle Bronze Equid Burials at Tell Jemmah and Reexamination of a Purportedly Hyksos Practice. in Oren, E.D. (ed) *The Hyksos: New Historical and Archaeological Perspectives*, Philadelphia, 335-368

Ward, W.A.
1978 Scarabs from the Montet jar, a Late 11th Dynasty Collection at Byblos. *Berytus* XXVI, 37-54
1991 Early Contacts Between Egypt, Canaan and Sinai : Remarks on the Paper by Amnon Ben-Tor. *BASOR* 281, 11-26
1992 The Present Status of Egyptian Chronology. *BASOR* 288, 47-52

Watson, P. and Tidmarsh, J.
1996 Pella/Tell al-Husn Excavations 1993 : The University of Sydney's Fifteenth Season. *ADAJ* 40, 293-314

Weill, R.
1920 *La Cite David : Fouilles de 1913-1914*, Paris
1947 *La Cite David : Fouilles de 1923-1924*, Paris

Weinstein, J.M.
1975 Egyptian Relations With Palestine in the Middle Kingdom. *BASOR* 217, 1-16
1984 The Significance of Tell Areini for Egyptian-Palestinian Relations at the Beginning of the Bronze Age. *BASOR* 256, 61-70
1991 Egypt and the MBIIC/LBIA Transition in Palestine. *Levant* XXIII, 105-115
1992 The Chronology of Palestine in the Early Second Millenium B.C.E. *BASOR* 288, 27-46
1997a Tell el-Far'ah South. in Meyers, E.M. (ed) *The Oxford Encyclopedia of Archaeology in the Near East*, Oxford
1997b Tell el-Yehudiyeh. in Meyers, E.M. (ed) *The Oxford Encyclopedia of Archaeology in the Near East*, Oxford

Weiss, H.
1983 Tell Leilan in the Third and Second Millenia B.C. *AAAS* XXXIII, 47-73

Welter, G.
1932 Review. *Archaologischer Anzeiger, Beiblatt zum Jahrbuch des Archaologischen Instituts* III-IV, 289-314

Wheeler, N.F.
1961 Diary of the Excavation of the Mirgissa Fort. *Kush* IX, 87-179

Wimmer, D.H.
1989 Safut (Tell). in Homes-Fredericq, D. and Hennessy, J.B. (eds) *Archaeology of Jordan, Volume* 2, 512-515
1997 Tell Safut. in Meyers, E.M. (ed) *The Oxford Encyclopedia of Archaeology in the Near East*, Oxford

Winnett, F.V.
1952 Excavations at Dibon in Moab, 1950-1951. *BASOR* 125, 7-19

Woolley, C.L.
1921 *Carchemish II*, Oxford
1952 *Carchemish III*, Oxford
1955 *Alalakh. An Account of the Excavations at Tell Atchana, 1937-1949*, Oxford

Wright, G.E.
1937 *The Pottery of Palestine From the Earliest Times to the End of the Early Bronze Age*, New Haven
1956 The First Campaign at Tell Balatah (Shechem). *BASOR* 144, 9-19
1957 The Second Campaign at Tell Balatah (Shechem). *BASOR* 148, 11-27
1961 The Archaeology of Palestine. in Wright, G.E. (ed) *The Bible and the Ancient Near East : Essays in Honor of William Foxwell Albright*, New York, 84-139
1968 Tell el-Yehudiyah and the Glacis. *ZDPV* 84, 1-17
1969 Iran and the Glacis. *ZDPV* 85.2, 24-34

Wright, G.R.H.
1985a *Ancient Building in South Syria and Palestine*, Leiden
1985b The City Gates at Shechem, Simple Reconstruction Drawings. *ZDPV* 101.1, 1-8
2000 *Ancient Building Technology*, Volume 1. Leiden

Yadin, Y.
1955 Hyksos Fortifications and the Battering-Ram. *BASOR* 137, 23-31
1958a Solomon's City Wall and Gate at Gezer. *IEJ* 8.2, 80-86
1958b *Hazor I*, Jerusalem
1960 *Hazor II*, Jerusalem
1973 The Tell Beit Mirsim G-F Alleged Fortifications. *BASOR* 212, 22-25
1978 The Nature of the Settlement During the MBIIA Period in Israel and the Problem of the Aphek Fortifications. *ZDPV* 94.1, 1-15
1989 *Hazor III-IV*, Jerusalem

Yakar, J.
1984a Regional and Local Schools of Metalwork in Early Bronze Age Anatolia. *AS* XXXIV, 59-86
1984b Tell Kinrot, 1983. *IEJ* 34.4, 191
1985 Regional and Local Schools of Metalwork in Early Bronze Age Anatolia. *AS* XXXV, 25-38

Yassine, K.
1974 City Planning of Tell el-Ajjul. *ADAJ* XIX,129-133

Yassine, K., Ibrahim, M. and Sauer, J.
1988a The East Jordan Valley Survey 1975. in Yassine, K. (ed) *Archaeology of Jordan : Essays and Reports*, Department of Archaeology, University of Jordan.
1988b The East Jordan Valley Survey 1975. in Yassine, K. (ed) *Archaeology of Jordan : Essays and Reports*. Department of Archaeology, University of Jordan

Yeivin, S.
1960 Early Contacts Between Canaan and Egypt. *IEJ* 10.4, 193-203

Yeivin, Z.
1974 Khirbet el-Mahruq. *IEJ* 24, 259

Zayadine, F.
1973 Recent Excavations on the Citadel of Amman. *ADAJ* 18, 17-36

Zayadine, F., Najjar, M. and Greene, J.A.
1987 Recent Excavations on the Citadel of Amman (Lower Terrace) : A Preliminary Report. *ADAJ* XXXI, 299-311

Zayadine, F., Humbert, J.B. and Najjar, M.
1989 The 1988 Excavations on the Citadel of Amman, Lower Terrace, Area A. *ADAJ* XXXIII, 357-363

Zeist, W. van
1985 Past and Present Environments of the Jordan Valley. *SHAJ* II, 199-204

Zohar, M.
1980 Tell Marjamah ('Ein Samiyeh). *IEJ* 30.3-4, 219

Zorn, J.R.
1997 An Inner and Outer Gate Complex at Tell en-Nasbeh. *BASOR* 307, 53-66

www.ingramcontent.com/pod-product-compliance
Lightning Source LLC
LaVergne TN
LVHW071216210926
848482LV00015B/827